'This articulate and engrossing book is as beautifully written as it is insightful'   *Sunday Times*

'A fascinating exploration of everything from industrial feedlots to how plants produce poisons to fend off potential predators'
                                                                                    Financial Times

'His exposition of organic agriculture is masterly'   *Guardian*

'I doubt that there is a book which succeeds more than *The Omnivore's Dilemma*—with its richness of information, eloquence of address and integrity of moral purpose—in rendering visible, and presenting for a "different" style of ethical reflection, that "profound engagement" with our world which eating represents'   *TLS*

'As riveting as a novel, and a passionate journey of the heart . . . this is the book he was born to write. He walks us through four meals that delineate the battlefield of dinner'   *Newsweek*

'He raises a lot of disturbing questions about how we eat and how we think about food today'   *Wall Street Journal*

'Michael Pollan has perfected a tone—one of gleeful irony and barely suppressed outrage—and a way of inserting himself into a narrative so that a subject comes alive through what he's feeling and thinking'
                                                                                Los Angeles Times

'Ambitious and thoroughly enjoyable . . . Pollan's account of hunting, cooking and serving a wild boar (along with wild mushrooms and bread baked from "wild" yeast), with which he ends his book, is entertaining and memorable . . . Readers of this intelligent and admirable book will almost certainly find their capacity to delight in food augmented rather than diminished.'   *San Francisco Chronicle*

MICHAEL POLLAN is the author of three previous books: *Second Nature*, *A Place of My Own*, and *The Botany of Desire*, a *New York Times* bestseller that was named a best book of the year by Borders, Amazon, and the American Booksellers Association. Pollan is a longtime contributing writer at the *New York Times Magazine* and teaches journalism at the University of California-Berkeley. He lives in Berkeley with his wife, the painter Judith Belzer, and their son, Isaac. To read more of his work, go to www.michaelpollan. com.

BY THE SAME AUTHOR

Second Nature

A Place of My Own

The Botany of Desire

In Defence of Food

Food Rules: An Eater's Manual

THE OMNIVORE'S DILEMMA

THE SEARCH FOR A PERFECT MEAL IN A FAST-FOOD WORLD

MICHAEL POLLAN

BLOOMSBURY

LONDON · NEW DELHI · NEW YORK · SYDNEY

First published in Great Britain 2006
This paperback published in 2011

Published by arrangement with Penguin Press,
a member of Penguin Group (USA) Inc.

Bloomsbury Publishing Plc, London, New Delhi, New York and Sydney

Bloomsbury Publishing Plc,
50 Bedford Square,
London WC1B 3DP

www.bloomsbury.com

A CIP catalogue record for this book is
available from the British Library

ISBN 978 1 4088 1218 1

10

Printed and bound in Great Britain by CPI Group (UK) Ltd, Croydon CR0 4YY

FOR JUDITH AND ISAAC

CONTENTS

III PERSONAL
THE FOREST

THE OMNIVORE'S DILEMMA

INTRODUCTION

OUR NATIONAL
EATING DISORDER

What should we have for dinner?

This book is a long and fairly involved answer to this seemingly simple question. Along the way, it also tries to figure out how such a simple question could ever have gotten so complicated. As a culture we seem to have arrived at a place where whatever native wisdom we may once have possessed about eating has been replaced by confusion and anxiety. Somehow this most elemental of activities—figuring out what to eat—has come to require a remarkable amount of expert help. How did we ever get to a point where we need investigative journalists to tell us where our food comes from and nutritionists to determine the dinner menu?

For me the absurdity of the situation became inescapable in the fall of 2002, when one of the most ancient and venerable staples of human life abruptly disappeared from the American dinner table. I'm talking of course about bread. Virtually overnight, Americans changed the way they eat. A collective spasm of what can only be described as carbophobia seized the country, supplanting an era of national lipophobia dating to the Carter administration. That was when, in 1977, a Senate committee had issued a set of "dietary goals" warning beef-loving Americans to lay off the red meat. And so we dutifully had done, until now.

What set off the sea change? It appears to have been a perfect media storm of diet books, scientific studies, and one timely magazine article. The new diet books, many of them inspired by the formerly discredited Dr. Robert C. Atkins, brought Americans the welcome news that they could eat more meat and lose weight just so long as they laid off the bread and pasta. These high-protein, low-carb diets found support in a handful of new epidemiological studies suggesting that the nutritional orthodoxy that had held sway in America since the 1970s might be wrong. It was not, as official opinion claimed, fat that made us fat, but the carbohydrates we'd been eating precisely in order to stay slim. So conditions were ripe for a swing of the dietary pendulum when, in the summer of 2002, the *New York Times Magazine* published a cover story on the new research entitled "What if Fat Doesn't Make You Fat?" Within months, supermarket shelves were restocked and restaurant menus rewritten to reflect the new nutritional wisdom. The blamelessness of steak restored, two of the most wholesome and uncontroversial foods known to man—bread and pasta—acquired a moral stain that promptly bankrupted dozens of bakeries and noodle firms and ruined an untold number of perfectly good meals.

So violent a change in a culture's eating habits is surely the sign of a national eating disorder. Certainly it would never have happened in a culture in possession of deeply rooted traditions surrounding food and eating. But then, such a culture would not feel the need for its most august legislative body to ever deliberate the nation's "dietary goals"—or, for that matter, to wage political battle every few years over the precise design of an official government graphic called the "food pyramid." A country with a stable culture of food would not shell out millions for the quackery (or common sense) of a new diet book every January. It would not be susceptible to the pendulum swings of food scares or fads, to the apotheosis every few years of one newly discovered nutrient and the demonization of another. It would not be apt to confuse protein bars and food supplements with meals or breakfast cereals with medicines. It probably would not eat a fifth of its meals in cars or feed fully a third of its children at a fast-food outlet every day. And it surely would not be nearly so fat.

Nor would such a culture be shocked to discover that there are other countries, such as Italy and France, that decide their dinner questions on the basis of such quaint and unscientific criteria as pleasure and tradition, eat all manner of "unhealthy" foods, and, lo and behold, wind up actually healthier and happier in their eating than we are. We show our surprise at this by speaking of something called the "French paradox," for how could a people who eat such demonstrably toxic substances as foie gras and triple crème cheese actually be slimmer and healthier than we are? Yet I wonder if it doesn't make more sense to speak in terms of an American paradox—that is, a notably unhealthy people obsessed by the idea of eating healthily.

To ONE DEGREE or another, the question of what to have for dinner assails every omnivore, and always has. When you can eat just about anything nature has to offer, deciding what you should eat will inevitably stir anxiety, especially when some of the potential foods on offer are liable to sicken or kill you. This is the omnivore's dilemma, noted long ago by writers like Rousseau and Brillat-Savarin and first given that name thirty years ago by a University of Pennsylvania research psychologist named Paul Rozin. I've borrowed his phrase for the title of this book because the omnivore's dilemma turns out to be a particularly sharp tool for understanding our present predicaments surrounding food.

In a 1976 paper called "The Selection of Foods by Rats, Humans, and Other Animals" Rozin contrasted the omnivore's existential situation with that of the specialized eater, for whom the dinner question could not be simpler. The koala doesn't worry about what to eat: If it looks and smells and tastes like a eucalyptus leaf, it must be dinner. The koala's culinary preferences are hardwired in its genes. But for omnivores like us (and the rat) a vast amount of brain space and time must be devoted to figuring out which of all the many potential dishes nature lays on are safe to eat. We rely on our prodigious powers of recognition and memory to guide us away from poisons (*Isn't that the mushroom that made me sick last week?*) and toward nutritious plants (*The red berries are the juicier, sweeter ones*). Our taste buds help too, predisposing us toward sweetness, which signals carbohydrate energy in nature, and away from bitterness, which is how many of the toxic alkaloids produced by plants taste. Our inborn sense of disgust keeps us from ingesting things that might infect us, such as rotten meat. Many anthropologists believe that the reason we evolved such big and intricate brains was precisely to help us deal with the omnivore's dilemma.

Being a generalist is of course a great boon as well as a challenge; it is what allows humans to successfully inhabit virtually every terrestrial environment on the planet. Omnivory offers the pleasures of variety, too. But the surfeit of choice brings with it a lot of stress and leads to a kind of Manichaean view of food, a division of nature into The Good Things to Eat, and The Bad.

The rat must make this all-important distinction more or less on its own, each individual figuring out for itself—and then remembering—which things will nourish and which will poison. The human omnivore has, in addition to his senses and memory, the incalculable advantage of a culture, which stores the experience and accumulated wisdom of countless human tasters before him. I don't need to experiment with the mushroom now called, rather helpfully, the "death cap," and it is common knowledge that that first intrepid lobster eater was on to something very good. Our culture codifies the rules of wise eating in an elaborate structure of taboos, rituals, recipes, manners, and culinary traditions that keep us from having to reenact the omnivore's dilemma at every meal.

One way to think about America's national eating disorder is as the return, with an almost atavistic vengeance, of the omnivore's dilemma. The cornucopia of the American supermarket has thrown us back on a bewildering food landscape where we once again have to worry that some of those tasty-looking morsels might kill us. (Perhaps not as quickly as a poisonous mushroom, but just as surely.) Certainly the extraordinary abundance of food in America complicates the whole problem of choice. At the same time, many of the tools with which people historically managed the omnivore's dilemma have lost their sharpness here—or simply failed. As a relatively new nation drawn from many different immigrant populations, each with its own culture of food, Americans have never had a single, strong, stable culinary tradition to guide us.

The lack of a steadying culture of food leaves us especially vulnerable to the blandishments of the food scientist and the marketer, for whom the omnivore's dilemma is not so much a dilemma as an opportunity. It is very much in the interest of the food industry to exacerbate our anxieties about what to eat, the better to then assuage them with new products. Our bewilderment in the supermarket is no accident; the return of the omnivore's dilemma has deep roots in the modern food industry, roots that, I found, reach all the way back to fields of corn growing in places like Iowa.

And so we find ourselves where we do, confronting in the supermarket or at the dinner table the dilemmas of omnivorousness, some of them ancient and others never before imagined. The organic apple or the conventional? And if the organic, the local one or the imported? The wild fish or the farmed? The trans fats or the butter or the "not butter"? Shall I be a carnivore or a vegetarian? And if a vegetarian, a lacto-vegetarian or a vegan? Like the hunter-gatherer picking a novel mushroom off the forest floor and consulting his sense memory to determine its edibility, we pick up the package in the supermarket and, no longer so confident of our senses, scrutinize the label, scratching our heads over the meaning of phrases like "heart healthy," "no trans fats," "cage-free," or "range-fed." What is "natural grill flavor" or TBHQ or xanthan gum? What is all this stuff, anyway, and where in the world did it come from?

My wager in writing The Omnivore's Dilemma was that the best way to answer the questions we face about what to eat was to go back to the very beginning, to follow the food chains that sustain us, all the way from the earth to the plate—to a small number of actual meals. I wanted to look at the getting and eating of food at its most fundamental, which is to say, as a transaction between species in nature, eaters and eaten. ("The whole of nature," wrote the English author William Ralph Inge, "is a conjugation of the verb to eat, in the active and passive.") What I try to do in this book is approach the dinner question as a naturalist might, using the long lenses of ecology and anthropology, as well as the shorter, more intimate lens of personal experience.

My premise is that like every other creature on earth, humans take part in a food chain, and our place in that food chain, or web, determines to a considerable extent what kind of creature we are. The fact of our omnivorousness has done much to shape our nature, both body (we possess the omnicompetent teeth and jaws of the omnivore, equally well suited to tearing meat and grinding seeds) and soul. Our prodigious powers of observation and memory, as well as our curious and experimental stance toward the natural world, owe much to the biological fact of omnivorousness. So do the various adaptations we've evolved to defeat the defenses of other creatures so that we might eat them, including our skills at hunting and cooking with fire. Some philosophers have argued that the very open-endedness of human appetite is responsible for both our savagery and civility, since a creature that could conceive of eating anything (including, notably, other humans) stands in particular need of ethical rules, manners, and rituals. We are not only what we eat, but how we eat, too.

Yet we are also different from most of nature's other eaters— markedly so. For one thing, we've acquired the ability to substantially modify the food chains we depend on, by means of such revolutionary technologies as cooking with fire, hunting with tools, farming, and food preservation. Cooking opened up whole new vistas of edibility by rendering various plants and animals more digestible, and overcoming many of the chemical defenses other species deploy against being eaten. Agriculture allowed us to vastly multiply the populations of a few favored food species, and therefore in turn our own. And, most recently, industry has allowed us to reinvent the human food chain, from the synthetic fertility of the soil to the microwaveable can of soup designed to fit into a car's cup holder. The implications of this last revolution, for our health and the health of the natural world, we are still struggling to grasp.

The Omnivore's Dilemma is about the three principal food chains that sustain us today: the industrial, the organic, and the hunter-gatherer. Different as they are, all three food chains are systems for doing more or less the same thing: linking us, through what we eat, to the fertility of the earth and the energy of the sun. It might be hard to see how, but even a Twinkie does this—constitutes an engagement with the natural world. As ecology teaches, and this book tries to show, it's all connected, even the Twinkie.

Ecology also teaches that all life on earth can be viewed as a competition among species for the solar energy captured by green plants and stored in the form of complex carbon molecules. A food chain is a system for passing those calories on to species that lack the plant's unique ability to synthesize them from sunlight. One of the themes of this book is that the industrial revolution of the food chain, dating to the close of World War II, has actually changed the fundamental rules of this game. Industrial agriculture has supplanted a complete reliance on the sun for our calories with something new under the sun: a food chain that draws much of its energy from fossil fuels instead. (Of course, even that energy originally came from the sun, but unlike sunlight it is finite and irreplaceable.) The result of this innovation has been a vast increase in the amount of food energy available to our species; this has been a boon to humanity (allowing us to multiply our numbers), but not an unalloyed one. We've discovered that an abundance of food does not render the omnivore's dilemma obsolete. To the contrary, abundance seems only to deepen it, giving us all sorts of new problems and things to worry about.

Each of this book's three parts follows one of the principal human food chains from beginning to end: from a plant, or group of plants, photosynthesizing calories in the sun, all the way to a meal at the dinner end of that food chain. Reversing the chronological order, I start with the industrial food chain, since that is the one that today involves and concerns us the most. It is also by far the biggest and longest. Since monoculture is the hallmark of the industrial food chain, this section focuses on a single plant: *Zea mays*, the giant tropical grass we call corn, which has become the keystone species of the industrial food chain, and so in turn of the modern diet. This section follows a bushel of commodity corn from the field in Iowa where it grew on its long, strange journey to its ultimate destination in a fast-food meal, eaten in a moving car on a highway in Marin County, California.

The book's second part follows what I call—to distinguish it from the industrial—the pastoral food chain. This section explores some of the alternatives to industrial food and farming that have sprung up in recent years (variously called "organic," "local," "biological," and "beyond organic"), food chains that might appear to be preindustrial but in surprising ways turn out in fact to be postindustrial. I set out thinking I could follow one such food chain, from a radically innovative farm in Virginia that I worked on one recent summer to an extremely local meal prepared from animals raised on its pastures. But I promptly discovered that no single farm or meal could do justice to the complex, branching story of alternative agriculture right now, and that I needed also to reckon with the food chain I call, oxymoronically, the "industrial organic." So the book's pastoral section serves up the natural history of two very different "organic" meals: one whose ingredients came from my local Whole Foods supermarket (gathered there from as far away as Argentina), and the other tracing its origins to a single polyculture of grasses growing at Polyface Farm in Swoope, Virginia.

The last section, titled Personal, follows a kind of neo-Paleolithic food chain from the forests of Northern California to a meal I prepared (almost) exclusively from ingredients I hunted, gathered, and grew myself. Though we twenty-first-century eaters still eat a handful of hunted and gathered food (notably fish and wild mushrooms), my interest in this food chain was less practical than philosophical: I hoped to shed fresh light on the way we eat now by immersing myself in the way we ate then. In order to make this meal I had to learn how to do some unfamiliar things, including hunting game and foraging for wild mushrooms and urban tree fruit. In doing so I was forced to confront some of the most elemental questions—and dilemmas—faced by the human omnivore: What are the moral and psychological implications of killing, preparing, and eating a wild animal? How does one distinguish between the delicious and the deadly when foraging in the woods? How do the alchemies of the kitchen transform the raw stuffs of nature into some of the great delights of human culture?

The end result of this adventure was what I came to think of as the Perfect Meal, not because it turned out so well (though in my humble opinion it did), but because this labor- and thought-intensive dinner, enjoyed in the company of fellow foragers, gave me the opportunity, so rare in modern life, to eat in full consciousness of everything involved in feeding myself: For once, I was able to pay the full karmic price of a meal.

Yet as different as these three journeys (and four meals) turned out to be, a few themes kept cropping up. One is that there exists a fundamental tension between the logic of nature and the logic of human industry, at least as it is presently organized. Our ingenuity in feeding ourselves is prodigious, but at various points our technologies come into conflict with nature's ways of doing things, as when we seek to maximize efficiency by planting crops or raising animals in vast monocultures. This is something nature never does, always and for good reasons practicing diversity instead. A great many of the health and environmental problems created by our food system owe to our attempts to oversimplify nature's complexities, at both the growing and the eating ends of our food chain. At either end of any food chain you find a biological system—a patch of soil, a human body—and the health of one is connected—literally—to the health of the other. Many of the problems of health and nutrition we face today trace back to things that happen on the farm, and behind those things stand specific government policies few of us know anything about.

I don't mean to suggest that human food chains have only recently come into conflict with the logic of biology; early agriculture and, long before that, human hunting proved enormously destructive. Indeed, we might never have needed agriculture had earlier generations of hunters not eliminated the species they depended upon. Folly in the getting of our food is nothing new. And yet the new follies we are per-petrating in our industrial food chain today are of a different order. By replacing solar energy with fossil fuel, by raising millions of food ani-mals in close confinement, by feeding those animals foods they never evolved to eat, and by feeding ourselves foods far more novel than we even realize, we are taking risks with our health and the health of the natural world that are unprecedented.

Another theme, or premise really, is that the way we eat represents our most profound engagement with the natural world. Daily, our eat-ing turns nature into culture, transforming the body of the world into our bodies and minds. Agriculture has done more to reshape the natu-ral world than anything else we humans do, both its landscapes and the composition of its flora and fauna. Our eating also constitutes a rela-tionship with dozens of other species—plants, animals, and fungi—with which we have coevolved to the point where our fates are deeply entwined. Many of these species have evolved expressly to gratify our desires, in the intricate dance of domestication that has allowed us and them to prosper together as we could never have prospered apart. But our relationships with the wild species we eat—from the mush-rooms we pick in the forest to the yeasts that leaven our bread—are no less compelling, and far more mysterious. Eating puts us in touch with all that we share with the other animals, and all that sets us apart. It de-fines us.

What is perhaps most troubling, and sad, about industrial eating is how thoroughly it obscures all these relationships and connections. To go from the chicken (*Gallus gallus*) to the Chicken McNugget is to leave this world in a journey of forgetting that could hardly be more costly, not only in terms of the animal's pain but in our pleasure, too. But for-getting, or not knowing in the first place, is what the industrial food chain is all about, the principal reason it is so opaque, for if we could see what lies on the far side of the increasingly high walls of our industrial agriculture, we would surely change the way we eat.

"Eating is an agricultural act," as Wendell Berry famously said. It is also an ecological act, and a political act, too. Though much has been done to obscure this simple fact, how and what we eat determines to a great extent the use we make of the world—and what is to become of it. To eat with a fuller consciousness of all that is at stake might sound like a burden, but in practice few things in life can afford quite as much satisfaction. By comparison, the pleasures of eating industrially, which is to say eating in ignorance, are fleeting. Many people today seem perfectly content eating at the end of an industrial food chain, without a thought in the world; this book is probably not for them. There are things in it that will ruin their appetites. But in the end this is a book about the pleasures of eating, the kinds of pleasure that are only deepened by knowing.

I

INDUSTRIAL

CORN

ONE

THE PLANT

Corn's Conquest

1. A NATURALIST IN THE SUPERMARKET

Air-conditioned, odorless, illuminated by buzzing fluorescent tubes, the American supermarket doesn't present itself as having very much to do with Nature. And yet what is this place if not a landscape (man-made, it's true) teeming with plants and animals?

I'm not just talking about the produce section or the meat counter, either—the supermarket's flora and fauna. Ecologically speaking, these are this landscape's most legible zones, the places where it doesn't take a field guide to identify the resident species. Over there's your eggplant, onion, potato, and leek; here your apple, banana, and orange. Spritzed with morning dew every few minutes, Produce is the only corner of the supermarket where we're apt to think "Ah, yes, the bounty of Nature!" Which probably explains why such a garden of fruits and vegetables (sometimes flowers, too) is what usually greets the shopper coming through the automatic doors.

Keep rolling, back to the mirrored rear wall behind which the butchers toil, and you encounter a set of species only slightly harder to identify—there's chicken and turkey, lamb and cow and pig. Though in Meat the creaturely character of the species on display does seem to be fading, as the cows and pigs increasingly come subdivided into boneless and bloodless geometrical cuts. In recent years some of this supermarket euphemism has seeped into Produce, where you'll now find formerly soil-encrusted potatoes cubed pristine white, and "baby" carrots machine-lathed into neatly tapered torpedoes. But in general here in flora and fauna you don't need to be a naturalist, much less a food scientist, to know what species you're tossing into your cart.

Venture farther, though, and you come to regions of the supermarket where the very notion of species seems increasingly obscure: the canyons of breakfast cereals and condiments; the freezer cases stacked with "home meal replacements" and bagged platonic peas; the broad expanses of soft drinks and towering cliffs of snacks; the unclassifiable Pop-Tarts and Lunchables; the frankly synthetic coffee whiteners and the Linnaeus-defying Twinkie. Plants? Animals?! Though it might not always seem that way, even the deathless Twinkie is constructed out of . . . well, precisely *what* I don't know offhand, but ultimately some sort of formerly living creature, i.e., a *species*. We haven't yet begun to synthesize our foods from petroleum, at least not directly.

If you do manage to regard the supermarket through the eyes of a naturalist, your first impression is apt to be of its astounding biodiversity. Look how many different plants and animals (and fungi) are represented on this single acre of land! What forest or prairie could hope to match it? There must be a hundred different species in the produce section alone, a handful more in the meat counter. And this diversity appears only to be increasing: When I was a kid, you never saw radicchio in the produce section, or a half dozen different kinds of mushrooms, or kiwis and passion fruit and durians and mangoes. Indeed, in the last few years a whole catalog of exotic species from the tropics has colonized, and considerably enlivened, the produce department. Over in fauna, on a good day you're apt to find—beyond beef—ostrich and quail and even bison, while in Fish you can catch not just salmon and shrimp but catfish and tilapia, too. Naturalists regard biodiversity as a measure of a landscape's health, and the modern supermarket's devotion to variety and choice would seem to reflect, perhaps even promote, precisely that sort of ecological vigor.

Except for the salt and a handful of synthetic food additives, every edible item in the supermarket is a link in a food chain that begins with a particular plant growing in a specific patch of soil (or, more seldom, stretch of sea) somewhere on earth. Sometimes, as in the produce section, that chain is fairly short and easy to follow: As the netted bag says, this potato was grown in Idaho, that onion came from a farm in Texas. Move over to Meat, though, and the chain grows longer and less comprehensible: The label doesn't mention that that rib-eye steak came from a steer born in South Dakota and fattened in a Kansas feedlot on grain grown in Iowa. Once you get into the processed foods you have to be a fairly determined ecological detective to follow the intricate and increasingly obscure lines of connection linking the Twinkie, or the nondairy creamer, to a plant growing in the earth someplace, but it can be done.

So what exactly would an ecological detective set loose in an American supermarket discover, were he to trace the items in his shopping cart all the way back to the soil? The notion began to occupy me a few years ago, after I realized that the straightforward question "What should I eat?" could no longer be answered without first addressing two other even more straightforward questions: "What am I eating? And where in the world did it come from?" Not very long ago an eater didn't need a journalist to answer these questions. The fact that today one so often does suggests a pretty good start on a working definition of industrial food: Any food whose provenance is so complex or obscure that it requires expert help to ascertain.

When I started trying to follow the industrial food chain—the one that now feeds most of us most of the time and typically culminates either in a supermarket or fast-food meal—I expected that my investigations would lead me to a wide variety of places. And though my journeys did take me to a great many states, and covered a great many miles, at the very end of these food chains (which is to say, at the very beginning), I invariably found myself in almost exactly the same place: a farm field in the American Corn Belt. The great edifice of variety and choice that is an American supermarket turns out to rest on a remarkably narrow biological foundation comprised of a tiny group of plants that is dominated by a single species: *Zea mays*, the giant tropical grass most Americans know as corn.

Corn is what feeds the steer that becomes the steak. Corn feeds the chicken and the pig, the turkey and the lamb, the catfish and the tilapia and, increasingly, even the salmon, a carnivore by nature that the fish farmers are reengineering to tolerate corn. The eggs are made of corn. The milk and cheese and yogurt, which once came from dairy cows that grazed on grass, now typically come from Holsteins that spend their working lives indoors tethered to machines, eating corn.

Head over to the processed foods and you find ever more intricate manifestations of corn. A chicken nugget, for example, piles corn upon corn: what chicken it contains consists of corn, of course, but so do most of a nugget's other constituents, including the modified corn starch that glues the thing together, the corn flour in the batter that coats it, and the corn oil in which it gets fried. Much less obviously, the leavenings and lecithin, the mono-, di-, and triglycerides, the attractive golden coloring, and even the citric acid that keeps the nugget "fresh" can all be derived from corn.

To wash down your chicken nuggets with virtually any soft drink in the supermarket is to have some corn with your corn. Since the 1980s virtually all the sodas and most of the fruit drinks sold in the supermarket have been sweetened with high-fructose corn syrup (HFCS)—after water, corn sweetener is their principal ingredient. Grab a beer for your beverage instead and you'd still be drinking corn, in the form of alcohol fermented from glucose refined from corn. Read the ingredients on the label of any processed food and, provided you know the chemical names it travels under, corn is what you will find. For modified or unmodified starch, for glucose syrup and maltodextrin, for crystalline fructose and ascorbic acid, for lecithin and dextrose, lactic acid and lysine, for maltose and HFCS, for MSG and polyols, for the caramel color and xanthan gum, read: corn. Corn is in the coffee whitener and Cheez Whiz, the frozen yogurt and TV dinner, the canned fruit and ketchup and candies, the soups and snacks and cake mixes, the frosting and gravy and frozen waffles, the syrups and hot sauces, the mayonnaise and mustard, the hot dogs and the bologna, the margarine and short-ening, the salad dressings and the relishes and even the vitamins. (Yes, it's in the Twinkie, too.) There are some forty-five thousand items in the average American supermarket and more than a quarter of them now contain corn. This goes for the nonfood items as well—everything from the toothpaste and cosmetics to the disposable diapers, trash bags, cleansers, charcoal briquettes, matches, and batteries, right down to the shine on the cover of the magazine that catches your eye by the checkout: corn. Even in Produce on a day when there's ostensibly no corn for sale you'll nevertheless find plenty of corn: in the vegetable wax that gives the cucumbers their sheen, in the pesticide responsible for the produce's perfection, even in the coating on the cardboard it was shipped in. Indeed, the supermarket itself—the wallboard and joint compound, the linoleum and fiberglass and adhesives out of which the building itself has been built—is in no small measure a manifestation of corn.

And us?

2. CORN WALKING

Descendents of the Maya living in Mexico still sometimes refer to them-selves as "the corn people." The phrase is not intended as metaphor. Rather, it's meant to acknowledge their abiding dependence on this miraculous grass, the staple of their diet for almost nine thousand years. Forty percent of the calories a Mexican eats in a day comes directly from corn, most of it in the form of tortillas. So when a Mexican says "I am maize" or "corn walking," it is simply a statement of fact: The very sub-stance of the Mexican's body is to a considerable extent a manifestation of this plant.

For an American like me, growing up linked to a very different food chain, yet one that is also rooted in a field of corn, not to think of himself as a corn person suggests either a failure of imagination or a triumph of capitalism. Or perhaps a little of both. It does take some imagination to recognize the ear of corn in the Coke bottle or the Big Mac. At the same time, the food industry has done a good job of persuading us that the forty-five thousand different items or SKUs (stock keeping units) in the supermarket—seventeen thousand new ones every year—represent genuine variety rather than so many clever rearrangements of molecules extracted from the same plant.

You are what you eat, it's often said, and if this is true, then what we mostly are is corn—or, more precisely, processed corn. This proposition is susceptible to scientific proof: The same scientists who glean the composition of ancient diets from mummified human remains can do the same for you or me, using a snip of hair or fingernail. The science works by identifying stable isotopes of carbon in human tissue that bear the signatures, in effect, of the different types of plants that originally took them from the air and introduced them into the food chain. The intricacies of this process are worth following, since they go some distance toward explaining how corn could have conquered our diet and, in turn, more of the earth's surface than virtually any other domesticated species, our own included.

Carbon is the most common element in our bodies—indeed, in all living things on earth. We earthlings are, as they say, a carbon life form. (As one scientist put it, carbon supplies life's quantity, since it is the main structural element in living matter, while much scarcer nitrogen supplies its quality—but more on that later.) Originally, the atoms of carbon from which we're made were floating in the air, part of a carbon dioxide molecule. The only way to recruit these carbon atoms for the molecules necessary to support life—the carbohydrates, amino acids, proteins, and lipids—is by means of photosynthesis. Using sunlight as a catalyst the green cells of plants combine carbon atoms taken from the air with water and elements drawn from the soil to form the simple organic compounds that stand at the base of every food chain. It is more than a figure of speech to say that plants create life out of thin air.

But corn goes about this procedure a little differently than most other plants, a difference that not only renders the plant more efficient than most, but happens also to preserve the identity of the carbon atoms it recruits, even after they've been transformed into things like Gatorade and Ring Dings and hamburgers, not to mention the human bodies nourished on those things. Where most plants during photosynthesis create compounds that have three carbon atoms, corn (along with a small handful of other species) make compounds that have four: hence "C-4," the botanical nickname for this gifted group of plants, which wasn't identified until the 1970s.

The C-4 trick represents an important economy for a plant, giving it an advantage, especially in areas where water is scarce and temperatures high. In order to gather carbon atoms from the air, a plant has to open its stomata, the microscopic orifices in the leaves through which plants both take in and exhaust gases. Every time a stoma opens to admit carbon dioxide precious molecules of water escape. It's as though every time you opened your mouth to eat you lost a quantity of blood. Ideally, you would open your mouth as seldom as possible, ingesting as much food as you could with every bite. This is essentially what a C-4 plant does. By recruiting extra atoms of carbon during each instance of photosynthesis, the corn plant is able to limit its loss of water and "fix"—that is, take from the atmosphere and link in a useful molecule—significantly more carbon than other plants.

At its most basic, the story of life on earth is the competition among species to capture and store as much energy as possible—either directly from the sun, in the case of plants, or, in the case of animals, by eating plants and plant eaters. The energy is stored in the form of carbon molecules and measured in calories. The calories we eat, whether in an ear of corn or a steak, represent packets of energy once captured by a plant. The C-4 trick helps explain the corn plant's success in this competition:

Few plants can manufacture quite as much organic matter (and calories) from the same quantities of sunlight and water and basic elements as corn. (Ninety-seven percent of what a corn plant is comes from the air, three percent from the ground.)

The trick doesn't yet, however, explain how a scientist could tell that a given carbon atom in a human bone owes its presence there to a photosynthetic event that occurred in the leaf of one kind of plant and not another—in corn, say, instead of lettuce or wheat. The scientist can do this because all carbon is not created equal. Some carbon atoms, called isotopes, have more than the usual complement of six protons and six neutrons, giving them a slightly different atomic weight. C-13, for example, has six protons and seven neutrons. (Hence "C-13.") For whatever reason, when a C-4 plant goes scavenging for its four-packs of carbon, it takes in more carbon 13 than ordinary—C-3—plants, which exhibit a marked preference for the more common carbon 12. Greedy for carbon, C-4 plants can't afford to discriminate among isotopes, and so end up with relatively more carbon 13. The higher the ratio of carbon 13 to carbon 12 in a person's flesh, the more corn has been in his diet—or in the diet of the animals he or she ate. (As far as we're concerned, it makes little difference whether we consume relatively more or less carbon 13.)

One would expect to find a comparatively high proportion of carbon 13 in the flesh of people whose staple food of choice is corn—Mexicans, most famously. Americans eat much more wheat than corn—114 pounds of wheat flour per person per year, compared to 11 pounds of corn flour. The Europeans who colonized America regarded themselves as wheat people, in contrast to the native corn people they encountered; wheat in the West has always been considered the most refined, or civilized, grain. If asked to choose, most of us would probably still consider ourselves wheat people (except perhaps the proud corn-fed Midwesterners, and they don't know the half of it), though by now the whole idea of identifying with a plant at all strikes us as a little old-fashioned. Beef people sounds more like it, though nowadays chicken people, which sounds not nearly so good, is probably closer to the truth of the matter. But carbon 13 doesn't lie, and researchers who have compared the isotopes in the flesh or hair of North Americans to those in the same tissues of Mexicans report that it is now we in the North who are the true people of corn. "When you look at the isotope ratios," Todd Dawson, a Berkeley biologist who's done this sort of research, told me, "we North Americans look like corn chips with legs." Compared to us, Mexicans today consume a far more varied carbon diet: the animals they eat still eat grass (until recently, Mexicans regarded feeding corn to livestock as a sacrilege); much of their protein comes from legumes; and they still sweeten their beverages with cane sugar.

So that's us: processed corn, walking.

3. THE RISE OF ZEA MAYS

How this peculiar grass, native to Central America and unknown to the Old World before 1492, came to colonize so much of our land and bodies is one of the plant world's greatest success stories. I say the plant world's success story because it is no longer clear that corn's triumph is such a boon to the rest of the world, and because we should give credit where credit is due. Corn is the hero of its own story, and though we humans played a crucial supporting role in its rise to world domination, it would be wrong to suggest we have been calling the shots, or acting always in our own best interests. Indeed, there is every reason to believe that corn has succeeded in domesticating us.

To some extent this holds true for all of the plants and animals that take part in the grand coevolutionary bargain with humans we call agriculture. Though we insist on speaking of the "invention" of agriculture as if it were our idea, like double-entry bookkeeping or the light-bulb, in fact it makes just as much sense to regard agriculture as a brilliant (if unconscious) evolutionary strategy on the part of the plants and animals involved to get us to advance their interests. By evolving certain traits we happen to regard as desirable, these species got themselves noticed by the one mammal in a position not only to spread their genes around the world, but to remake vast swaths of that world in the image of the plants' preferred habitat. No other group of species gained more from its association with humans than the edible grasses, and no grass has reaped more from agriculture than *Zea mays*, today the world's most important cereal crop.

Corn's success might seem fated in retrospect, but it was not something anyone would have predicted on that day in May 1493 when Columbus first described the botanical oddity he had encountered in the New World to Isabella's court. He told of a towering grass with an ear as thick as a man's arm, to which grains were "affixed by nature in a wondrous manner and in form and size like garden peas, white when young." Wondrous, perhaps, yet this was, after all, the staple food of a people that would shortly be vanquished and all but exterminated.

By all rights, maize should have shared the fate of that other native species, the bison, which was despised and targeted for elimination precisely because it was "the Indians' commissary," in the words of General Philip Sheridan, commander of the armies of the West. Exterminate the species, Sheridan advised, and "[t]hen your prairies can be covered with speckled cattle and the festive cowboy." In outline Sheridan's plan was the plan for the whole continent: The white man brought his own "associate species" with him to the New World—cattle and apples, pigs and wheat, not to mention his accustomed weeds and microbes—and wherever possible helped them to displace the native plants and animals allied with the Indian. More even than the rifle, it was this biotic army that did the most to defeat the Indians.

But corn enjoyed certain botanical advantages that would allow it to thrive even as the Native Americans with whom it had coevolved were being eliminated. Indeed, maize, the one plant without which the American colonists probably would never have survived, let alone prospered, wound up abetting the destruction of the very people who had helped develop it. In the plant world at least, opportunism trumps gratitude. Yet in time, the plant of the vanquished would conquer even the conquerors.

Squanto taught the Pilgrims how to plant maize in the spring of

1621, and the colonists immediately recognized its value: No other plant could produce quite as much food quite as fast on a given patch of New World ground as this Indian corn. (Originally "corn" was a generic English word for any kind of grain, even a grain of salt—hence "corned beef"; it didn't take long for *Zea mays* to appropriate the word for itself, at least in America.) The fact that the plant was so well adapted to the climate and soils of North America gave it an edge over European grains, even if it did make a disappointingly earthbound bread. Centuries before the Pilgrims arrived the plant had already spread north from central Mexico, where it is thought to have originated, all the way to New England, where Indians were probably cultivating it by 1000. Along the way, the plant—whose prodigious genetic variability allows it to adapt rapidly to new conditions—made itself at home in virtually every microclimate in North America; hot or cold, dry or wet, sandy soil or heavy, short day or long, corn, with the help of its Native American allies, evolved whatever traits it needed to survive and flourish.

Lacking any such local experience, wheat struggled to adapt to the continent's harsh climate, and yields were often so poor that the settlements that stood by the old world staple often perished. Planted, a single corn seed yielded more than 150 fat kernels, often as many as 300, while the return on a seed of wheat, when all went well, was something less than 50:1. (At a time when land was abundant and labor scarce, agricultural yields were calculated on a per-seed-sown basis.)

Corn won over the wheat people because of its versatility, prized especially in new settlements far from civilization. This one plant supplied settlers with a ready-to-eat vegetable and a storable grain, a source of fiber and animal feed, a heating fuel and an intoxicant. Corn could be eaten fresh off the cob ("green") within months after planting, or dried on the stalk in fall, stored indefinitely, and ground into flour as needed. Mashed and fermented, corn could be brewed into beer or distilled into whiskey; for a time it was the only source of alcohol on the frontier. (Whiskey and pork were both regarded as "concentrated corn," the latter a concentrate of its protein, the former of its calories; both had the virtue of reducing corn's bulk and raising its price.) No part of the big grass went to waste: The husks could be woven into rugs and twine; the leaves and stalks made good silage for livestock; the shelled cobs were burned for heat and stacked by the privy as a rough substitute for toilet paper. (Hence the American slang term "corn hole.")

"Corn was the means that permitted successive waves of pioneers to settle new territories," writes Arturo Warman, a Mexican historian. "Once the settlers had fully grasped the secrets and potential of corn, they no longer needed the Native Americans." Squanto had handed the white man precisely the tool he needed to dispossess the Indian. Without the "fruitfulness" of Indian corn, the nineteenth-century English writer William Cobbett declared, the colonists would never have been able to build "a powerful nation." Maize, he wrote, was "the greatest blessing God ever gave to man."

Valuable as corn is as a means of subsistence, the kernel's qualities make it an excellent means of accumulation as well. After the crop has supplied its farmer's needs, he can go to market with any surplus, dried corn being the perfect commodity: easy to transport and virtually indestructible. Corn's dual identity, as food and commodity, has allowed many of the peasant communities that have embraced it to make the leap from a subsistence to a market economy. The dual identity also made corn indispensable to the slave trade: Corn was both the currency traders used to pay for slaves in Africa and the food upon which slaves subsisted during their passage to America. Corn is the protocapitalist plant.

4. MARRIED TO MAN

But while both the new and the native Americans were substantially dependent on corn, the plant's dependence on the Americans had become total. Had maize failed to find favor among the conquerors, it would have risked extinction, because without humans to plant it every spring, corn would have disappeared from the earth in a matter of a few years. The novel cob-and-husk arrangement that makes corn such a convenient grain for us renders the plant utterly dependent for its survival on an animal in possession of the opposable thumb needed to remove the husk, separate the seeds, and plant them.

Plant a whole corncob and watch what happens: If any of the kernels manage to germinate, and then work their way free of the smothering husk, they will invariably crowd themselves to death before their second set of leaves has emerged. More than most domesticated plants (a few of whose offspring will usually find a way to grow unassisted), corn completely threw its lot in with humanity when it evolved its peculiar husked ear. Several human societies have seen fit to worship corn, but perhaps it should be the other way around: For corn, we humans are the contingent beings. So far, this reckless-seeming act of evolutionary faith in us has been richly rewarded.

It is tempting to think of maize as a human artifact, since the plant is so closely linked to us and so strikingly different from any wild species. There are in fact no wild maize plants, and teosinte, the weedy grass from which corn is believed to have descended (the word is Nahuatl for "mother of corn"), has no ear, bears its handful of tiny naked seeds on a terminal rachis like most other grasses, and generally looks nothing whatsoever like maize. The current thinking among botanists is that several thousand years ago teosinte underwent an abrupt series of mutations that turned it into corn; geneticists calculate that changes on as few as four chromosomes could account for the main traits that distinguish teosinte from maize. Taken together, these mutations amounted to (in the words of botanist Hugh Iltis) a "catastrophic sexual transmutation": the transfer of the plant's female organs from the top of the grass to a monstrous sheathed ear in the middle of the stalk. The male organs stayed put, remaining in the tassel.

It is, for a grass, a bizarre arrangement with crucial implications: The ear's central location halfway down the stalk allows it to capture far more nutrients than it would up top, so suddenly producing hundreds of gigantic seeds becomes metabolically feasible. Yet because those seeds are now trapped in a tough husk, the plant has lost its ability to reproduce itself—hence the catastrophe in teosinte's sex change. A mutation this freakish and maladaptive would have swiftly brought the plant to an evolutionary dead end had one of these freaks not happened to catch the eye of a human somewhere in Central America who, looking for something to eat, peeled open the husk to free the seeds. What would have been an unheralded botanical catastrophe in a world without humans became an incalculable evolutionary boon. If you look hard enough, you can still find teosinte growing in certain Central American highlands; you can find maize, its mutant offspring, anywhere you find people.

5. CORN SEX

Maize is self-fertilized and wind-pollinated, botanical terms that don't begin to describe the beauty and wonder of corn sex. The tassel at the top of the plant houses the male organs, hundreds of pendant anthers that over the course of a few summer days release a superabundance of powdery yellow pollen: 14 million to 18 million grains per plant, 20,000 for every potential kernel. ("Better safe than sorry" or "more is more" being nature's general rule for male genes.) A meter or so below await the female organs, hundreds of minuscule flowers arranged in tidy rows along a tiny, sheathed cob that juts upward from the stalk at the crotch of a leaf midway between tassel and earth. That the male anthers resemble flowers and the female cob a phallus is not the only oddity in the sex life of corn.

Each of the four hundred to eight hundred flowers on a cob has the potential to develop into a kernel—but only if a grain of pollen can find its way to its ovary, a task complicated by the distance the pollen has to travel and the intervening husk in which the cob is tightly wrapped. To surmount this last problem, each flower sends out through the tip of the husk a single, sticky strand of silk (technically its "style") to snag its own grain of pollen. The silks emerge from the husk on the very day the tassel is set to shower its yellow dust.

What happens next is very strange. After a grain of pollen has fallen through the air and alighted on the moistened tip of silk, its nucleus divides in two, creating a pair of twins, each with the same set of genes but a completely different role to perform in the creation of the kernel. The first twin's job is to tunnel a microscopic tube down through the center of the silk thread. That accomplished, its clone slides down through the tunnel, past the husk, and into the waiting flower, a journey of between six and eight inches that takes several hours to complete. Upon arrival in the flower the second twin fuses with the egg to form the embryo—the germ of the future kernel. Then the first twin follows, entering the now fertilized flower, where it sets about forming the endosperm—the big, starchy part of the kernel. Every kernel of corn is the product of this intricate ménage à trois; the tiny, stunted kernels you often see at the narrow end of a cob are flowers whose silk no pollen grain ever penetrated. Within a day of conception, the now superfluous silk dries up, eventually turning reddish brown; fifty or so days later, the kernels are mature.*

The mechanics of corn sex, and in particular the great distance over open space corn pollen must travel to complete its mission, go a long way toward accounting for the success of maize's alliance with humankind. It's a simple matter for a human to get between a corn plant's pollen and its flower, and only a short step from there to deliberately crossing one corn plant with another with an eye to encouraging specific traits in the offspring. Long before scientists understood hybridization, Native Americans had discovered that by taking the pollen from the tassel of one corn plant and dusting it on the silks of another, they could create new plants that combined the traits of both parents. American Indians were the world's first plant breeders, developing literally thousands of distinct cultivars for every conceivable environment and use.

Looked at another way, corn was the first plant to involve humans so intimately in its sex life. For a species whose survival depends on how well it can gratify the ever shifting desires of its only sponsor, this has proved to be an excellent evolutionary strategy. More even than other domesticated species, many of which can withstand a period of human

---

*My account of the sex life of corn is drawn from Betty Fussell's *The Story of Corn* (1992) and Frederick Sargent's *Corn Plants* (1901).

neglect, it pays for corn to be obliging—and to be so quick about it. The usual way a domesticated species figures out what traits its human ally will reward is through the slow and wasteful process of Darwinian trial and error. Hybridization represents a far swifter and more efficient means of communication, or feedback loop, between plant and human; by allowing humans to arrange its marriages, corn can discover in a single generation precisely what qualities it needs to prosper.

It is by being so obliging that corn have won itself as much human attention and habitat as it has. The plant's unusual sexual arrangements, so amenable to human intervention, have allowed it to adapt to the very different worlds of Native Americans (and to their very different worlds, from southern Mexico to New England), of colonists and settlers and slaves, and of all the other corn-eating societies that have come and gone since the first human chanced upon that first teosinte freak.

But of all the human environments to which corn has successfully adapted since then, the adaptation to our own—the world of industrial consumer capitalism; the world, that is, of the supermarket and fast-food franchise—surely represents the plant's most extraordinary evolutionary achievement to date. For to prosper in the industrial food chain to the extent it has, corn had to acquire several improbable new tricks. It had to adapt itself not just to humans but to their machines, which it did by learning to grow as upright, stiff-stalked, and uniform as soldiers. It had to multiply its yield by an order of magnitude, which it did by learning to grow shoulder to shoulder with other corn plants, as many as thirty thousand to the acre. It had to develop an appetite for fossil fuel (in the form of petrochemical fertilizer) and a tolerance for various synthetic chemicals. But even before it could master these tricks and make a place for itself in the bright sunshine of capitalism, corn first had to turn itself into something never before seen in the plant world: a form of intellectual property.

The free corn sex I've described allowed people to do virtually anything they wanted with the genetics of corn except own them—a big problem for a would-be capitalist plant. If I crossed two corn plants to create a variety with an especially desirable trait, I could sell you my special seeds, but only once, since the corn you grew from my special seeds would produce lots more special seeds, for free and forever, putting me out business in short order. It's difficult to control the means of production when the product you're selling can reproduce itself endlessly. This is one of the ways in which the imperatives of biology are difficult to mesh with the imperatives of business.

Difficult, but not impossible. Early in the twentieth century American corn breeders figured out how to bring corn reproduction under firm control and to protect the seed from copiers. The breeders discovered that when they crossed two corn plants that had come from inbred lines—from ancestors that had themselves been exclusively self-pollinated for several generations—the hybrid offspring displayed some highly unusual characteristics. First, all the seeds in that first generation (F-1, in the plant breeder's vocabulary) produced genetically identical plants—a trait that, among other things, facilitates mechanization. Second, those plants exhibited heterosis, or hybrid vigor— better yields than either of their parents. But most important of all, they found that the seeds produced by these seeds did not "come true"—the plants in the second (F-2) generation bore little resemblance to the plants in the first. Specifically, their yields plummeted by as much as a third, making their seeds virtually worthless.

Hybrid corn now offered its breeders what no other plant at that time could: the biological equivalent of a patent. Farmers now had to buy new seeds every spring; instead of depending upon their plants to reproduce themselves, they now depended on a corporation. The corporation, assured for the first time of a return on its investment in breeding, showered corn with attention—R&D, promotion, advertising—and the plant responded, multiplying its fruitfulness year after year. With the advent of the F-1 hybrid, a technology with the power to remake nature in the image of capitalism, Zea mays entered the industrial age and, in time, it brought the whole American food chain with it.

TWO

THE FARM

1. ONE FARMER, 129 EATERS

To take the wheel of a clattering 1975 International Harvester tractor, pulling a spidery eight-row planter through an Iowa cornfield during the first week of May, is like trying to steer a boat through a softly rolling sea of dark chocolate. The hard part is keeping the thing on a straight line, that and hearing the shouted instructions of the farmer sitting next to you when you both have wads of Kleenex jammed into your ears to muffle the diesel roar. Driving a boat, you try to follow the compass heading or aim for a landmark on shore; planting corn, you try to follow the groove in the soil laid down on the previous pass by a rolling disk at the end of a steel arm attached to the planter behind us. Deviate from the line and your corn rows will wobble, overlapping or drifting away from one another. Either way, it'll earn you a measure of neighborly derision and hurt your yield. And yield, measured in bushels per acre, is the measure of all things here in corn country.

The tractor I was driving belonged to George Naylor, who bought it new back in the midseventies, when, as a twenty-seven-year-old, he returned to Greene County, Iowa, to farm his family's 470 acres. Naylor is a big man with a moon face and a scraggly gray beard. On the phone his gravelly voice and incontrovertible pronouncements ("That is just the biggest bunch of bullshit! Only the *New York Times* would be dumb enough to believe the Farm Bureau still speaks for American farmers!") led me to expect someone considerably more ornery than the shy fellow who climbed down from his tractor cab to greet me in the middle of a field in the middle of a slate-gray day threatening rain. Naylor had on the farmer's standard-issue baseball cap, a yellow chamois shirt, and overalls—the stripy blue kind favored by railroad workers, about as un-intimidating an article of clothing as has ever been donned by a man. My first impression was more shambling Gentle Ben than fiery prairie populist, but I would discover that Naylor can be either fellow, the mere mention of "Cargill" or "Earl Butz" supplying the transformational trigger.

This part of Iowa has some of the richest soil in the world, a layer of cakey alluvial loam nearly two feet thick. The initial deposit was made by the retreat of the Wisconsin glacier ten thousand years ago, and then compounded at the rate of another inch or two every decade by prairie grasses—big bluestem, foxtail, needlegrass, and switchgrass. Tall-grass prairie is what this land was until the middle of the nine-teenth century, when the sod was first broken by the settler's plow. George's grandfather moved his family to Iowa from Derbyshire, Eng-land, in the 1880s, a coal miner hoping to improve his lot in life. The sight of such soil, pushing up and then curling back down behind the blade of his plow like a thick black wake behind a ship, must have stoked his confidence, and justifiably so: It's gorgeous stuff, black gold as deep as you can dig, as far as you can see. What you can't see is all the soil that's no longer here, having been blown or washed away since the sod was broken; the two-foot crust of topsoil here probably started out closer to four.

The story of the Naylor farm since 1919, when George's grandfather bought it, closely tracks the twentieth-century story of American agriculture, its achievements as well as its disasters. It begins with a farmer supporting a family on a dozen different species of plants and animals. There would have been a fair amount of corn then too, but also fruits and other vegetables, as well as oats, hay, and alfalfa to feed the pigs, cattle, chickens, and horses—horses being the tractors of that time. One of every four Americans lived on a farm when Naylor's grandfather arrived here in Churdan; his land and labor supplied enough food to feed his family and twelve other Americans besides. Less than a century after, fewer than 2 million Americans still farm—and they grow enough to feed the rest of us. What that means is that Naylor's grandson, raising nothing but corn and soybeans on a fairly typical Iowa farm, is so astoundingly productive that he is, in effect, feeding some 129 Americans. Measured in terms of output per worker, American farmers like Naylor are the most productive humans who have ever lived.

Yet George Naylor is all but going broke—and he's doing better than many of his neighbors. (Partly because he's still driving that 1975 tractor.) For though this farm might feed 129, it can no longer support the four who live on it: The Naylor farm survives by the grace of Peggy Naylor's paycheck (she works for a social services agency in Jefferson) and an annual subsidy payment from Washington, D.C. Nor can the Naylor farm literally feed the Naylor family, as it did in grandfather Naylor's day. George's crops are basically inedible—they're commodities that must be processed or fed to livestock before they can feed people. Water, water, everywhere and not a drop to drink: Like most of Iowa, which now imports 80 percent of its food, George's farm (apart from his garden, his laying hens, and his fruit trees) is basically a food desert.

The 129 people who depend on George Naylor for their sustenance are all strangers, living at the far end of a food chain so long, intricate, and obscure that neither producer nor consumer has any reason to know the first thing about the other. Ask one of those eaters where their steak or soda comes from and she'll tell you "the supermarket." Ask George Naylor whom he's growing all that corn for and he'll tell you "the military-industrial complex." Both are partly right.

I came to George Naylor's farm as an unelected representative of the Group of 129, curious to learn whom, and what, I'd find at the far end of the food chain that keeps me alive. There's no way of knowing whether George Naylor is literally growing the corn that feeds the steer that becomes my steak, or that sweetened my son's soft drink, or that supplied the dozen or so corn-derived ingredients from which his chicken nugget is constructed. But given the complexly ramifying fate of a bushel of commodity corn, the countless forking paths followed by its ninety thousand kernels as they're dispersed across the nation's sprawling food system, the odds are good that at least one of the kernels grown on the Naylor farm has, like the proverbial atom from Caesar's dying breath, made its way to me. And if not me, then certainly you. This Iowa cornfield (and all the others just like it) is the place most of our food comes from.

2. PLANTING THE CITY OF CORN

The day I showed up was supposed to be the only dry one all week, so George and I spent most of it in the cab of his tractor, trying to get acquainted and get his last 160 acres of corn planted at the same time; a week or two later he'd start in on the soybeans. The two crops take turns in these fields year after year, in what has been the classic Corn Belt rotation since the 1970s. (Since that time soybeans have become the second leg supporting the industrial food system: It too is fed to livestock and now finds its way into two-thirds of all processed foods.) For most of the afternoon I sat on a rough cushion George had fashioned for me from crumpled seed bags, but after a while he let me take the wheel.

Back and forth and back again, a half a mile in each direction, planting corn feels less like planting, or even driving, than stitching an interminable cloak, or covering a page with the same sentence over and over again. The monotony, compounded by the roar of a diesel engine well past its prime, is hypnotic after a while. Every pass across this field, which is almost but not quite dead flat, represents another acre of corn planted, another thirty thousand seeds tucked into one of the eight furrows being simultaneously etched into the soil by pairs of stainless steel disks; a trailing roller then closes the furrows over the seed.

The seed we were planting was Pioneer Hi-Bred's 34H31, a strain that the catalog described as "an adaptable hybrid with solid agronomics and yield potential." The lack of hype, notable for a seed catalog, probably reflects the fact that 34H31 does not contain the "YieldGard gene," the Monsanto-developed line of genetically engineered corn that Pioneer is currently pushing: The genetically modified 34B98, on the same page, promises "outstanding yield potential." Despite the promises, Naylor, unlike many of his neighbors, doesn't plant GMOs (genetically modified organisms). He has a gut distrust of the technology ("They're messing with three billion years of evolution") and doesn't think it's worth the extra twenty-five dollars a bag (in technology fees) they cost. "Sure, you might get a yield bump, but whatever you make on the extra corn goes right back to cover the premium for the seed. I fail to see why I should be laundering money for Monsanto." As Naylor sees it, GMO seed is just the latest chapter in an old story: Farmers eager to increase their yields adopt the latest innovation, only to find that it's the companies selling the innovations who reap the most from the gain in the farmer's productivity.

Even without the addition of transgenes for traits like insect resistance, the standard F-1 hybrids Naylor plants are technological marvels, capable of coaxing 180 bushels of corn from an acre of Iowa soil. One bushel holds 56 pounds of kernels, so that's slightly more than ten thousand pounds of food per acre; the field George and I planted that day would produce 1.8 million pounds of corn. *Not bad for a day's work sitting down,* I thought to myself that afternoon, though of course there'd be several more days of work between now and the harvest in October.

One way to tell the story of this farm is by following the steady upward arc in the yield of corn. Naylor has no idea how many bushels of corn per acre his grandfather could produce, but the average back in

1920 was about twenty bushels per acre—roughly the same yields historically realized by Native Americans. Corn then was planted in widely spaced bunches in a checkerboard pattern so farmers could easily cultivate between the stands in either direction. Hybrid seed came on the market in the late the 1930s, when his father was farming. "You heard stories," George shouted over the din of the tractor. "How they talked him into raising an acre or two of the new hybrid, and by god when the old corn fell over, the hybrid stood straight up. Doubled Dad's yields, till he was getting seventy to eighty an acre in the fifties." George has doubled that yet again, some years getting as much as two hundred bushels of corn per acre. The only other domesticated species ever to have multiplied its productivity by such a factor is the Holstein cow.

"High yield" is a fairly abstract concept, and I wondered what it meant at the level of the plant: more cobs per stalk? more kernels per cob? Neither of the above, Naylor explained. The higher yield of modern hybrids stems mainly from the fact that they can be planted so close together, thirty thousand to the acre instead of eight thousand in his father's day. Planting the old open-pollinated (nonhybrid) varieties so densely would result in stalks grown spindly as they jostled each other for sunlight; eventually the plants would topple in the wind. Hybrids have been bred for thicker stalks and stronger root systems, the better to stand upright in a crowd and withstand mechanical harvesting. Basically, modern hybrids can tolerate the corn equivalent of city life, growing amid the multitudes without succumbing to urban stress.

You would think that competition among individuals would threaten the tranquility of such a crowded metropolis, yet the modern field of corn forms a most orderly mob. This is because every plant in it, being an F-1 hybrid, is genetically identical to every other. Since no individual plant has inherited any competitive edge over any other, precious resources like sunlight, water, and soil nutrients are shared equitably. There are no alpha corn plants to hog the light or fertilizer. The true socialist utopia turns out to be a field of F-1 hybrid plants.

Iowa begins to look a little different when you think of its sprawling fields as cities of corn, the land, in its own way, settled as densely as

Manhattan for the very same purpose: to maximize real estate values. There may be little pavement out here, but this is no middle landscape. Though by any reasonable definition Iowa is a rural state, it is more thoroughly developed than many cities: A mere 2 percent of the state's land remains what it used to be (tall-grass prairie), every square foot of the rest having been completely remade by man. The only thing missing from this man-made landscape is . . . man.

3. VANISHING SPECIES

A case can be made that the corn plant's population explosion in places like Iowa is responsible for pushing out not only other plants but the animals and then finally the people, too. When Naylor's grandfather arrived in America the population of Greene County was near its peak: 16,467 people. In the most recent census it had fallen to 10,366. There are many reasons for the depopulation of the American Farm Belt, but the triumph of corn deserves a large share of the blame—or the credit, depending on your point of view.

When George Naylor's grandfather was farming, the typical Iowa farm was home to whole families of different plant and animal species, corn being only the fourth most common. Horses were the first, because every farm needed working animals (there were only 225 tractors in all of America in 1920), followed by cattle, chickens, and then corn. After corn came hogs, apples, hay, oats, potatoes, and cherries; many Iowa farms also grew wheat, plums, grapes, and pears. This diversity allowed the farm not only to substantially feed itself—and by that I don't mean feed only the farmers, but also the soil and the livestock—but to withstand a collapse in the market for any one of those crops. It also produced a completely different landscape than the Iowa of today.

"You had fences everywhere," George recalled, "and of course pastures. Everyone had livestock, so large parts of the farm would be green most of the year. The ground never used to be this bare this long." For much of the year, from the October harvest to the emergence of the corn in mid-May, Greene County is black now, a great tarmac only slightly more hospitable to wildlife than asphalt. Even in May the only green you see are the moats of lawn surrounding the houses, the narrow strips of grass dividing one farm from another, and the roadside ditches. The fences were pulled up when the animals left, beginning in the fifties and sixties, or when they moved indoors, as Iowa's hogs have more recently done; hogs now spend their lives in aluminum sheds perched atop manure pits. Greene County in the spring has become a monotonous landscape, vast plowed fields relieved only by a dwindling number of farmsteads, increasingly lonesome islands of white wood and green grass marooned in a sea of black. Without the fences and hedgerows to slow it down, Naylor says, the winds blow more fiercely in Iowa today than they once did.

Corn isn't solely responsible for remaking this landscape: It was the tractor, after all, that put the horses out of work, and with the horses went the fields of oats and some of the pasture. But corn was the crop that put cash in the farmer's pocket, so as corn yields began to soar at midcentury, the temptation was to give the miracle crop more and more land. Of course, every other farmer in America was thinking the same way (having been encouraged to do so by government policies), with the inevitable result that the price of corn declined. One might think falling corn prices would lead farmers to plant less of it, but the economics and psychology of agriculture are such that exactly the opposite happened.

Beginning in the fifties and sixties, the flood tide of cheap corn made it profitable to fatten cattle on feedlots instead of on grass, and to raise chickens in giant factories rather than in farmyards. Iowa livestock farmers couldn't compete with the factory-farmed animals their own cheap corn had helped spawn, so the chickens and cattle disappeared from the farm, and with them the pastures and hay fields and fences. In their place the farmers planted more of the one crop they could grow more of than anything else: corn. And whenever the price of corn slipped they planted a little more of it, to cover expenses and stay even. By the 1980s the diversified family farm was history in Iowa, and corn was king.

(Planting corn on the same ground year after year brought down the predictable plagues of insects and disease, so beginning in the 1970s Iowa farmers started alternating corn with soybeans, a legume. Recently, though, bean prices having fallen and bean diseases having risen, some farmers are going back to a risky rotation of "corn on corn.")

With the help of its human and botanical allies (i.e., farm policy and soybeans), corn had pushed the animals and their feed crops off the land, and steadily expanded into their paddocks and pastures and fields. Now it proceeded to push out the people. For the radically simplified farm of corn and soybeans doesn't require nearly as much human labor as the old diversified farm, especially when the farmer can call on sixteen-row planters and chemical weed killers. One man can handle a lot more acreage by himself when it's planted in monoculture, and without animals to care for he can take the weekend off, and even think about spending the winter in Florida.

"Growing corn is just riding tractors and spraying," Naylor told me; the number of riding and spraying days it takes to raise five hundred acres of industrial corn can probably be counted in weeks. So the farms got bigger, and eventually the people, whom the steadily falling price of corn could no longer support anyway, went elsewhere, ceding the field to the monstrous grass.

Today Churdan is virtually a ghost town, much of its main street shuttered. The barbershop, a food market, and the local movie theater have all closed in recent years; there's a café and one sparsely stocked little market somehow still hanging on, but most people drive the ten miles to Jefferson to buy their groceries or pick up milk and eggs when they're getting gas at the Kum & Go. The middle school can no longer field a baseball team or put together a band, it has so few students left, and it takes four local high schools to field a single football team: the Jefferson-Scranton-Paton-Churdan Rams. Just about the only going concern left standing in Churdan is the grain elevator, rising at the far end of town like a windowless concrete skyscraper. It endures because, people or no people, the corn keeps coming, more of it every year.

4. THERE GOES THE SUN

I've oversimplified the story a bit; corn's rapid rise is not quite as self-propelled as I've made it sound. As in so many other "self-made" American successes, the closer you look the more you find the federal government lending a hand—a patent, a monopoly, a tax break—to our hero at a critical juncture. In the case of corn, the botanical hero I've depicted as plucky and ambitious was in fact subsidized in crucial ways, both economically and biologically. There's a good reason I met farmers in Iowa who don't respect corn, who will tell you in disgust that the plant has become "a welfare queen."

The great turning point in the modern history of corn, which in turn marks a key turning point in the industrialization of our food, can be dated with some precision to the day in 1947 when the huge munitions plant at Muscle Shoals, Alabama, switched over to making chemical fertilizer. After the war the government had found itself with a tremendous surplus of ammonium nitrate, the principal ingredient in the making of explosives. Ammonium nitrate also happens to be an excellent source of nitrogen for plants. Serious thought was given to spraying America's forests with the surplus chemical, to help out the timber industry. But agronomists in the Department of Agriculture had a better idea: Spread the ammonium nitrate on farmland as fertilizer. The chemical fertilizer industry (along with that of pesticides, which are based on poison gases developed for the war) is the product of the government's effort to convert its war machine to peacetime purposes. As the Indian farmer activist Vandana Shiva says in her speeches, "We're still eating the leftovers of World War II."

Hybrid corn turned out to be the greatest beneficiary of this conversion. Hybrid corn is the greediest of plants, consuming more fertilizer than any other crop. For though the new hybrids had the genes to survive in teeming cities of corn, the richest acre of Iowa soil could never have fed thirty thousand hungry corn plants without promptly bankrupting its fertility. To keep their land from getting "corn sick" farmers in Naylor's father's day would carefully rotate their crops with legumes (which add nitrogen to the soil), never growing corn more than twice in the same field every five years; they would also recycle nutrients by spreading their cornfields with manure from their livestock. Before synthetic fertilizers the amount of nitrogen in the soil strictly limited the amount of corn an acre of land could support. Though hybrids were introduced in the thirties, it wasn't until they made the acquaintance of chemical fertilizers in the 1950s that corn yields exploded.

The discovery of synthetic nitrogen changed everything—not just for the corn plant and the farm, not just for the food system, but also for the way life on earth is conducted. All life depends on nitrogen; it is the building block from which nature assembles amino acids, proteins, and nucleic acids; the genetic information that orders and perpetuates life is written in nitrogen ink. (This is why scientists speak of nitrogen as supplying life's quality, while carbon provides the quantity.) But the supply of usable nitrogen on earth is limited. Although earth's atmosphere is about 80 percent nitrogen, all those atoms are tightly paired, nonreactive, and therefore useless; the nineteenth-century chemist Justus von Liebig spoke of atmospheric nitrogen's "indifference to all other substances." To be of any value to plants and animals, these self-involved nitrogen atoms must be split and then joined to atoms of hydrogen. Chemists call this process of taking atoms from the atmosphere and combining them into molecules useful to living things "fixing" that element. Until a German Jewish chemist named Fritz Haber figured out how to turn this trick in 1909, all the usable nitrogen on earth had at one time been fixed by soil bacteria living on the roots of leguminous plants (such as peas or alfalfa or locust trees) or, less commonly, by the shock of electrical lightning, which can break nitrogen bonds in the air, releasing a light rain of fertility.

Vaclav Smil, a geographer who has written a fascinating book about Fritz Haber called *Enriching the Earth*, pointed out that "there is no way to grow crops and human bodies without nitrogen." Before Fritz Haber's invention the sheer amount of life earth could support—the size of crops and therefore the number of human bodies—was limited by the amount of nitrogen that bacteria and lightning could fix. By 1900, European scientists recognized that unless a way was found to augment this naturally occurring nitrogen, the growth of the human population would soon grind to a very painful halt. The same recognition by Chinese scientists a few decades later is probably what compelled China's opening to the West: After Nixon's 1972 trip the first major order the Chinese government placed was for thirteen massive fertilizer factories. Without them, China would probably have starved.

This is why it may not be hyperbole to claim, as Smil does, that the Haber-Bosch process (Carl Bosch gets the credit for commercializing Haber's idea) for fixing nitrogen is the most important invention of the twentieth century. He estimates that two of every five humans on earth today would not be alive if not for Fritz Haber's invention. We can easily imagine a world without computers or electricity, Smil points out, but without synthetic fertilizer billions of people would never have been born. Though, as these numbers suggest, humans may have struck something of a Faustian bargain with nature when Fritz Haber gave us the power to fix nitrogen.

Fritz Haber? No, I'd never heard of him either, even though he was awarded the Nobel Prize in 1920 for "improving the standards of agriculture and the well-being of mankind." But the reason for his obscurity has less to do with the importance of his work than the ugly twist of his biography, which recalls the dubious links between modern warfare and industrial agriculture. During World War I, Haber threw himself into the German war effort, and his chemistry kept alive Germany's hopes for victory. After Britain choked off Germany's supply of nitrates from Chilean mines, an essential ingredient in the manufacture of explosives, Haber's technology allowed Germany to continue making bombs from synthetic nitrate. Later, as the war became mired in the trenches of France, Haber put his genius for chemistry to work developing poison gases—ammonia, then chlorine. (He subsequently developed Zyklon B, the gas used in Hitler's concentration camps.) On April 22, 1915, Smil writes, Haber was "on the front lines directing the first gas attack in military history." His "triumphant" return to Berlin was ruined a few days later when his wife, a fellow chemist sickened by her husband's contribution to the war effort, used Haber's army pistol to kill herself. Though Haber later converted to Christianity, his Jewish background forced him to flee Nazi Germany in the thirties; he died, broken, in a Basel hotel room in 1934. Perhaps because the history of science gets written by the victors, Fritz Haber's story has been all but written out of the twentieth century. Not even a plaque marks the site of his great discovery at the University of Karlsruhe.

Haber's story embodies the paradoxes of science: the double edge to our manipulations of nature, the good and evil that can flow not only from the same man but the same knowledge. Haber brought a vital new source of fertility and an awful new weapon into the world; as his biographer wrote, "[I]t's the same science and the same man doing both." Yet this dualism dividing the benefactor of agriculture from the chemical weapons maker is far too pat, for even Haber's benefaction has proven decidedly to be a mixed blessing.

When humankind acquired the power to fix nitrogen, the basis of soil fertility shifted from a total reliance on the energy of the sun to a new reliance on fossil fuel. For the Haber-Bosch process works by combining nitrogen and hydrogen gases under immense heat and pressure in the presence of a catalyst. The heat and pressure are supplied by prodigious amounts of electricity, and the hydrogen is supplied by oil, coal, or, most commonly today, natural gas—fossil fuels. True, these fossil fuels were at one time billions of years ago created by the sun, but they are not renewable in the same way that the fertility created by a legume nourished by sunlight is. (That nitrogen is actually fixed by a bacterium living on the roots of the legume, which trades a tiny drip of sugar for the nitrogen the plant needs.)

On the day in the 1950s that George Naylor's father spread his first load of ammonium nitrate fertilizer, the ecology of his farm underwent a quiet revolution. What had been a local, sun-driven cycle of fertility, in which the legumes fed the corn which fed the livestock which in turn (with their manure) fed the corn, was now broken. Now he could plant corn every year and on as much of his acreage as he chose, since he had no need for the legumes or the animal manure. He could buy fertility in a bag, fertility that had originally been produced a billion years ago halfway around the world.

Liberated from the old biological constraints, the farm could now be managed on industrial principles, as a factory transforming inputs of raw material—chemical fertilizer—into outputs of corn. Since the farm no longer needs to generate and conserve its own fertility by maintaining a diversity of species, synthetic fertilizer opens the way to monoculture, allowing the farmer to bring the factory's economies of scale and mechanical efficiency to nature. If, as has sometimes been said, the discovery of agriculture represented the first fall of man from the state of nature, then the discovery of synthetic fertility is surely a second precipitous fall. Fixing nitrogen allowed the food chain to turn from the logic of biology and embrace the logic of industry. Instead of eating exclusively from the sun, humanity now began to sip petroleum.

Corn adapted brilliantly to the new industrial regime, consuming prodigious quantities of fossil fuel energy and turning out ever more prodigious quantities of food energy. More than half of all the synthetic nitrogen made today is applied to corn, whose hybrid strains can make better use of it than any other plant. Growing corn, which from a biological perspective had always been a process of capturing sunlight to turn it into food, has in no small measure become a process of converting fossil fuels into food. This shift explains the color of the land: The reason Greene County is no longer green for half the year is because the farmer who can buy synthetic fertility no longer needs cover crops to capture a whole year's worth of sunlight; he has plugged himself into a new source of energy. When you add together the natural gas in the fertilizer to the fossil fuels it takes to make the pesticides, drive the tractors, and harvest, dry, and transport the corn, you find that every bushel of industrial corn requires the equivalent of between a quarter and a third of a gallon of oil to grow it—or around fifty gallons of oil per acre of corn. (Some estimates are much higher.) Put another way, it takes more than a calorie of fossil fuel energy to produce a calorie of food; before the advent of chemical fertilizer the Naylor farm produced more than two calories of food energy for every calorie of energy invested. From the standpoint of industrial efficiency, it's too bad we can't simply drink the petroleum directly.

Ecologically this is a fabulously expensive way to produce food—but "ecologically" is no longer the operative standard. As long as fossil fuel energy is so cheap and available, it makes good economic sense to produce corn this way. The old way of growing corn—using fertility drawn from the sun—may have been the biological equivalent of a free lunch, but the service was much slower and the portions were much skimpier. In the factory time is money, and yield is everything.

One problem with factories, as compared to biological systems, is that they tend to pollute. Hungry for fossil fuel as hybrid corn is, farmers still feed it far more than it can possibly eat, wasting most of the fertilizer they buy. Maybe it's applied at the wrong time of year; maybe it runs off the fields in the rain; maybe the farmer puts down extra just to play it safe. "They say you only need a hundred pounds per acre. I don't know. I'm putting on up to two hundred. You don't want to err on the side of too little," Naylor explained to me, a bit sheepishly. "It's a form of yield insurance."

But what happens to the one hundred pounds of synthetic nitrogen that Naylor's corn plants don't take up? Some of it evaporates into the air, where it acidifies the rain and contributes to global warming. (Ammonium nitrate is transformed into nitrous oxide, an important greenhouse gas.) Some seeps down to the water table. When I went to pour myself a glass of water in the Naylors' kitchen, Peggy made sure I drew it from a special faucet connected to a reverse-osmosis filtration system in the basement. As for the rest of the excess nitrogen, the spring rains wash it off Naylor's fields, carrying it into drainage ditches that eventually spill into the Raccoon River. From there it flows into the Des Moines River, down to the city of Des Moines—which drinks from the Des Moines River. In spring, when nitrogen runoff is at its heaviest, the city issues "blue baby alerts," warning parents it's unsafe to give children water from the tap. The nitrates in the water convert to nitrite, which binds to hemoglobin, compromising the blood's ability to carry oxygen to the brain. So I guess I was wrong to suggest we don't sip fossil fuels directly; sometimes we do.

It has been less than a century since Fritz Haber's invention, yet already it has changed the earth's ecology. More than half of the world's supply of usable nitrogen is now man-made. (Unless you grew up on organic food, most of the kilo or so of nitrogen in your body was fixed by the Haber-Bosch process.) "We have perturbed the global nitrogen cycle," Smil wrote, "more than any other, even carbon." The effects may be harder to predict than the effects of the global warming caused by our disturbance of the carbon cycle, but they may be no less momentous. The flood of synthetic nitrogen has fertilized not just the farm fields but the forests and the oceans too, to the benefit of some species (corn and algae being two of the biggest beneficiaries), and to the detriment of countless others. The ultimate fate of the nitrates that George Naylor spreads on his cornfield in Iowa is to flow down the Mississippi into the Gulf of Mexico, where their deadly fertility poisons the marine ecosystem. The nitrogen tide stimulates the wild growth of algae, and the algae smother the fish, creating a "hypoxic," or dead, zone as big as the state of New Jersey—and still growing. By fertilizing the world, we alter the planet's composition of species and shrink its biodiversity.

5. A PLAGUE OF CHEAP CORN

The day after George Naylor and I finished planting his corn, the rains came, so we spent most of it around his kitchen table, drinking coffee and talking about what farmers always talk about: lousy commodity prices; benighted farm policies; making ends meet in a dysfunctional farm economy. Naylor came back to the farm in what would turn out to be the good old days in American agriculture: Corn prices were at an all-time high, and it looked as though it might actually be possible to make a living growing it. But by the time Naylor was ready to take his first crop to the elevator, the price of a bushel of corn had dropped from three dollars to two dollars, the result of a bumper crop. So he held his corn off the market, storing it in the hope that the price would rebound. But the price kept falling all through that winter and into the following spring and, if you factor in inflation, it has pretty much been falling ever since. These days the price of a bushel of corn is about a dollar beneath the true cost of growing it, a boon for everyone but the corn farmer. What I was hoping George Naylor could help me understand is, if there's so much corn being grown in America today that the market won't pay the cost of producing it, then why would any farmer in his right mind plant another acre of it?

The answer is complicated, as I would learn, but it has something to do with the perverse economics of agriculture, which would seem to defy the classical laws of supply and demand; a little to do with the psychology of farmers; and everything to do with farm policies, which underwent a revolution right around the time George Naylor was buying his first tractor. Government farm programs once designed to limit production and support prices (and therefore farmers) were quietly rejiggered to increase production and drive down prices. Put another way, instead of supporting farmers, during the Nixon administration the government began supporting corn at the expense of farmers. Corn, already the recipient of a biological subsidy in the form of synthetic nitrogen, would now receive an economic subsidy too, ensuring its final triumph over the land and the food system.

NAYLOR'S PERSPECTIVE on farm policy was shaped by a story his dad used to tell him. It takes place during the winter of 1933, in the depths of the farm depression. "That's when my father hauled corn to town and found out that the price of corn had been ten cents a bushel the day before, but on that day the elevator wasn't even buying." The price of corn had fallen to zero. "Tears always came to his eyes when he recounted the neighbors losing their farms in the 1920s and '30s," Naylor told me. America's farm policy was forged during the Depression not, as many people seem to think, to encourage farmers to produce more food for a hungry nation, but to rescue farmers from the disastrous effects of growing too much food—far more than Americans could afford to buy.

For as long as people have been farming, fat years have posed almost as stiff a challenge as lean, since crop surpluses collapse prices and bankrupt farmers who will be needed again when the inevitable lean years return. When it comes to food, nature can make a mockery of the classical economics of supply and demand—nature in the form of good or bad weather, of course, but also the nature of the human body, which can consume only so much food no matter how plentiful the supply. So, going back to the Old Testament, communities have devised various strategies to even out the destructive swings of agricultural production. The Bible's recommended farm policy was to establish a grain reserve. Not only did this ensure that when drought or pestilence ruined a harvest there'd still be food to eat, but it kept farmers whole by taking food off the market when the harvest was bountiful.

This is more or less what New Deal farm programs attempted to do. For storable commodities such as corn, the government established a target price based on the cost of production, and whenever the market price dropped below that target, the farmer was given a choice. Instead of dumping corn onto a weak market (thereby weakening it further), the farmer could take out a loan from the government—using his crop as collateral—that allowed him to store his grain until prices recovered. At that point, he sold the corn and paid back the loan; if corn prices stayed low, he could elect to keep the money he'd borrowed and, in repayment, give the government his corn, which would then go into something that came to be called, rather quaintly, the "Ever-Normal Granary." Other New Deal programs, such as those administered by the Soil Conservation Service, sought to avert overproduction (and soil erosion) by encouraging farmers to idle their most environmentally sensitive land.

The system, which remained in place more or less until shortly before George Naylor came back to the farm in the 1970s, did a fairly good job of keeping corn prices from collapsing in the face of the twentieth century's rapid gains in yield. Surpluses were held off the market by the offer of these "nonrecourse loans," which cost the government relatively little, since most of the loans were eventually repaid. And when prices climbed, as a result of bad weather, say, the government sold corn from its granary, which helped both to pay for the farm programs and smooth out the inevitable swings in price.

I say this system remained in place "more or less" until the 1970s because, beginning in the 1950s, a campaign to dismantle the New Deal farm programs took root, and with every new farm bill since then another strut was removed from the structure of support. Almost from the start, the policy of supporting prices and limiting production had collected powerful enemies: exponents of laissez-faire economics, who didn't see why farming should be treated differently than any other economic sector; food processors and grain exporters, who profited from overproduction and low crop prices; and a coalition of political and business leaders who for various reasons thought America had far too many farmers for her (or at least their) own good.

America's farmers had long been making political trouble for Wall Street and Washington; in the words of historian Walter Karp, "since the Civil War at least, the most unruly, the most independent, the most republican of American citizens have been the small farmers." Beginning with the populist revolt of the 1890s, farmers had made common cause with the labor movement, working together to check the power of corporations. Rising agricultural productivity handed a golden opportunity to the farmers' traditional adversaries. Since a smaller number of farmers could now feed America, the moment had come to "rationalize" agriculture by letting the market force prices down and farmers off the land. So Wall Street and Washington sought changes in farm policies that would loose "a plague of cheap corn" (in the words of George Naylor, a man very much in the old rural-populist mold) on the nation, the effects of which are all around us—indeed, in us.

6. THE SAGE OF PURDUE

Earl "Rusty" Butz, Richard Nixon's second secretary of agriculture, probably did more than any other single individual to orchestrate George Naylor's plague of cheap corn. In every newspaper article about him, and there were scores, the name of Earl Butz, a blustering, highly quotable agricultural economist from Purdue University, is invariably accompanied by the epithet "colorful." Butz's plainspoken manner and barnyard humor persuaded many people he must be a friend to the farmer, but his presence on the board of Ralston Purina probably offered a more reliable guide to his sympathies. Though chiefly remembered outside agriculture for the racist joke that cost him his job during the 1976 election, Butz revolutionized American agriculture, helping to shift the food chain onto a foundation of cheap corn.

Butz took over the Department of Agriculture during the last period in American history that food prices climbed high enough to generate real political heat; his legacy would be to make sure that never happened again. In the fall of 1972 Russia, having suffered a series of disastrous harvests, purchased 30 million tons of American grain. Butz had helped arrange the sale, in the hopes of giving a boost to crop prices in order to bring restive farmers tempted to vote for George McGovern into the Republican fold. The plan worked all too well: The unexpected surge in demand, coinciding with a spell of bad weather in the Farm Belt, drove grain prices to historic heights. These were the corn prices that persuaded George Naylor he could make a go of it on his family's farm.

The 1972 Russian grain sale and the resulting spike in farm income that fall helped Nixon nail down the farm vote for his reelection, but by the following year those prices had reverberated through the food chain, all the way to the supermarket. By 1973 the inflation rate for groceries reached an all-time high, and housewives were organizing protests at supermarkets. Farmers were killing chicks because they couldn't afford to buy feed, and the price of beef was slipping beyond the reach of middle-class consumers. Some foods became scarce; horse meat began showing up in certain markets. "Why a Food Scare in a Land of Plenty?" was a headline in *U.S. News and World Report* that summer. Nixon had a consumer revolt on his hands, and he dispatched Earl Butz to quell it. The Sage of Purdue set to work reengineering the American food system, driving down prices and vastly increasing the output of American farmers. What had long been the dream of agribusiness (cheaper raw materials) and the political establishment (fewer restive farmers) now became official government policy.

Butz made no secret of his agenda: He exhorted farmers to plant their fields "fencerow to fencerow" and advised them to "get big or get out." Bigger farms were more productive, he believed, so he pushed farmers to consolidate ("adapt or die" was another of his credos) and to regard themselves not as farmers but as "agribusinessmen." Somewhat less noisily, Butz set to work dismantling the New Deal farm regime of price supports, a job made easier by the fact that prices at the time were so high. He abolished the Ever-Normal Granary and, with the 1973 farm bill, began replacing the New Deal system of supporting prices through loans, government grain purchases, and land idling with a new system of direct payments to farmers.

The change from loans to direct payments hardly seems momentous—either way, the government pledges to make sure the farmer receives some target price for a bushel of corn when prices are weak. But in fact paying farmers directly for the shortfall in the price of corn was revolutionary, as its proponents surely must have understood. They had removed the floor under the price of grain. Instead of keeping corn out of a falling market, as the old loan programs and federal granary had done, the new subsidies encouraged farmers to sell their corn at any price, since the government would make up the difference. Or, as it turned out, make up *some* of the difference, since just about every farm bill since has lowered the target price in order, it was claimed, to make American grain more competitive in world markets. (Beginning in the 1980s, big buyers of grain like Cargill and Archer Daniels Midland [ADM] took a hand in shaping the farm bills, which predictably came to reflect their interests more closely than those of farmers.) Instead of supporting farmers, the government was now sub-sidizing every bushel of corn a farmer could grow—and American farmers pushed to go flat out could grow a hell of a lot of corn.

7. THE NAYLOR CURVE

It's not at all clear that very many American farmers know exactly what hit them, even now. The rhetoric of competitiveness and free trade per-suaded many of them that cheap corn would be their salvation, and several putative farmers' organizations have bought into the virtues of cheap corn. But since the heyday of corn prices in the early seventies, farm income has steadily declined along with corn prices, forcing mil-lions of farmers deeper into debt and thousands of them into bank-ruptcy every week. Exports, as a percentage of the American corn harvest, have barely budged from around 20 percent, even as prices have fallen. Iowa State University estimates that it costs roughly $2.50 to grow a bushel of Iowa corn; in October 2005 Iowa grain elevators were paying $1.45, so the typical Iowa farmer is selling corn for a dol-lar less than it costs him to grow it. Yet the corn keeps coming, more of it every year.

How can this possibly be?

George Naylor has studied this question, and he has come up with a convincing answer. He's often asked to speak at meetings on the farm crisis, and to testify at hearings about farm policy, where he often pre-sents a graph he's drawn to explain the mystery. He calls it the Naylor Curve. ("Remember the Laffer curve? Well, this one looks a little like that one, only it's true.") Basically it purports to show why falling farm prices force farmers to increase production in defiance of all rational economic behavior.

"Farmers facing lower prices have only one option if they want to be able to maintain their standard of living, pay their bills, and service their debt, and that is to produce more." A farm family needs a certain amount of cash flow every year to support itself, and if the price of corn falls, the only way to stay even is to sell more corn. Naylor says that farmers desperate to boost yield end up degrading their land, plowing and planting marginal land, applying more nitrogen—anything to squeeze a few more bushels from the soil. Yet the more bushels each farmer produces, the lower prices go, giving another turn to the perverse spiral of overproduction. Even so, corn farmers persist in measuring their success in bushels per acre, a measurement that improves even as they go broke.

"The free market has never worked in agriculture and it never will. The economics of a family farm are very different than a firm's: When prices fall, the firm can lay off people, idle factories, and make fewer widgets. Eventually the market finds a new balance between supply and demand. But the demand for food isn't elastic; people don't eat more just because food is cheap. And laying off farmers doesn't help to reduce supply. You can fire me, but you can't fire my land, because some other farmer who needs more cash flow or thinks he's more efficient than I am will come in and farm it. Even if I go out of business this land will keep producing corn."

But why corn and not something else? "We're on the bottom rung of the industrial food chain here, using this land to produce energy and protein, mostly to feed animals. Corn is the most efficient way to produce energy, soybeans the most efficient way to produce protein." The notion of switching to some other crop Naylor gruffly dismisses. "What am I going to grow here, broccoli? Lettuce? We've got a long-term investment in growing corn and soybeans; the elevator is the only buyer in town, and the elevator only pays me for corn and soybeans. The market is telling me to grow corn and soybeans, period." As is the government, which calculates his various subsidy payments based on his yield of corn.

So the plague of cheap corn goes on, impoverishing farmers (both here and in the countries to which we export it), degrading the land, polluting the water, and bleeding the federal treasury, which now spends up to $5 billion a year subsidizing cheap corn. But though those subsidy checks go to the farmer (and represent nearly half of net farm income today), what the Treasury is really subsidizing are the buyers of all that cheap corn. "Agriculture's always going to be organized by the government; the question is, organized for whose benefit? Now it's for Cargill and Coca-Cola. It's certainly not for the farmer."

Early that afternoon, after George and I had been talking agricultural policy for longer than I ever thought possible, the phone rang; his neighbor Billy needed a hand with a balky corn planter. On the drive over Naylor told me a little about Billy. "He's got all the latest toys: the twelve-row planter, Roundup Ready seed, the new John Deere combine." George rolled his eyes. "Billy's in debt up to his eyeballs." George believes he's managed to survive on the farm by steering clear of debt, nursing along his antique combine and tractor, and avoiding the trap of expansion.

A blockish fellow in his fifties, with a seed cap perched over a graying crew cut, Billy seemed cheerful enough, especially considering he'd just blown his morning fiddling with a broken tractor cable. While he and George were working on it I checked out the shed full of state-of-the-art farm equipment and asked him what he thought about the Bt corn he was planting—corn genetically engineered to produce its own pesticide. Billy thought the seed was the greatest. "I'm getting 220 bushels an acre on that seed," he boasted. "How's that compare, George?"

George owned he was getting something just south of two hundred, but he was too polite to say what he knew, which was that he was almost certainly clearing more money per acre growing less corn more cheaply. But in Iowa, bragging rights go to the man with the biggest yield, even if it's bankrupting him.

In a shed across the way I noticed the shiny chrome prow of a tractor trailer poking out and asked Billy about it. He explained he'd had to take on long-distance hauling work to keep the farm afloat. "Have to drive the big rig to pay for all my farm toys," he chuckled.

George tossed me a look, as if to say, kind of pathetic, isn't it? Poignant seemed more like it, to think what this farmer had to do to hold on to his farm. I was reminded of Thoreau's line: "Men have become the tools of their tools." And I wondered if Billy gave much thought, in those late-night hours rolling up the miles on Interstate 80, to how he got to this point, and about who he was really working for now. The bank? John Deere? Monsanto? Pioneer? Cargill? Two hundred and twenty bushels of corn is an astounding accomplishment, yet it didn't do Billy nearly as much good as it did those companies.

And then of course there's the corn itself, which if corn could form an opinion would surely marvel at the absurdity of it all—and at its great good fortune. For corn has been exempted from the usual rules of nature and economics, both of which have rough mechanisms to check any such wild, uncontrolled proliferation. In nature, the population of a species explodes until it exhausts its supply of food; then it crashes. In the market, an oversupply of a commodity depresses prices until either the surplus is consumed or it no longer makes sense to produce any more of it. In corn's case, humans have labored mightily to free it from either constraint, even if that means going broke growing it, and consuming it just as fast as we possibly can.

THREE

THE ELEVATOR

On the spring afternoon I visited the grain elevator in Jefferson, Iowa, where George Naylor hauls his corn each October, the sky was a soft gray, drizzling lightly. Grain elevators, the only significant verticals for miles around in this part of Iowa, resemble tight clusters of window-less concrete office towers, but this day the cement sky had robbed them of contrast, rendering the great cylinders nearly invisible. What stood out as my car rumbled across the railroad tracks and passed the green and white "Iowa Farmers Cooperative" sign was a bright yellow pyramid the size of a circus tent pitched near the base of the elevator: an immense pile of corn left out in the rain.

The previous year's had been a bumper crop in this part of the Mid-west; the pile represented what was left of the millions of bushels of corn that had overflowed the elevators last October. Even now, seven months later, there was still a surfeit of corn, and I watched a machine that looked like a portable escalator pour several tons of it over the lip of a railroad car. As I circumnavigated the great pile, I started to see the golden kernels everywhere, ground into the mud by tires and boots, floating in the puddles of rainwater, pancaked on the steel rails. Most of this grain is destined for factory farms and processing plants, so no one worries much about keeping it particularly clean. Even so, it was hard not to register something deeply amiss in the sight of so much food lying around on the wet ground.

In Ames the following afternoon I met a Mexican American agronomist named Ricardo Salvador, a professor at Iowa State University, who told me he'd had a similar reaction the first time he'd seen kernels littering Iowa roads in October; farmers haul their corn to town in big open wagons that fishtail across the county highways, scattering a light rain of yellow kernels as they go. "To be honest, I felt a revulsion. In Mexico, even today, you do not let corn lay on the ground; it is considered almost sacrilegious." He sent me to a passage from a sixteenth-century writer, Friar Sahagún, who had chronicled the Aztecs' reverence for maize:

> If they saw dry grains of maize scattered on the ground, they
> quickly gathered them up, saying "Our Sustenance suffereth, it
> lieth weeping. If we should not gather it up, it would accuse us
> before our Lord. It would say, 'O, Our Lord, this vassal picked
> me not up when I lay scattered upon the ground. Punish him!'
> Or perhaps we should starve."

The agronomist's reaction, like mine, owes something to our confusion of corn-the-food with corn-the-commodity, which turn out to be two subtly but crucially different things. What George Naylor grows, and what the pile by the elevator consists of, is "number 2 field corn," an internationally recognized commodity grown everywhere (and nowhere in particular), fungible, traded in and speculated upon and accepted as a form of capital all over the world. And while number 2 field corn certainly *looks* like the corn you would eat, and is directly descended from the maize Friar Sahagún's Aztecs worshipped as the source of life, it is less a food than an industrial raw material—and an abstraction. The kernels are hard to eat, but if you soak them in water for several hours you'll find they taste less like corn than lightly corn-flavored starch.

Actually there are many different kinds of corn heaped together in this pile: George Naylor's Pioneer Hi-Bred 34H31 mixed in with his neighbor Billy's genetically modified 33P67; corn grown with atrazine mixed with corn grown with metolachlor. Number 2 corn is a lowest common denominator; all the designation tells you is that the moisture content of this corn is no more than 14 percent, and that fewer than 5 percent of the kernels exhibit insect damage. Other than that, this is the corn without qualities; quantity is really the only thing that counts. Such corn is not something to feel reverent or even sentimental about, and nobody in Iowa, save the slightly embarrassed agronomist, does.

Commodity corn, which is as much an economic abstraction as it is a biological fact, was invented in Chicago in the 1850s.* Before then corn was bought and sold in burlap sacks. More often than not the sacks bore the name of the farm where the corn had been grown. You could follow a sack from a farm in Iowa to the mill in Manhattan where it was ground into meal, or to the dairy in Brooklyn where it was fed to a cow. This made a difference. For most of history farmers have had to think about the buyers of their crops, to worry about making sure their corn found its way to the right place at the right time, before it spoiled or got waylaid or its price collapsed. Farmers had to worry, too, about the quality of their corn, since customers didn't pay before sampling what was in the sack. In America before the 1850s a farmer owned his sacks of corn up to the moment when a buyer took delivery, and so bore the risk for anything that went wrong between farm and table or trough. For better or worse that burlap sack linked a corn buyer anywhere in America with a particular farmer cultivating a particular patch of the earth.

With the coming of the railroads and the invention of the grain elevator (essentially a great vertical warehouse filled by conveyor belt and

---

*I'm drawing on the excellent account of the invention of agricultural commodities in William Cronon's *Nature's Metropolis: Chicago and the Great West* (1991).

emptied by spigot) the sacks suddenly became a problem. Now it made sense to fill railroad cars and elevators by conveyor, to treat corn less as a certain number of discrete packages someone had to haul and more like an unbounded liquid that could be pumped, in effect, by machine. Mix it all together in a great golden river. The river of corn would flow from the farms to the Chicago market and then out from there to buyers anywhere in the world. But before buyers would accept this new, nonspecific, trackless corn they would have to have some assurance of its quality.

The breakthrough came in 1856, when the Chicago Board of Trade instituted a grading system. Now any number 2 corn was guaranteed to be as good as any other number 2 corn. So there was no longer any reason for anyone to care where the corn came from or who grew it, as long as it met the board's standard. Since this standard was fairly minimal (specifying acceptable levels of insect damage, dirt and extraneous matter, and moisture) growers and breeders were now free to train their energies on producing bigger and bigger harvests. Before the commodity system farmers prided themselves on a panoply of qualities in their crop: big ears, plump kernels, straight rows, various colors; even the height of their corn plants became a point of pride. Now none of these distinctions mattered; "bushels per acre" became the only boast you heard. No one could foresee it at the time, but the Chicago Board of Trade's decision redirected the evolution of *Zea mays*. From that moment on the trajectory of the species' descent was guided by a single quality: yield. Which is to say, by the quality of sheer quantity.

The invention of commodity grain severed any link between the producer of a foodstuff and its ultimate consumer. A commodity is like a filter, stripping qualities and histories from the harvest of a particular farm and farmer. When George Naylor delivers his wagonload to the elevator in Jefferson, which at the height of the harvest runs twenty-four hours a day seven days a week, his corn is weighed and graded, his account is credited with that day's posted price per bushel, and Naylor's worries about his crop—his responsibility for it, indeed his whole relationship to it—are over for another year.

Within hours Naylor's corn joins the streams of corn coming off his neighbors' farms; later, that tributary flows from Jefferson County into the river of commodity corn flowing mostly east and south from Iowa into the tremendous maw of the American food system. (Today much of it flows farther south, into Mexico.) Watching a pile of corn stream over the lip of a hopper car painted with Cargill's blue-and-yellow logo, a car destined to join a train more than a mile long and holding 440,000 bushels of corn, I began to see what George Naylor was getting at when he'd told me whom it was he grew his corn for: "the military-industrial complex."

The immense pyramid of corn I stood before in Jefferson is of course only a tiny part of an infinitely more immense mountain of corn dispersed over thousands of grain elevators across the American Corn Belt every autumn. That mountain is the product of the astounding efficiency of American corn farmers, who—with their technology, machinery, chemicals, hybrid genetics, and sheer skill—can coax five tons of corn from an acre of Iowa soil. All this you can see with your own eyes, hanging around during the harvest. What is much harder to see is that all this corn is also the product of government policies, which have done more than anything else to raise that mountain and shrink the price of each bushel in it.

The Iowa Farmers Cooperative does not write the only check George Naylor will receive for his corn crop this fall. He gets a second check from the U.S. Department of Agriculture (USDA)—about twenty-eight cents a bushel no matter what the market price of corn is, and considerably more should the price of corn drop below a certain threshold. Let's say the price of a bushel falls to $1.45, as it most recently did in October 2005. Since the official target price (called the "loan rate") in Greene County stands at $1.87, the government would then send farmers another $0.42 in "deficiency payments," for a total of $0.70 for every bushel of corn they can grow. Taken together these federal payments account for nearly half the income of the average Iowa corn farmer and represent roughly a quarter of the $19 billion U.S. taxpayers spend each year on payments to farmers.

This is a system designed to keep production high and prices low. In fact, it's designed to drive prices ever lower, since handing farmers deficiency payments (as compared to the previous system of providing loans to support prices) encourages them to produce as much corn as they possibly can, and then to dump it all on the market no matter what the price—a practice that inevitably pushes prices even lower. And as prices decline, the only way a farmer like George Naylor can keep his income from declining is by producing still more corn. So the mountain grows, from 4 billion bushels in 1970 to 10 billion bushels today. Moving that mountain of cheap corn—finding the people and animals to consume it, the cars to burn it, the new products to absorb it, and the nations to import it—has become the principal task of the industrial food system, since the supply of corn vastly exceeds the demand.

Another way to look at this 10-billion-bushel pile of commodity corn—a naturalist's way of looking at it*—is that industrial agriculture has introduced a vast new stock of biomass to the environment, creating what amounts to an imbalance—a kind of vacuum in reverse. Ecology teaches that whenever an excess of organic matter arises anywhere in nature, creatures large and small inevitably step forward to consume it, sometimes creating whole new food chains in the process. In this case the creatures feasting on the surplus biomass are both metaphorical and real: There are the agribusiness corporations, foreign markets, and whole new industries (such as ethanol), and then there are the food scientists, livestock, and human eaters, as well as the usual array of microorganisms (such as *E. coli* O157:H7).

What's involved in absorbing all this excess biomass goes a long way toward explaining several seemingly unconnected phenomena, from the rise of factory farms and the industrialization of our food, to the epidemic of obesity and prevalence of food poisoning in America, to the fact that in the country where *Zea mays* was originally domesticated, *campesinos* descended from those domesticators are losing their

---

*See Manning (2004).

farms because imported corn, flooding in from the North, has become too cheap. Such is the protean, paradoxical nature of the corn in that pile that getting rid of it could contribute to obesity and to hunger both.

My plan when I came to Iowa was to somehow follow George Naylor's corn on its circuitous path to our plates and into our bodies. I should have known that tracing any single bushel of commodity corn is as impossible as tracing a bucket of water after it's been poured into a river. Making matters still more difficult, the golden river of American commodity corn, wide though it is, passes through a tiny number of corporate hands. Though the companies won't say, it has been estimated that Cargill and ADM together probably buy somewhere near a third of all the corn grown in America.

These two companies now guide corn's path at every step of the way: They provide the pesticide and fertilizer to the farmers; operate most of America's grain elevators (Naylor's member-owned cooperative is an exception); broker and ship most of the exports; perform the wet and dry milling; feed the livestock and then slaughter the corn-fattened animals; distill the ethanol; and manufacture the high-fructose corn syrup and the numberless other fractions derived from number 2 field corn. Oh, yes—and help write many of the rules that govern this whole game, for Cargill and ADM exert considerable influence over U.S. agricultural policies. More even than the farmers who receive the checks (and the political blame for cashing them), these companies are the true beneficiaries of the "farm" subsidies that keep the river of cheap corn flowing. Cargill is the biggest privately held corporation in the world.

Cargill and ADM together comprise the vanishingly narrow sluice gate through which the great corn river passes every year. That gate is also virtually invisible. Neither company sells products directly to consumers, so they have little to gain from cooperating with journalists— and seldom do. Both companies declined to let me follow the corn river as it passes through their elevators, pipes, vats, tankers, freighters, feedlots, mills, and laboratories on its complex and increasingly obscure path to our bodies. The reason this segment of our food chain is essentially off-limits, they explained, is "food security."

Even so, it is possible to follow a bushel of George Naylor's corn, provided you are willing to regard it as the commodity it is—that is, treat it not as a specific physical entity you can hold in your hands but as a generic, fungible quantity, no different from any other bushel of number 2 field corn boarding that Cargill train or any other. Since Naylor's corn is mixed in with all the other corn grown this year, the destinations of the kernels in any one of his bushels will mirror, more or less precisely, the ultimate destinations of the crop as a whole—export, livestock, high-fructose corn syrup, etc.

So where do those ninety thousand generic kernels wind up? After they've been milled and fractionated, processed and exported and passed through the guts of cows and chickens and pigs, what sort of meal do they make? And—at the risk of employing a word that might sound extreme attached to something as wholesome and all-American as corn—what sort of havoc can those ninety thousand kernels wreak?

THE PLACE where most of those kernels wind up—about three of every five—is on the American factory farm, a place that could not exist without them. Here, hundreds of millions of food animals that once lived on family farms and ranches are gathered together in great commissaries, where they consume as much of the mounting pile of surplus corn as they can digest, turning it into meat. Enlisting the cow in this undertaking has required particularly heroic efforts, since the cow is by nature not a corn eater. But Nature abhors a surplus, and the corn must be consumed.

Enter the corn-fed American steer.

FOUR

THE FEEDLOT

Making Meat

(54,000 KERNELS)

1. CATTLE METROPOLIS

The landscape that corn has made in the American Middle West is unmistakable: It forms a second great American lawn, unfurling through the summer like an absurdly deep-pile carpet of green across the vast lands drained by the Mississippi River. Corn the plant has colonized some 125,000 square miles of the American continent, an area twice the size of New York State; even from outer space you can't miss it. It takes a bit more looking, however, to see some of the other landscapes that corn-the-commodity has created, in obscure places like Garden City, Kansas. Here in the high plains of western Kansas is where America's first feedlots were built, beginning in the early fifties.

You'll be speeding down one of Finney County's ramrod roads when the empty, dun-colored January prairie suddenly turns black and geometric, an urban grid of steel-fenced rectangles as far as the eye can see—which in Kansas is really far. I say "suddenly" but in fact the swiftly rising odor—an aroma whose Proustian echoes are decidedly more bus station men's room than cows in the country—has been heralding the feedlot's approach for more than a mile. And then it's upon you: Poky Feeders, population, thirty-seven thousand. A sloping subdivision of cattle pens stretches to the horizon, each one home to a hundred or so animals standing dully or lying around in a grayish mud that, it eventually dawns on you, isn't mud at all. The pens line a network of unpaved roads that loop around vast waste lagoons on their way to the feedyard's thunderously beating heart and dominating landmark: a rhythmically chugging feed mill that rises, soaring and silvery in the early morning light, like an industrial cathedral in the midst of a teeming metropolis of meat. As it does twelve hours a day seven days a week, the mill is noisily converting America's river of corn into cattle feed.

I'd traveled to Poky early one January with the slightly improbable notion of visiting one particular resident, though as I nosed my rental car through the feedlot's rolling black sea of bovinity, I began to wonder if this was realistic. I was looking for a young black steer with three white blazes on his face that I'd met the previous fall on a ranch in Vale, South Dakota, five hundred miles due north of here. In fact, the steer I hoped to find belonged to me: I'd purchased him as an eight-month-old calf from the Blair Ranch for $598. I was paying Poky Feeders $1.60 a day for his room and board (all the corn he could eat) and meds.

My interest in this steer was not strictly financial, or even gustatory. No, my primary interest in this animal was educational. I wanted to learn how the industrial food chain transforms bushels of corn into steaks. How do you enlist so unlikely a creature—for the cow is a herbivore by nature—to help dispose of America's corn surplus? By far the biggest portion of a bushel of American commodity corn (about 60 percent of it, or some fifty-four thousand kernels) goes to feeding livestock, and much of that goes to feeding America's 100 million beef cattle—cows and bulls and steers that in times past spent most of their lives grazing on grasses out on the prairie.

America's food animals have undergone a revolution in lifestyle in the years since World War II. At the same time as much of America's human population found itself leaving the city for the suburbs, our food animals found themselves traveling in the opposite direction, leaving widely dispersed farms in places like Iowa to live in densely populated new animal cities. These places are so different from farms and ranches that a new term was needed to denote them: CAFO—Concentrated Animal Feeding Operation. The new animal and human landscapes were both products of government policy. The postwar suburbs would never have been built if not for the interstate highway system, as well as the G.I. Bill and federally subsidized mortgages. The urbanization of America's animal population would never have taken place if not for the advent of cheap, federally subsidized corn.

Corn itself profited from the urbanization of livestock twice. As the animals left the farm, more of the farm was left for corn, which rapidly colonized the paddocks and pastures and even the barnyards that had once been the animals' territory. The animals left because the farmers simply couldn't compete with the CAFOs. It cost a farmer more to grow feed corn than it cost a CAFO to buy it, for the simple reason that commodity corn now was routinely sold for less than it cost to grow. Corn profited again as the factory farms expanded, absorbing increasing amounts of its surplus. Corn found its way into the diet of animals that never used to eat very much of it (like cattle) or any corn at all, like the farmed salmon now being bred to tolerate grain. All that excess biomass has to go somewhere.

The economic logic of gathering so many animals together to feed them cheap corn in CAFOs is hard to argue with; it has made meat, which used to be a special occasion in most American homes, so cheap and abundant that many of us now eat it three times a day. Not so compelling is the biological logic behind this cheap meat. Already in their short history CAFOs have produced more than their share of environmental and health problems: polluted water and air, toxic wastes, novel and deadly pathogens.

Raising animals on old-fashioned mixed farms such as the Naylors' used to make simple biological sense: You can feed them the waste products of your crops, and you can feed their waste products to your crops. In fact, when animals live on farms the very idea of waste ceases to exist; what you have instead is a closed ecological loop—what in retrospect you might call a solution. One of the most striking things that animal feedlots do (to paraphrase Wendell Berry) is to take this elegant solution and neatly divide it into two new problems: a fertility problem on the farm (which must be remedied with chemical fertilizers) and a pollution problem on the feedlot (which seldom is remedied at all).

This biological absurdity, characteristic of all CAFOs, is compounded in the cattle feedyard by a second absurdity. Here animals exquisitely adapted by natural selection to live on grass must be adapted by us—at considerable cost to their health, to the health of the land, and ultimately to the health of their eaters—to live on corn, for no other reason than it offers the cheapest calories around and because the great pile must be consumed. This is why I decided to follow the trail of industrial corn through a single steer rather than, say, a chicken or a pig, which can get by just fine on a diet of grain: The short, unhappy life of a corn-fed feedlot steer represents the ultimate triumph of industrial thinking over the logic of evolution.

2. PASTORAL: VALE, SOUTH DAKOTA

The Blair Ranch occupies fifty-five hundred acres of rolling short-grass prairie a few miles outside Sturgis, South Dakota, and directly in the shadow of Bear Butte. The Bismarck-Deadwood trail crossed its land just to the north of the butte, which rises dramatically from the plains like a chubby ten-story exclamation mark. You can still make out ruts in the turf dug by stagecoaches and cattle drives the century before last. The turf itself in November, when I visited, forms a luxuriant pelt of grass oscillating yellow and gold in the constant wind and sprinkled with perambulating black dots: Angus cows and calves, grazing.

Ed and Rich Blair run what's called a "cow-calf" operation, the first stage in the production of a hamburger and the stage least changed by the modern industrialization of meat. While the pork and chicken industries have consolidated the life cycle of those animals under a single roof, beef cattle still get born on hundreds of thousands of independently owned ranches scattered mainly across the West. Although a mere four giant meatpacking companies (Tyson subsidiary IBP, Cargill subsidiary Excel, Swift & Company, and National) now slaughter and market four of every five beef cattle born in this country, that concentration represents the narrow end of a funnel that starts out as wide as the Great Plains. These corporations have concluded that it takes so much land (and therefore capital) to produce a calf ready for the feedlot—ten acres per head at a minimum—that they're better off leaving the ranching (and the risk) to the ranchers.

Steer number 534 spent his first six months in these lush pastures alongside his mother, 9534. The number signifies she was the thirty-fourth cow born in 1995; since none of her male offspring stick around long enough to meet, they're all named 534. His father was a registered Angus by the name of Gar Precision 1680, a bull distinguished by the size and marbling of his offsprings' rib-eye steaks. Gar Precision's only contact with 9534 came by way of a fifteen-dollar mail-order straw of his semen.

Born on March 13, 2001, in the birthing shed across the road, 534 and his mother were turned out on pasture just as soon as the eighty-pound calf stood up and began nursing. Within a few weeks the calf began supplementing his mother's milk by nibbling on a salad bar of mostly native grasses: western wheatgrass, little bluestem, buffalo grass, green needlegrass.

Apart from the trauma of the Saturday in April when he was branded and castrated, one could imagine 534 looking back on those six months as the good old days. It might be foolish for us to presume to know what a cow experiences, yet we can say that a calf grazing on grass is at least doing what he has been supremely well suited by evolution to do. Oddly enough, though, eating grass is something that after October my steer will never have the opportunity to do again.

THE COEVOLUTIONARY RELATIONSHIP between cows and grass is one of nature's underappreciated wonders; it also happens to be the key to understanding just about everything about modern meat. For the grasses, which have evolved to withstand the grazing of ruminants, the cow maintains and expands their habitat by preventing trees and shrubs from gaining a foothold and hogging the sunlight; the animal also spreads grass seed, plants it with his hooves, and then fertilizes it with his manure. In exchange for these services the grasses offer ruminants a plentiful and exclusive supply of lunch. For cows (like sheep, bison, and other ruminants) have evolved the special ability to convert grass—which single-stomached creatures like us can't digest—into high-quality protein. They can do this because they possess what is surely the most highly evolved digestive organ in nature: the rumen. About the size of a medicine ball, the organ is essentially a forty-five-gallon fermentation tank in which a resident population of bacteria dines on grass. Living their unseen lives at the far end of the food chain that culminates in a hamburger, these bacteria have, like the grasses, coevolved with the cow, whom they feed.

Truly this is an excellent system for all concerned: for the grasses, for the bacteria, for the animals, and for us, the animals' eaters. While it is true that overgrazing can do ecological harm to a grassland, in recent years ranchers have adopted rotational grazing patterns that more closely mimic the patterns of the bison, a ruminant that sustainably grazed these same grasses for thousands of years before the cow displaced it. In fact, a growing number of ecologists now believe the rangelands are healthier with cattle on them, provided they're moved frequently. Today the most serious environmental harm associated with the cattle industry takes place on the feedlot.

In fact, growing meat on grass makes superb ecological sense: It is a sustainable, solar-powered food chain that produces food by transforming sunlight into protein. Row crops could accomplish this trick too, but not around here: In places like western South Dakota the land is far too arid, thin, and hilly to grow crops without large amounts of irrigation, chemicals, and erosion. "My cattle can take low-quality forage and convert it into a pretty desirable product," Rich Blair pointed out. "If you didn't have ruminant animals, all this"—he gestures to the high plains rolling out from his ranch in every direction—"would be the great American desert."

So then why is it that steer number 534 hasn't tasted a blade of prairie grass since October? Speed, in a word, or, in the industry's preferred term, "efficiency." Cows raised on grass simply take longer to reach slaughter weight than cows raised on a richer diet, and for half a century now the industry has devoted itself to shortening a beef animal's allotted span on earth. "In my grandfather's time, cows were four or five years old at slaughter," Rich explained. "In the fifties, when my father was ranching, it was two or three years old. Now we get there at fourteen to sixteen months." Fast food, indeed. What gets a steer from 80 to 1,100 pounds in fourteen months is tremendous quantities of corn, protein and fat supplements, and an arsenal of new drugs.

Weaning marks the fateful moment when the natural, evolutionary logic represented by a ruminant grazing on grass bumps up against the industrial logic that will propel the animal on the rest of its swift journey to a wholesale box of beef. This industrial logic is rational and even irresistible—after all, it has succeeded in making beef everyday fare for millions of people for whom it once represented a luxury. And yet the further you follow it, the more likely you are to begin wondering if that rational logic might not also be completely mad.

In October, two weeks before I made his acquaintance, steer number 534 was weaned from his mother. Weaning is perhaps the most traumatic time on a ranch for animals and ranchers alike; cows separated from their calves will mope and bellow for days, and the calves, stressed by the change in circumstance and diet, are prone to getting sick. Calves are weaned for a couple of reasons: to free their mothers to have more calves (9534 had already been inseminated again in June), and to get the animals, now five or six hundred pounds, ready for life on the feedlot.

The animals are rounded up and herded into a "backgrounding" pen, where they'll spend a couple of months before boarding the truck for Poky Feeders. Think of backgrounding as prep school for feedlot life: The animals are, for the first time in their lives, confined to a pen, "bunk broken"—taught to eat from a trough—and gradually accustomed to eating what is for them a new and unnatural diet. Here is where the rumen first encounters corn.

It was in the backgrounding pen that I first made the acquaintance of 534. Before coming to Vale I'd told the Blairs I wanted to follow one of their steers through the life cycle; Ed Blair, the older of the brothers, suggested only half in jest that I might as well go whole hog and buy the animal, if I really wanted to appreciate the challenges of ranching. This immediately struck me as a promising idea.

Ed and Rich told me what to look for: a broad straight back and thick shoulders—basically, a sturdy frame on which to hang a lot of meat. I was also looking for a memorable face in this black Angus sea, one that I could pick out of the crowd at the feedlot. Almost as soon as I began surveying the ninety or so animals in the pen, 534 moseyed up to the railing and made eye contact. He had a wide stout frame and was brockle-faced—he had three easy-to-spot white blazes. Here was my boy.

3. INDUSTRIAL: GARDEN CITY, KANSAS

Traveling from the ranch to the feedyard, as 534 and I both did (in separate vehicles) the first week of January, feels a lot like going from the country to the big city. A feedlot is very much a premodern city, however, teeming and filthy and stinking, with open sewers, unpaved roads, and choking air rendered visible by dust.

The urbanization of the world's livestock being a fairly recent historical development, it makes a certain sense that cow towns like Poky

Feeders would recall human cities centuries ago, in the days before modern sanitation. As in fourteenth-century London, say, the workings of the metropolitan digestion remain vividly on display, the foodstuffs coming in, the streams of waste going out. The crowding into tight quarters of recent arrivals from all over, together with the lack of sanitation, has always been a recipe for disease. The only reason contemporary animal cities aren't as plague-ridden or pestilential as their medieval human counterparts is a single historical anomaly: the modern antibiotic.

I spent the better part of a day at Poky Feeders, walking the streets, cattle watching, looking up my steer, and touring local landmarks like the towering feed mill. In any city it's easy to lose track of nature—of the transactions between various species and the land on which everything ultimately depends. Back on the ranch the underlying ecological relationship could not have been more legible: It is a local food chain built upon grass and the ruminants that can digest grass, and it draws its energy from the sun. But what about here?

As the long shadow of the mill suggests, the feedlot is a city built upon America's mountain of surplus corn—or rather, corn plus the various pharmaceuticals a ruminant must have if it is to tolerate corn. Yet, having started out from George Naylor's farm, I understood that the corn on which this place runs is implicated in a whole other set of ecological relationships powered by a very different source of energy— the fossil fuel it takes to grow all that corn. So if the modern CAFO is a city built upon commodity corn, it is a city afloat on an invisible sea of petroleum. How this peculiar state of affairs came to seem sensible is a question I spent my day at Poky trying to answer.

IT WAS ONLY NATURAL that I start my tour at the feed mill, the feedlot's thundering hub, where three meals a day for thirty-seven thousand animals are designed and mixed by computer. A million pounds of feed pass through the mill each day. Every hour of every day a tractor trailer pulls up to the loading dock to deliver another fifty tons of corn. The driver opens a valve in the belly of the truck and a golden stream of grain—one thin rivulet of the great corn river coursing out of the Middle West—begins to flow, dropping down a chute into the bowels of the mill. Around to the other side of the building, tanker trucks back up to silo-shaped tanks into which they pump thousands of gallons of liquefied fat and protein supplements. In a shed attached to the mill sit vats of liquid vitamins and synthetic estrogen beside pallets stacked with fifty-pound sacks of antibiotics—Rumensin and Tylosin. Along with alfalfa hay and silage (for roughage), all these ingredients will be automatically blended and then piped into the parade of dump trucks that three times a day fan out from here to keep Poky's eight and a half miles of trough filled.

The feed mill's pulsing din is the sound of two giant steel rollers turning against one another twelve hours a day, crushing steamed corn kernels into warm and fragrant flakes. (Flaking the corn makes it easier for cattle to digest it.) This was the only feed ingredient I sampled, and it wasn't half bad; not as crisp as a Kellogg's flake, but with a cornier flavor. I passed on the other ingredients: the liquefied fat (which on today's menu is beef tallow, trucked in from one of the nearby slaughterhouses), and the protein supplement, a sticky brown goop consisting of molasses and urea. The urea is a form of synthetic nitrogen made from natural gas, similar to the fertilizer spread on George Naylor's fields.

Before being put on this highly concentrated diet, new arrivals to the feedyard are treated to a few days of fresh long-stemmed hay. (They don't eat on the long ride and can lose up to one hundred pounds, so their rumens need to be carefully restarted.) Over the next several weeks they'll gradually step up to a daily ration of thirty-two pounds of feed, three-quarters of which is corn—nearly a half bushel a day.

What got corn onto the menu at this and almost every other American feedlot is price, of course, but also USDA policy, which for decades has sought to help move the mountain of surplus corn by passing as much of it as possible through the digestive tracts of food animals, who can convert it into protein.

We've come to think of "corn-fed" as some kind of old-fashioned virtue, which it may well be when you're referring to Midwestern children, but feeding large quantities of corn to cows for the greater part of their lives is a practice neither particularly old nor virtuous. Its chief advantage is that cows fed corn, a compact source of caloric energy, get fat quickly; their flesh also marbles well, giving it a taste and texture American consumers have come to like. Yet this corn-fed meat is demonstrably less healthy for us, since it contains more saturated fat and less omega-3 fatty acids than the meat of animals fed grass. A growing body of research suggests that many of the health problems associated with eating beef are really problems with corn-fed beef. (Modern-day hunter-gatherers who subsist on wild meat don't have our rates of heart disease.) In the same way ruminants are ill adapted to eating corn, humans in turn may be poorly adapted to eating ruminants that eat corn.

Yet the USDA's grading system has been designed to reward marbling (a more appealing term than "intramuscular fat," which is what it is) and thus the feeding of corn to cattle. Indeed, corn has become so deeply ingrained in the whole system of producing beef in America that whenever I raised any questions about it among ranchers or feedlot operators or animal scientists, people looked at me as if I'd just arrived from another planet. (Or perhaps from Argentina, where excellent steaks are produced on nothing but grass.)

The economic logic behind corn is unassailable, and on a factory farm there is no other kind. Calories are calories, and corn is the cheapest, most convenient source of calories on the market. Of course, it was the same industrial logic—protein is protein—that made feeding rendered cow parts back to cows seem like a sensible thing to do, until scientists figured out that this practice was spreading bovine spongiform encephalopathy (BSE), more commonly known as mad cow disease. Rendered bovine meat and bonemeal represented the cheapest, most convenient way of satisfying a cow's protein requirement (never mind these animals were herbivores by evolution) and so appeared on the daily menus of Poky and most other feedyards until the Food and Drug Administration (FDA) banned the practice in 1997.

We now understand that while at a reductive, molecular level protein may indeed be protein, at an ecological or species level, this isn't quite true. As cannibal tribes have discovered, eating the flesh of one's own species carries special risks of infection. Kuru, a disease bearing a striking resemblance to BSE, spread among New Guinea tribesmen who ritually ate the brains of their dead kin. Some evolutionary biologists believe that evolution selected against cannibalism as a way to avoid such infections; animals' aversion to their own feces, and the carcasses of their species, may represent a similar strategy. Through natural selection animals have developed a set of hygiene rules, functioning much like taboos. One of the most troubling things about factory farms is how cavalierly they flout these evolutionary rules, forcing animals to overcome deeply ingrained aversions. We make them trade their instincts for antibiotics.

Though the industrial logic that made feeding cattle to cattle seem like a good idea has been thrown into doubt by mad cow disease, I was surprised to learn it hadn't been discarded. The FDA ban on feeding ruminant protein to ruminants makes an exception for blood products and fat; my steer will probably dine on beef tallow recycled from the very slaughterhouse he's heading to in June. ("Fat is fat," the feedlot manager shrugged, when I raised an eyebrow.) Though Poky doesn't do it, the rules still permit feedlots to feed nonruminant animal protein to ruminants. Feather meal and chicken litter (that is, bedding, feces, and discarded bits of feed) are accepted cattle feeds, as are chicken, fish, and pig meal. Some public health experts worry that since the bovine meat and bonemeal that cows used to eat is now being fed to chickens, pigs, and fish, infectious prions could find their way back into cattle when they're fed the protein of the animals that have been eating them.

Before mad cow disease remarkably few people in the cattle business, let alone the general public, comprehended the strange new semicircular food chain that industrial agriculture had devised for the beef animal—and so, in turn, for the beef eater. When I mentioned to Rich

Blair how surprised I'd been to learn cattle were eating cattle, he said, "To tell you the truth, it was kind of a shock to me, too."

COMPARED TO ALL the other things we feed cattle these days, corn seems positively wholesome. And yet it too violates the biological or evolutionary logic of bovine digestion. During my day at Poky I spent a few hours with Dr. Mel Metzin, the staff veterinarian, learning more than any beef eater really should know about the gastrointestinal life of the modern cow. Dr. Mel, as he's known at Poky, oversees a team of eight cowboys who spend their days riding the yard's dusty streets, spotting sick animals and bringing them into Poky's three "hospitals" for treatment. Most of the health problems that afflict feedlot cattle can be traced either directly or indirectly to their diet. "They're made to eat forage," Dr. Metzin explained, "and we're making them eat grain.

"It's not that they can't adjust," he continues, "and now we're breeding cattle to do well in a feedyard." One way to look at the breeding work going on at ranches like the Blairs' is that the contemporary beef cow is being selected for the ability to eat large quantities of corn and efficiently convert it to protein without getting too sick. (These, after all, are precisely the genes prized in 534's father, Gar Precision 1680.) The species is evolving, in other words, to help absorb the excess biomass coming off America's cornfields. But the cow's not there quite yet, and a great many feedlot cattle—virtually all of them to one degree or another, according to several animal scientists I talked to—are simply sick.

Bloat is perhaps the most serious thing that can go wrong with a ruminant on corn. The fermentation in the rumen produces copious amounts of gas, which is normally expelled by belching during rumination. But when the diet contains too much starch and too little roughage, rumination all but stops, and a layer of foamy slime forms in the rumen that can trap the gas. The rumen inflates like a balloon until it presses against the animal's lungs. Unless action is taken promptly to relieve the pressure (usually by forcing a hose down the animal's esoph-agus), the animal suffocates.

A concentrated diet of corn can also give a cow acidosis. Unlike our own highly acid stomachs, the normal pH of a rumen is neutral. Corn ren-ders it acidic, causing a kind of bovine heartburn that in some cases can kill the animal, but usually just makes him sick. Acidotic animals go off their feed, pant and salivate excessively, paw and scratch their bellies, and eat dirt. The condition can lead to diarrhea, ulcers, bloat, rumenitis, liver disease, and a general weakening of the immune system that leaves the animal vulnerable to the full panoply of feedlot diseases—pneumonia, coccid-iosis, enterotoxemia, feedlot polio. Much like modern humans, modern cattle are susceptible to a set of relatively new diseases of civilization—assuming, that is, we're willing to put the modern feedlot under the rubric of civilization.

Cattle rarely live on feedlot diets for more than 150 days, which might be about as much as their systems can tolerate. "I don't know how long you could feed them this ration before you'd see problems," Dr. Metzin said; another vet told me the diet would eventually "blow out their livers" and kill them. Over time the acids eat away at the ru-men wall, allowing bacteria to enter the animal's bloodstream. These microbes wind up in the liver, where they form abscesses and impair the liver's function. Between 15 percent and 30 percent of feedlot cows are found at slaughter to have abscessed livers; Dr. Mel told me that in some pens the figure runs as high as 70 percent.

What keeps a feedlot animal healthy—or healthy enough—are anti-biotics. Rumensin buffers acidity in the rumen, helping to prevent bloat and acidosis, and Tylosin, a form of erythromycin, lowers the in-cidence of liver infection. Most of the antibiotics sold in America today end up in animal feed, a practice that, it is now generally acknowledged (except in agriculture), is leading directly to the evolution of new antibiotic-resistant superbugs. In the debate over the use of antibiotics in agriculture, a distinction is usually made between their clinical and nonclinical uses. Public health advocates don't object to treating sick animals with antibiotics; they just don't want to see the drugs lose their effectiveness because factory farms are feeding them to healthy animals to promote growth. But the use of antibiotics in feedlot cattle confounds this distinction. Here the drugs are plainly being used to treat sick animals, yet the animals probably wouldn't be sick if not for the diet of grain we feed them.

I asked Dr. Mel what would happen if drugs like Rumensin and Tylosin were banned from cattle feed, as some public health experts advocate. "We'd have a high death rate [it's currently about 3 percent, matching the industry average] and poorer performing cattle. We just couldn't feed them as hard." The whole system would have to change—and slow down.

"Hell, if you gave them lots of grass and space, I wouldn't have a job."

MY FIRST IMPRESSION of pen 63, where my steer is spending his last five months, was, *Not a bad little piece of real estate, all considered.* The pen is far enough from the feed mill to be fairly quiet and it has a water view of what I thought was a pond or reservoir until I noticed the brown scum. The body of water is what is known, in the geography of CAFOs, as a manure lagoon. I asked the feedlot manager why they didn't just spray the liquefied manure on neighboring farms. The farmers don't want it, he explained. The nitrogen and phosphorus levels are so high that spraying the crops would kill them. He didn't say that feedlot wastes also contain heavy metals and hormone residues, persistent chemicals that end up in waterways downstream, where scientists have found fish and amphibians exhibiting abnormal sex characteristics. CAFOs like Poky transform what at the proper scale would be a precious source of fertility—cow manure—into toxic waste.

The pen 534 lives in is surprisingly spacious, about the size of a hockey rink, with a concrete feed bunk along the road, and a fresh water trough out back. I climbed over the railing and joined the ninety steers, which, en masse, retreated a few lumbering steps, and then stopped to see what I would do.

I had on the same carrot-colored sweater I'd worn to the ranch in South Dakota, hoping to elicit some glint of recognition from my steer. I couldn't find him at first; all the faces staring at me were either completely black or bore an unfamiliar pattern of white marks. And then I spotted him—the three white blazes—way off in the back. As I gingerly stepped toward him the quietly shuffling mass of black cowhide between us parted, and there stood 534 and I, staring dumbly at one another. Glint of recognition? None, none whatsoever. I told myself not to take it personally; 534 and his pen mates have been bred for their marbling, after all, not their ability to form attachments.

I noticed that 534's eyes looked a little bloodshot. Dr. Metzin had told me that some animals are irritated by feedlot dust. The problem is especially serious in the summer months, when the animals kick up clouds of the stuff and workers have to spray the pens with water to keep it down. I had to remind myself that this is not ordinary dirt dust, inasmuch as the dirt in a feedyard is not ordinary dirt; no, this is fecal dust. But apart from the air quality, how did feedlot life seem to be agreeing with 534? I don't know enough about the emotional life of a steer to say with confidence that 534 was miserable, bored, or indifferent, but I would not say he looked happy.

He's clearly eating well, though. My steer had put on a couple hundred pounds since we'd last met, and he looked it: thicker across the shoulder and round as a barrel through the middle. He carried himself more like a steer now than a calf, even though his first birthday was still two months away. Dr. Metzin complimented me on his size and conformation. "That's a handsome-looking beef you got there." (Shucks.)

If I stared at my steer hard enough, I could imagine the white lines of the butcher's chart dissecting his black hide: rump roast, flank steak, standing rib, tenderloin, brisket. One way of looking at 534—the feedlot way, the industrial way—was as a most impressive machine for turning number 2 field corn into cuts of beef. Every day between now and his slaughter in six months, 534 will convert thirty-two pounds of feed into four pounds of gain—new muscle, fat, and bone. This at least is how 534 appears in the computer program I'd seen at the mill: the ratio of feed to gain that determines his efficiency. (Compared to other food animals, cattle are terribly inefficient: The ratio of feed to flesh in chicken, the most efficient animal by this measure, is two pounds of corn to one of meat, which is why chicken costs less than beef.) Poky Feeders is indeed a factory, transforming—as fast as bovinely possible—cheap raw materials into a less cheap finished product, through the mechanism of bovine metabolism.

Yet metaphors of the factory and the machine obscure as much as they reveal about the creature standing before me. He has, of course, another, quite different identity—as an animal, I mean, connected as all animals must be to certain other animals and plants and microbes, as well as to the earth and the sun. He's a link in a food chain, a thread in a far-reaching web of ecological relationships. Looked at from this perspective, everything going on in this cattle pen appears quite different, and not nearly as far removed from our world as this manure-encrusted patch of ground here in Nowhere, Kansas, might suggest.

For one thing, the health of these animals is inextricably linked to our own by that web of relationships. The unnaturally rich diet of corn that undermines a steer's health fattens his flesh in a way that undermines the health of the humans who will eat it. The antibiotics these animals consume with their corn at this very moment are selecting, in their gut and wherever else in the environment they end up, for new strains of resistant bacteria that will someday infect us and withstand the drugs we depend on to treat that infection. We inhabit the same microbial ecosystem as the animals we eat, and whatever happens in it also happens to us.

Then there's the deep pile of manure on which I stand, in which 534 sleeps. We don't know much about the hormones in it—where they will end up, or what they might do once they get there—but we do know something about the bacteria, which can find their way from the manure on the ground to his hide and from there into our hamburgers. The speed at which these animals will be slaughtered and processed—four hundred an hour at the plant where 534 will go—means that sooner or later some of the manure caked on these hides gets into the meat we eat. One of the bacteria that almost certainly resides in the manure I'm standing in is particularly lethal to humans. *Escherichia coli* O157:H7 is a relatively new strain of the common intestinal bacteria (no one had seen it before 1980) that thrives in feedlot cattle, 40 percent of which carry it in their gut. Ingesting as few as ten of these microbes can cause a fatal infection; they produce a toxin that destroys human kidneys.

Most of the microbes that reside in the gut of a cow and find their way into our food get killed off by the strong acids in our stomachs, since they evolved to live in the neutral pH environment of the rumen. But the rumen of a corn-fed feedlot steer is nearly as acidic as our own stomachs, and in this new, man-made environment new acid-resistant strains of *E. coli*, of which O157:H7 is one, have evolved—yet another creature recruited by nature to absorb the excess biomass coming off the Farm Belt. The problem with these bugs is that they can shake off the acid bath in our stomachs—and then go on to kill us. By acidifying the rumen with corn we've broken down one of our food chain's most important barriers to infection. Yet another solution turned into a problem.

We've recently discovered that this process of acidification can be reversed, and that doing so can greatly diminish the threat from *E. coli* O157:H7. Jim Russell, a USDA microbiologist on the faculty at Cornell, has found that switching a cow's diet from corn to grass or hay for a few days prior to slaughter reduces the population of *E. coli* O157:H7 in the animal's gut by as much as 80 percent. But such a solution (*Grass?!*) is considered wildly impractical by the cattle industry and (therefore) by the USDA. Their preferred solution for dealing with bacterial contamination is irradiation—essentially, to try to sterilize the manure getting into the meat.

So much comes back to corn, this cheap feed that turns out in so many ways to be not cheap at all. While I stood in pen 63 a dump truck pulled up alongside the feed bunk and released a golden stream of feed. The black mass of cowhide moved toward the trough for lunch. The $1.60 a day I'm paying for three meals a day here is a bargain only by the narrowest of calculations. It doesn't take into account, for example, the cost to the public health of antibiotic resistance or food poisoning by E. coli O157:H7. It doesn't take into account the cost to taxpayers of the farm subsidies that keep Poky's raw materials cheap. And it certainly doesn't take into account all the many environmental costs incurred by cheap corn.

I stood alongside 534 as he lowered his big head into the stream of fresh grain. How absurd, I thought, the two of us standing hock-deep in manure in this godforsaken place, overlooking a manure lagoon in the middle of nowhere somewhere in Kansas. Godforsaken perhaps, and yet not apart, I realized, as I thought of the other places connected to this place by the river of commodity corn. Follow the corn from this bunk back to the fields where it grows and I'd find myself back in the middle of that 125,000-mile-square monoculture, under a steady rain of pesticide and fertilizer. Keep going, and I could follow the nitrogen runoff from that fertilizer all the way down the Mississippi into the Gulf of Mexico, adding its poison to an eight-thousand-square-mile zone so starved of oxygen nothing but algae can live in it. And then go farther still, follow the fertilizer (and the diesel fuel and the petrochemical pesticides) needed to grow the corn all the way to the oil fields of the Persian Gulf.

I don't have a sufficiently vivid imagination to look at my steer and see a barrel of oil, but petroleum is one of the most important ingredients in the production of modern meat, and the Persian Gulf is surely a link in the food chain that passes through this (or any) feedlot. Steer 534 started his life part of a food chain that derived all of its energy from the sun, which nourished the grasses that nourished him and his mother. When 534 moved from ranch to feedlot, from grass to corn, he joined an industrial food chain powered by fossil fuel—and therefore defended by the U.S. military, another never-counted cost of cheap food. (One-fifth of America's petroleum consumption goes to producing and transporting our food.) After I got home from Kansas, I asked an economist who specializes in agriculture and energy if it might be possible to calculate precisely how much petroleum it will take to grow my steer to slaughter weight. Assuming 534 continues to eat twenty-five pounds of corn a day and reaches a weight of twelve hundred pounds, he will have consumed in his lifetime the equivalent of thirty-five gallons of oil—nearly a barrel.

So this is what commodity corn can do to a cow: industrialize the miracle of nature that is a ruminant, taking this sunlight- and prairie grass–powered organism and turning it into the last thing we need: another fossil fuel machine. This one, however, is able to suffer.

Standing there in the pen alongside my steer, I couldn't imagine ever wanting to eat the flesh of one of these protein machines. Hungry was the last thing I felt. Yet I'm sure that after enough time goes by, and the stink of this place is gone from my nostrils, I will eat feedlot beef again. Eating industrial meat takes an almost heroic act of not knowing or, now, forgetting. But I left Poky determined to follow this meat to a meal on a table somewhere, to see this food chain at least that far. I was curious to know what feedlot beef would taste like now, if I could taste the corn or even, since taste is as much a matter of what's in the head as it is about molecules dancing on the tongue, some hint of the petroleum. "You are what you eat" is a truism hard to argue with, and yet it is, as a visit to a feedlot suggests, incomplete, for you are what what you eat eats, too. And what we are, or have become, is not just meat but number 2 corn and oil.

FIVE

THE PROCESSING PLANT

Making Complex Foods

(18,000 KERNELS)

1. TAKING THE KERNEL APART: THE MILL

One of the truly odd things about the 10 billion bushels of corn harvested each year is how little of it we eat. Sure, we grind some of it to make cornmeal, but most of the corn we eat *as corn*—whether on the cob, flaked, or baked into muffins or tortillas or chips—comes from varieties other than number 2: usually sweet corn or white corn. These uses represent a tiny fraction of the harvest—less than a bushel per person per year—which is probably why we don't think of ourselves as big corn eaters. And yet each of us is personally responsible for consuming a ton of the stuff every year.

Much of the rest of that per capita ton does enter our bodies, but not before it has been heavily processed, broken down into simple compounds either by animals like steer 534 or a processing plant, and then reassembled either as beef, chicken, or pork, or as soft drinks, breakfast cereals, or snacks. What doesn't pass through the gut of a food animal to become meat will pass through one of America's twenty-five "wet mills" on its way to becoming one of the innumerable products food science has figured out how to tease from a kernel of corn. (These mills are called wet to distinguish them from the traditional mills where corn is simply ground into dry meal for things like tortillas.)

About a fifth of the corn river flowing out from the elevators at the Iowa Farmers Cooperative travels to a wet milling plant, usually by train. There it diverges into a great many slender branching tributaries, only to converge much later on a plate or in a cup. For what the wet mill does to a bushel of corn is to turn it into the building blocks from which companies like General Mills, McDonald's, and Coca-Cola assemble our processed foods.

The first rough breakdown of all that corn begins with the subdivision of the kernel itself: Its yellow skin will be processed into various vitamins and nutritional supplements; the tiny germ (the dark part nearest the cob, which holds the embryo of the potential future corn plant) will be crushed for its oil; and the biggest part, the endosperm, will be plundered for its rich cache of complex carbohydrates.

This oversized packet of starch is corn's most important contribution to the industrial food chain: an abundance of carbohydrate molecules in long chains that chemists have learned to break down and then rearrange into hundreds of different organic compounds—acids, sugars, starches, and alcohols. The names of many of these compounds will be familiar to anyone who's studied the ingredient label on a package of processed food: citric and lactic acid; glucose, fructose, and maltodextrin; ethanol (for alcoholic beverages as well as cars), sorbitol, mannitol, and xanthan gum; modified and unmodified starches; as well as dextrins and cyclodextrins and MSG, to name only a few.

To watch the stream of corn coming off of George Naylor's farm proceed to divide, subdivide, and ultimately branch off into a molecule of fructose destined to sweeten a soda is not as easy as following it to a feedlot into a cut of meat. For one thing, the two companies who wet mill most of America's corn (Cargill and ADM) declined to let me watch them do it. For another, the process is largely invisible, since it takes place inside a series of sealed vats, pipes, fermentation tanks, and filters. Even so, I would have liked to follow my bushel of corn through ADM's plant in Decatur, Illinois (the unofficial capital of corn processing in America), or to Cargill's mill in Iowa City (the likely destination of the train I saw being loaded at the elevator in Jefferson), but the industrial food chain goes underground, in effect, as it passes through these factories on its path to our plates.

The closest I got to following corn through a mill was at the Center for Crops Utilization Research at Iowa State University, in Ames, forty-five miles from the farmers cooperative elevator in Jefferson. After my visit to George Naylor's farm, I spent a couple of days on the Ames campus, which really should be called the University of Corn. Corn is the hero of the most prominent sculptures and murals on campus, and the work of the institution is dedicated in large part to the genetics, culture, history, and uses of this plant, though the soybean, Iowa's second crop, gets its share of attention too. The Center for Crops Utilization Research is charged with developing new uses for America's corn and soybean surplus, and to this end operates a scaled-down wet milling operation, a Rube Goldberg contraption of stainless steel tubes, pipes, valves, vents, drying tables, centrifuges, filters, and tanks that Larry Johnson, the center's director, was more than happy to show me.

To hear Johnson describe it, the wet milling process is essentially an industrial version of digestion: A food is broken down through a series of steps that includes the application of physical pressure, acids, and enzymes. The order of the steps is different in industrial digestion—the acids come before the mechanical chewing, for instance—but the results are much the same: A complex food is reduced to simple molecules, mostly sugars.

"First we separate the corn into its botanical parts—embryo, endosperm, fiber—and then into its chemical parts," Johnson explained as we began our tour of the plant. When a shipment of corn arrives at the mill, it is steeped for thirty-six hours in a bath of water containing a small amount of sulphur dioxide. The acid bath swells the kernels and frees the starch from the proteins that surround it.

After the soak, the swollen kernels are ground in a mill. "By now the germ is rubbery and it pops right off," Johnson explained. "We take the slurry to a hydroclone"—basically a centrifuge for liquids—"where the germ floats off. After it's dried, we squeeze it for corn oil." Corn oil can be used as a cooking or salad oil, or hydrogenated for use in margarine and other processed foods: Atoms of hydrogen are forced into the fat molecules to make them solid at room temperature. (Though originally designed as a healthy substitute for animal fats, medical researchers now think these trans fats are actually worse for our arteries than butter.)

Once the germ has been removed and the kernels crushed, what's left is a white mush of protein and starch called "mill starch." To draw off as much of the protein as possible, the mill starch undergoes a progressively finer series of grindings and filterings and centrifuges. The extracted protein, called gluten, is used in animal feed. At each step more fresh water is added—it takes about five gallons to process a bushel of corn, and prodigious amounts of energy. Wet milling is an energy-intensive way to make food; for every calorie of processed food it produces, another ten calories of fossil fuel energy are burned.

At this point the process has yielded a white slurry that's poured out onto a stainless steel table and dried to a fine, superwhite powder—cornstarch. Cornstarch comprised wet milling's sole product when the industry got its start in the 1840s. At first the laundry business was its biggest customer, but cooks and early food processors soon began adding cornstarch to as many recipes as they could: It offered the glamour of modernity, purity, and absolute whiteness. By 1866, corn refiners had learned how to use acids to break down cornstarch into glucose, and sweeteners quickly became—as they remain today—the industry's most important product. Corn syrup (which is mostly glucose or dextrose—the terms are interchangeable) became the first cheap domestic substitute for cane sugar.

I remember an elementary school science experiment in which we were instructed to chew—and chew and chew—a cracker until the slurry of starch turned suddenly sweet on our tongues. The teacher explained that the enzymes in our saliva had broken the long starch molecules into shorter molecules of glucose. Much the same process—it's called "enzyme hydrolysis"—revolutionized corn refining in the 1940s. As enzymes replaced acids, refiners were able to produce progressively sweeter sweeteners from corn. Yet none were quite as sweet as sugar (or, to be more precise, sucrose). That threshold wasn't crossed until the late 1960s, when Japanese chemists "broke the sweetness barrier," in the words of the Corn Refiners Association's official history of high-fructose corn sweetener. They discovered that an enzyme called glucose isomerase could transform glucose into the much sweeter sugar molecule called fructose. By the 1970s the process of refining corn into fructose had been perfected, and high-fructose corn syrup—which is a blend of 55 percent fructose and 45 percent glucose that tastes exactly as sweet as sucrose—came onto the market. Today it is the most valuable food product refined from corn, accounting for 530 million bushels every year. (A bushel of corn yields thirty-three pounds of fructose.)

But if the pipe marked "HFCS" leads to the fattest spigot at the far end of a corn refinery's bewildering tangle of pipes and valves, it is by no means the only spigot you'll find back there. There are dozens of other "output streams." At various points along its way through the mill some portion of the thick white slurry of starch is diverted to another purpose or, in the refiner's jargon, another "fraction." The starch itself is capable of being modified into spherical, crystalline, or highly branched molecules, each suitable for a different use: adhesives, coatings, sizings, and plastics for industry; stabilizers, thickeners, gels, and "viscosity-control agents" for food.

What remains in the slurry is "saccharified"—treated with enzymes that turn it into dextrose syrup. A portion of this dextrose is siphoned off for use as corn syrup; other fractions are recruited to become sugars like maltodextrin and maltose. The largest portion of the corn syrup stream is piped into a tank where it is exposed to glucose isomerase enzymes and then passed through ion exchange filters, emerging eventually as fructose. Now what's left of the dextrose stream is piped into a fermentation tank, where yeasts or amino acids go to work eating the sugars, in several hours yielding an alcoholic brew. This itself is fractionated into various alcohols, ethanol chief among them, our gas tanks being the ultimate destination of a tenth of the corn crop. The fermented brew can also be refined into a dozen different organic and amino acids for use in food processing or the manufacture of plastic.

And then that's about it: There's no corn left, and not much of anything else either, except for some dirty water. (Though even some of this "steep water" is used to make animal feeds.) The primary difference between the industrial digestion of corn and an animal's is that in this case there is virtually no waste at the end of it.

Step back for a moment and behold this great, intricately piped stainless steel beast: This is the supremely adapted creature that has evolved to help eat the vast surplus biomass coming off America's farms, efficiently digesting the millions of bushels of corn fed to it each day by the trainload. Go around back of this beast and you'll see a hundred different spigots, large and small, filling tanker cars of other trains with HFCS, ethanol, syrups, starches, and food additives of every description. The question now is, Who or what (besides our cars) is going to consume and digest all this freshly fractionated biomass—the sugars and starches, the alcohols and acids, the emulsifiers and stabilizers and viscosity-control agents? This is where we come in. It takes a certain kind of eater—an industrial eater—to consume these fractions of corn, and we are, or have evolved into, that supremely adapted creature: the eater of processed food.

2. PUTTING IT BACK TOGETHER AGAIN: PROCESSED FOODS

The dream of liberating food from nature is as old as eating. People began processing food to keep nature from taking it back: What is spoilage, after all, if not nature, operating through her proxy microorganisms, repossessing our hard-won lunch? So we learned to salt and dry and cure and pickle in the first age of food processing, and to can, freeze, and vacuum-pack in the second. These technologies were blessings, freeing people from nature's cycles of abundance and scarcity, as well as from the tyranny of the calendar or locale: Now a New Englander could eat sweet corn, or something reminiscent of it, in January, and taste a pineapple for the first time in his life. As Massimo Montanari, an Italian food historian, points out, the fresh, local, and seasonal food we prize today was for most of human history "a form of slavery," since it left us utterly at the mercy of the local vicissitudes of nature.

Even after people had learned the rudiments of preserving food, however, the dream of liberating food from nature continued to flourish—indeed, to expand in ambition and confidence. In the third age of food processing, which begins with the end of World War II, merely preserving the fruits of nature was deemed too modest: The goal now was to improve on nature. The twentieth-century prestige of technology and convenience combined with advances in marketing to push aside butter to make shelf space for margarine, replace fruit juice with juice drinks and then entirely juice-free drinks like Tang, cheese with Cheez Whiz, and whipped cream with Cool Whip.

Corn, a species that had been a modest beneficiary of the first two ages of food processing (having taken well to the can and the freezer), really came into its own during the third. You would never know it without reading the ingredient label (a literary genre unknown until the third age), but corn is the key constituent of all four of these processed foods. Along with the soybean, its rotational partner in the field, corn has done more than any other species to help the food industry realize the dream of freeing food from nature's limitations and seducing the omnivore into eating more of a single plant than anyone would ever have thought possible.

In fact, you would be hard-pressed to find a late-model processed food that isn't made from corn or soybeans. In the typical formulation, corn supplies the carbohydrates (sugars and starches) and soy the protein; the fat can come from either plant. (Remember what George Naylor said about the real produce of his farm: not corn and soybeans but "energy and protein.") The longer the ingredient label on a food, the more fractions of corn and soybeans you will find in it. They supply the essential building blocks, and from those two plants (plus a handful of synthetic additives) a food scientist can construct just about any processed food he or she can dream up.

A FEW YEARS AGO, in the days when "food security" meant something very different than it does today, I had the chance to visit one of the small handful of places where this kind of work is done. The Bell Institute, a leafy corporate campus on the outskirts of Minneapolis, is the research-and-development laboratory for General Mills, the sixth-largest food company in the world. Here nine hundred food scientists spend their days designing the future of food—its flavor, texture, and packaging.

Much of their work is highly secretive, but nowhere more so than in the cereals area. Deep in the heart of the heart of the Bell Institute, down in the bowels of the laboratory, you come to a warren of windowless rooms called, rather grandly, the Institute of Cereal Technology. I was permitted to pass through a high-security conference room furnished with a horseshoe-shaped table that had a pair of headphones at every seat. This was the institute's inner sanctum, the cereal situation room, where General Mills executives gather to hear briefings about new products.

The secrecy surrounding the successor to Cocoa Pebbles struck me as laughable, and I said so. But as an executive explained to me, "Recipes are not intellectual property; you can't patent a new cereal. All you can hope for is to have the market to yourself for a few months to establish your brand before a competitor knocks off the product. So we're very careful not to show our hand." For the same reason, the institute operates its own machine shop, where it designs and builds the machines that give breakfast cereals their shapes, making it that much harder for a competitor to knock off, say, a new marshmallow bit shaped to resemble a shooting star. In the interests of secrecy, the food scientists would not talk to me about current projects, only past failures, like the breakthrough cereal in the shapes of little bowling pins and balls. "In focus group the kids loved it," the product's rueful inventor told me, "but the mothers didn't like the idea of kids bowling their breakfast across the table." Which is why bowling pin cereal never showed up in your supermarket.

In many ways breakfast cereal is the prototypical processed food: four cents' worth of commodity corn (or some other equally cheap grain) transformed into four dollars' worth of processed food. What an alchemy! Yet it is performed straightforwardly enough: by taking several of the output streams issuing from a wet mill (corn meal, corn starch, corn sweetener, as well as a handful of tinier chemical fractions) and then assembling them into an attractively novel form. Further value is added in the form of color and taste, then branding and packaging. Oh yes, and vitamins and minerals, which are added to give the product a sheen of healthfulness and to replace the nutrients that are lost whenever whole foods are processed. On the strength of this alchemy the cereals group generates higher profits for General Mills than any other division. Since the raw materials in processed foods are so abundant and cheap (ADM and Cargill will gladly sell them to all comers) protecting whatever is special about the value you add to them is imperative.

I think it was at General Mills that I first heard the term "food system." Since then, I've seen in the pages of Food Technology, the monthly bible of the food-processing industry, that this term seems to be taking over from plain old "food." Food system is glossier and more high-tech than food, I guess; it also escapes some of the negative connotations that got attached to "processed food" during the sixties. It's probably as good a term as any when you're describing, as that magazine routinely does, new edible materials constructed from "textured vegetable protein," or a nutraceutical breakfast cereal so fortified with green tea, grape seed extract, and antioxidants that it's not even called a cereal but a "healthy heart system."

Exactly what corn is doing in such food systems has less to do with nutrition or taste than with economics. For the dream of liberating food from nature, which began as a dream of the eaters (to make it less perishable), is now primarily a dream of the feeders—of the corporations that sell us our food. No one was clamoring for synthetic cheese, or a cereal shaped like a bowling pin; processed food has become largely a supply-driven business—the business of figuring out clever ways to package and market the glut of commodities coming off the farm and out of the wet mills. Today the great advantages of processing food redound to the processors themselves. For them, nature is foremost a problem—not so much of perishable food (though that's always a concern when your market is global) as of perishable profits.

Like every other food chain, the industrial food chain is rooted at either end in a natural system: the farmer's field at one end, and the human organism at the other. From the capitalist's point of view, both of these systems are less than ideal.

The farm, being vulnerable to the vicissitudes of weather and pests, is prone to crises of over- and underproduction, both of which can hurt business. Rising raw material prices cut into profits, obviously enough. Yet the potential boon of falling raw material prices—which should allow you to sell a lot more of your product at a lower price— can't be realized in the case of food because of the special nature of your consumer, who can eat only so much food, no matter how cheap it gets. (Food industry executives used to call this the problem of the "fixed stomach"; economists speak of "inelastic demand.") Nature has cursed the companies working the middle of the food chain with a recipe for falling rates of profits.

The growth of the American food industry will always bump up against this troublesome biological fact: Try as we might, each of us can eat only about fifteen hundred pounds of food a year. Unlike many other products—CDs, say, or shoes—there's a natural limit to how much food we can each consume without exploding. What this means for the food industry is that its natural rate of growth is somewhere around 1 percent per year—1 percent being the annual growth rate of the American population. The problem is that won't tolerate such an anemic rate of growth.

This leaves companies like General Mills and McDonald's with two options if they hope to grow faster than the population: figure out how to get people to spend more money for the same three-quarters of a ton of food, or entice them to actually eat more than that. The two strategies are not mutually exclusive, of course, and the food industry energetically pursues them both at the same time. Which is good news indeed for the hero of our story, for it happens that turning cheap corn into complex food systems is an excellent way to achieve both goals.

BUILDING PROCESSED FOOD out of a commodity like corn doesn't completely cushion you from the vicissitudes of nature, but it comes close. The more complex your food system, the more you can practice "substitutionism" without altering the taste or appearance of the product. So if the price of hydrogenated fat or lecithin derived from corn spikes one day, you simply switch to fat or lecithin from soy, and the consumer will never know the difference. (This is why ingredient labels says things like "Contains one or more of the following: corn, soybean, or sunflower oil.") As a management consultant once advised his food industry clients, "The further a product's identity moves from a specific raw material—that is, the more processing steps involved—the less vulnerable is its processor" to the variability of nature.

In fact, there are lots of good reasons to complicate your product—or, as the industry prefers to say, to "add value" to it. Processing food can add months, even years, to its shelf life, allowing you to market globally. Complicating your product also allows you to capture more of the money a consumer spends on food. Of a dollar spent on a whole food such as eggs, $0.40 finds its way back to the farmer. By comparison, George Naylor will see only $0.04 of every dollar spent on corn sweeteners; ADM and Coca-Cola and General Mills capture most of the rest. (Every farmer I've ever met eventually gets around to telling the story about the food industry executive who declared, "There's money to be made in food, unless you're trying to grow it.") When Tyson food scientists devised the chicken nugget in 1983, a cheap bulk commodity—chicken—overnight became a high-value-added product, and most of the money Americans spend on chicken moved from the farmer's pocket to the processor's.

As Tyson understood, you want to be selling something more than a commodity, something more like a service: novelty, convenience, status, fortification, lately even medicine. The problem is, a value-added product made from a cheap commodity can itself become a commodity, so cheap and abundant are the raw materials. That lesson runs straight through the history of a company like General Mills, which started out in 1926 as a mill selling whole wheat flour: ground wheat. When that product became a cheap commodity, the company kept ahead of the competition by processing the grain a bit more, creating bleached and then "enriched" flour. Now they were adding value, selling not just wheat but an idea of purity and health, too. In time, however, even enriched white flour became a commodity, so General Mills took another step away from nature—from the farm and the plants in question—by inventing cake mixes and sweetened breakfast cereals. Now they were selling convenience, with a side of grain and corn sweetener, and today they're beginning to sell cereals that sound an awful lot like medicines. And so it goes, the rushing stream of ever cheaper agricultural commodities driving food companies to figure out new and ever more elaborate ways to add value and so induce us to buy more.

When I was in Minneapolis I spoke to a General Mills vice president who was launching a new line of organic TV dinners, a product that at first blush sounded like an oxymoron. The ingredient list went on forever, brimming with additives and obscure fractions of corn: maltodextrin, corn starch, xanthan gum. It seems that even organic food has succumbed to the economic logic of processing. The executive patiently explained that selling unprocessed or minimally processed whole foods will always be a fool's game, since the price of agricultural commodities tends to fall over time, whether they're organic or not. More food coming off the farm leads to either falling profits—or more processing.

The other problem with selling whole foods, he explained, is that it will always be hard to distinguish one company's corn or chickens or apples from any other company's. It makes much more sense to turn the corn into a brand-name cereal, the chicken into a TV dinner, and the apples into a component in a nutraceutical food system.

This last is precisely what one company profiled in a recent issue of *Food Technology* has done. TreeTop has developed a "low-moisture, naturally sweetened apple piece infused with a red-wine extract." Just eighteen grams of these apple pieces have the same amount of cancer-fighting "flavonoid phenols as five glasses of wine and the dietary fiber equivalent of one whole apple." Remember the sixties dream of an entire meal served in a pill, like the Jetsons? We've apparently moved from the meal-in-a-pill to the pill-in-a-meal, which is to say, not very far at all. Either way, the message is: We need food scientists to feed us. Of course, it was fortified breakfast cereal that first showed the way, by supplying more vitamins and minerals than any mere grain could hope to. Nature, these products implied, was no match for food science.

The news of TreeTop's breakthrough came in a recent *Food Technology* trend story titled "Getting More Fruits and Vegetables into Food." I had thought fruits and vegetables were *already* foods, and so didn't need to be gotten into them, but I guess that just shows I'm stuck in the food past. Evidently we're moving into the fourth age of food processing, in which the processed food will be infinitely better (i.e., contain more of whatever science has determined to be the good stuff) than the whole foods on which they're based. The food industry has gazed upon nature and found it wanting—and has gotten to work improving it.

Back in the seventies, a New York food additive manufacturer called International Flavors & Fragrances used its annual report to defend itself against the rising threat of "natural foods" and explain why we were better off eating synthetics. Natural ingredients, the company pointed out rather scarily, are a "wild mixture of substances created by plants and animals for completely non-food purposes—their survival and reproduction." These dubious substances "came to be consumed by humans at their own risk."

Now, thanks to the ingenuity of modern food science, we had a choice. We could eat things designed by humans for the express purpose of being eaten by people—or eat "substances" designed by natu-

ral selection for its own purposes: to, say, snooker a bee or lift a wing
or (eek!) make a baby. The meal of the future would be fabricated "in
the laboratory out of a wide variety of materials," as one food historian
wrote in 1973, including not only algae and fungi but also petrochemicals. Protein would be extracted directly from petroleum and then
"spun and woven into 'animal' muscle—long, wrist-thick tubes of 'filet
steak.'" (Come to think of it, agribusiness has long since mastered this
trick of turning petroleum into steak, though it still needs corn and cattle to do it.)

All that's really changed since the high-tech food future of the sixties is that the laboratory materials out of which these meals will be
constructed are nominally natural—the relative prestige of nature and
modern chemistry having traded places in the years since the rise of environmentalism. And besides, why go to the trouble and expense of
manufacturing food from petroleum when there is such a flood of
cheap carbon coming off the farm? So instead of creating foods whole
cloth from completely synthetic materials, the industry is building
them from fortified apple bits, red-wine extract, flavor fractions derived
from oranges, isoflavones from soy, meat substitutes fashioned from
mycoprotein, and resistant starches derived from corn. ("Natural raspberry flavor" doesn't mean the flavor came from a raspberry; it may well
have been derived from corn, just not from something synthetic.) But
the underlying reductionist premise—that a food is nothing more than
the sum of its nutrients—remains undisturbed. So we break down the
plants and animals into their component parts and then reassemble
them into high-value-added food systems. The omnivore's predilection
to eat a variety of species is tricked by this protean plant, and even the
biological limit on his appetite is overcome.

Resistant starch, the last novelty on that list of ingredients, has the
corn refiners particularly excited today. They've figured out how to
tease a new starch from corn that is virtually indigestible. You would
not think this is a particularly good thing for a food to be, unless of
course your goal is to somehow get around the biological limit on how
much each of us can eat in a year. Since the body can't break down re-

sistant starch, it slips through the digestive track without ever turning into calories of glucose—a particular boon, we're told, for diabetics. When fake sugars and fake fats are joined by fake starches, the food industry will at long last have overcome the dilemma of the fixed stomach: whole meals you can eat as often or as much of as you like, since this food will leave no trace. Meet the ultimate—the utterly elastic!—industrial eater.

SIX

THE CONSUMER

A Republic of Fat

In the early years of the nineteenth century, Americans began drinking more than they ever had before or since, embarking on a collective bender that confronted the young republic with its first major public health crisis—the obesity epidemic of its day. Corn whiskey, suddenly superabundant and cheap, became the drink of choice, and in 1820 the typical American was putting away half a pint of the stuff every day. That comes to more than five gallons of spirits a year for every man, woman, and child in America. The figure today is less than one.

As the historian W. J. Rorabaugh tells the story in *The Alcoholic Republic*, we drank the hard stuff at breakfast, lunch, and dinner, before work and after and very often during. Employers were expected to supply spirits over the course of the workday; in fact, the modern coffee break began as a late-morning whiskey break called "the elevenses." (Just to pronounce it makes you sound tipsy.) Except for a brief respite Sunday morning in church, Americans simply did not gather—whether for a barn raising or quilting bee, corn husking or political rally—without passing the whiskey jug. Visitors from Europe—hardly models of sobriety themselves—marveled at the free flow of American spirits. "Come on then, if you love toping," the journalist William Cobbett wrote his fellow Englishmen in a dispatch from America. "For here you may drink yourself blind at the price of sixpence."

The results of all this toping were entirely predictable: a rising tide of public drunkenness, violence, and family abandonment, and a spike in alcohol-related diseases. Several of the Founding Fathers—including George Washington, Thomas Jefferson, and John Adams—denounced the excesses of "the Alcoholic Republic," inaugurating an American quarrel over drinking that would culminate a century later in Prohibition.

But the outcome of our national drinking binge is not nearly as relevant to our own situation as its underlying cause. Which, put simply, was this: American farmers were producing far too much corn. This was particularly true in the newly settled regions west of the Appalachians, where fertile, virgin soils yielded one bumper crop after another. A mountain of surplus corn piled up in the Ohio River Valley. Much as today, the astounding productivity of American farmers proved to be their own worst enemy, as well as a threat to public health. For when yields rise, the market is flooded with grain, and its price collapses. What happens next? The excess biomass works like a vacuum in reverse: Sooner or later, clever marketers will figure out a way to induce the human omnivore to consume the surfeit of cheap calories.

As it is today, the clever thing to do with all that cheap corn was to process it—specifically, to distill it into alcohol. The Appalachian range made it difficult and expensive to transport surplus corn from the lightly settled Ohio River Valley to the more populous markets of the East, so farmers turned their corn into whiskey—a more compact and portable, and less perishable, value-added commodity. Before long the price of whiskey plummeted to the point that people could afford to drink it by the pint. Which is precisely what they did.

The Alcoholic Republic has long since given way to the Republic of Fat; we're eating today much the way we drank then, and for some of the same reasons. According to the surgeon general, obesity today is officially an epidemic; it is arguably the most pressing public health problem we face, costing the health care system an estimated $90 billion a year. Three of every five Americans are overweight; one of every five is obese. The disease formerly known as adult-onset diabetes has had to be renamed Type II diabetes since it now occurs so frequently in children. A recent study in the *Journal of the American Medical Association* predicts that a child born in 2000 has a one-in-three chance of developing diabetes. (An African American child's chances are two in five.) Because of diabetes and all the other health problems that accompany obesity, today's children may turn out to be the first generation of Americans whose life expectancy will actually be shorter than that of their parents. The problem is not limited to America: The United Nations reported that in 2000 the number of people suffering from overnutrition—a billion— had officially surpassed the number suffering from malnutrition—800 million.

You hear plenty of explanations for humanity's expanding waistline, all of them plausible. Changes in lifestyle (we're more sedentary; we eat out more). Affluence (more people can afford a high-fat Western diet). Poverty (healthier whole foods cost more). Technology (fewer of us use our bodies in our work; at home, the remote control keeps us pinned to the couch). Clever marketing (supersized portions; advertising to children). Changes in diet (more fats; more carbohydrates; more processed foods).

All these explanations are true, as far as they go. But it pays to go a little further, to search for the cause behind the causes. Which, very simply, is this: When food is abundant and cheap, people will eat more of it and get fat. Since 1977 an American's average daily intake of calories has jumped by more than 10 percent. Those two hundred calories have to go somewhere, and absent an increase in physical activity (which hasn't happened), they end up being stored away in fat cells in our bodies. But the important question is, Where, exactly, did all those extra calories come from in the first place? And the answer to that question takes us back to the source of almost all calories: the farm.

Most researchers trace America's rising rates of obesity to the 1970s.

This was, of course, the same decade that America embraced a cheap-food farm policy and began dismantling forty years of programs designed to prevent overproduction. Earl Butz, you'll recall, sought to drive up agricultural yields in order to drive down the price of the industrial food chain's raw materials, particularly corn and soybeans. It worked: The price of food is no longer a political issue. Since the Nixon administration, farmers in the United States have managed to produce 500 additional calories per person every day (up from 3,300, already substantially more than we need); each of us is, heroically, managing to put away 200 of those surplus calories at the end of their trip up the food chain. Presumably the other 300 are being dumped overseas, or turned (once again!) into ethyl alcohol: ethanol for our cars.

The parallels with the alcoholic republic of two hundred years ago are hard to miss. Before the changes in lifestyle, before the clever marketing, comes the mountain of cheap corn. Corn accounts for most of the surplus calories we're growing and most of the surplus calories we're eating. As then, the smart thing to do with all that surplus grain is to process it, transform the cheap commodity into a value-added consumer product—a denser and more durable package of calories. In the 1820s the processing options were basically two: You could turn your corn into pork or alcohol. Today there are hundreds of things a processor can do with corn: They can use it to make everything from chicken nuggets and Big Macs to emulsifiers and nutraceuticals. Yet since the human desire for sweetness surpasses even our desire for intoxication, the cleverest thing to do with a bushel of corn is to refine it into thirty-three pounds of high-fructose corn syrup.

That at least is what we're doing with about 530 million bushels of the annual corn harvest—turning it into 17.5 billion pounds of high-fructose corn syrup. Considering that the human animal did not taste this particular food until 1980, for HFCS to have become the leading source of sweetness in our diet stands as a notable achievement on the part of the corn-refining industry, not to mention this remarkable plant. (But then, plants have always known that one of the surest paths to evolutionary success is by gratifying the mammalian omnivore's innate desire for sweetness.) Since 1985, an American's annual consumption of HFCS has gone from forty-five pounds to sixty-six pounds. You might think that this growth would have been offset by a decline in sugar consumption, since HFCS often replaces sugar, but that didn't happen: During the same period our consumption of refined sugar actually went up by five pounds. What this means is that we're eating and drinking all that high-fructose corn syrup *on top* of the sugars we were already consuming. In fact, since 1985 our consumption of all added sugars—cane, beet, HFCS, glucose, honey, maple syrup, whatever—has climbed from 128 pounds to 158 pounds per person.

This is what makes high-fructose corn syrup such a clever thing to do with a bushel of corn: By inducing people to consume more calories than they otherwise might, it gets them to really chomp through the corn surplus. Corn sweetener is to the republic of fat what corn whiskey was to the alcoholic republic. Read the food labels in your kitchen and you'll find that HFCS has insinuated itself into every corner of the pantry: not just into our soft drinks and snack foods, where you would expect to find it, but into the ketchup and mustard, the breads and cereals, the relishes and crackers, the hot dogs and hams.

But it is in soft drinks that we consume most of our sixty-six pounds of high-fructose corn syrup, and to the red-letter dates in the natural history of *Zea mays*—right up there with teosinte's catastrophic sexual mutation, Columbus's introduction of maize to the court of Queen Isabella in 1493, and Henry Wallace's first F-1 hybrid seed in 1927—we must now add the year 1980. That was the year corn first became an ingredient in Coca-Cola. By 1984, Coca-Cola and Pepsi had switched over entirely from sugar to high-fructose corn syrup. Why? Because HFCS was a few cents cheaper than sugar (thanks in part to tariffs on imported sugarcane secured by the corn refiners) and consumers didn't seem to notice the substitution.

The soft drink makers' switch should have been a straightforward, zero-sum trade-off between corn and sugarcane (both, incidentally, C-4 grasses), but it wasn't: We soon began swilling a lot more soda and therefore corn sweetener. The reason isn't far to seek: Like corn whiskey in the 1820s, the price of soft drinks plummeted. Note, however, that Coca-Cola and Pepsi did not simply cut the price of a bottle of cola. That would only have hurt profit margins, for how many people are going to buy a second soda just because it cost a few cents less? The companies had a much better idea: They would supersize their sodas. Since a soft drink's main raw material—corn sweetener—was now so cheap, why not get people to pay just a few pennies more for a substantially bigger bottle? Drop the price per ounce, but sell a lot more ounces. So began the transformation of the svelte eight-ounce Coke bottle into the chubby twenty-ouncer dispensed by most soda machines today.

But the soda makers don't deserve credit for the invention of supersizing. That distinction belongs to a man named David Wallerstein. Until his death in 1993, Wallerstein served on the board of directors at McDonald's, but in the fifties and sixties he worked for a chain of movie theaters in Texas, where he labored to expand sales of soda and popcorn—the high-markup items that theaters depend on for their profitability. As the story is told in John Love's official history of McDonald's, Wallerstein tried everything he could think of to goose up sales—two-for-one deals, matinee specials—but found he simply could not induce customers to buy more than one soda and one bag of popcorn. He thought he knew why: Going for seconds makes people feel piggish.

Wallerstein discovered that people *would* spring for more popcorn and soda—a lot more—as long as it came in a single gigantic serving. Thus was born the two-quart bucket of popcorn, the sixty-four-ounce Big Gulp, and, in time, the Big Mac and the jumbo fries, though Ray Kroc himself took some convincing. In 1968, Wallerstein went to work for McDonald's, but try as he might, he couldn't convince Kroc, the company's founder, of supersizing's magic powers.

"If people want more fries," Kroc told him, "they can buy two bags." Wallerstein patiently explained that McDonald's customers did want more but were reluctant to buy a second bag. "They don't want to look like gluttons."

Kroc remained skeptical, so Wallerstein went looking for proof. He began staking out McDonald's outlets in and around Chicago, observing how people ate. He saw customers noisily draining their sodas, and digging infinitesimal bits of salt and burnt spud out of their little bags of French fries. After Wallerstein presented his findings, Kroc relented and approved supersized portions, and the dramatic spike in sales confirmed the marketer's hunch. Deep cultural taboos against gluttony—one of the seven deadly sins, after all—had been holding us back. Wallerstein's dubious achievement was to devise the dietary equivalent of a papal dispensation: Supersize it! He had discovered the secret to expanding the (supposedly) fixed human stomach.

One might think that people would stop eating and drinking these gargantuan portions as soon as they felt full, but it turns out hunger doesn't work that way. Researchers have found that people (and animals) presented with large portions will eat up to 30 percent more than they would otherwise. Human appetite, it turns out, is surprisingly elastic, which makes excellent evolutionary sense: It behooved our hunter-gatherer ancestors to feast whenever the opportunity presented itself, allowing them to build up reserves of fat against future famine. Obesity researchers call this trait the "thrifty gene." And while the gene represents a useful adaptation in an environment of food scarcity and unpredictability, it's a disaster in an environment of fast-food abundance, when the opportunity to feast presents itself 24/7. Our bodies are storing reserves of fat against a famine that never comes.

But if evolution has left the modern omnivore vulnerable to the blandishments of supersizing, the particular nutrients he's most likely to encounter in those supersized portions—lots of added sugar and fat—make the problem that much worse. Like most other warm-blooded creatures, humans have inherited a preference for energy-dense foods, a preference reflected in the sweet tooth shared by most mammals. Natural selection predisposed us to the taste of sugar and fat (its texture as well as taste) because sugars and fats offer the most energy (which is what a calorie is) per bite. Yet in nature—in whole foods—we seldom encounter these nutrients in the concentrations we now find them in in processed foods: You won't find a fruit with anywhere near the amount of fructose in a soda, or a piece of animal flesh with quite as much fat as a chicken nugget.

You begin to see why processing foods is such a good strategy for getting people to eat more of them. The power of food science lies in its ability to break foods down into their nutrient parts and then re-assemble them in specific ways that, in effect, push our evolutionary buttons, fooling the omnivore's inherited food selection system. Add fat or sugar to anything and it's going to taste better on the tongue of an animal that natural selection has wired to seek out energy-dense foods. Animal studies prove the point: Rats presented with solutions of pure sucrose or tubs of pure lard—goodies they seldom encounter in nature—will gorge themselves sick. Whatever nutritional wisdom the rats are born with breaks down when faced with sugars and fats in un-natural concentrations—nutrients ripped from their natural context, which is to say, from those things we call foods. Food systems can cheat by exaggerating their energy density, tricking a sensory apparatus that evolved to deal with markedly less dense whole foods.

It is the amped-up energy density of processed foods that gets om-nivores like us into trouble. Type II diabetes typically occurs when the body's mechanism for managing glucose simply wears out from over-use. Just about everything we eat sooner or later winds up in the blood as molecules of glucose, but sugars and simple starches turn to glucose faster than anything else. Type II diabetes and obesity are exactly what you would expect to see in a mammal whose environment has over-whelmed its metabolism with energy-dense foods.

This begs the question of why the problem has gotten so much worse in recent years. It turns out the price of a calorie of sugar or fat has plummeted since the 1970s. One reason that obesity and diabetes become more prevalent the further down the socioeconomic scale you look is that the industrial food chain has made energy-dense foods the cheapest foods in the market, when measured in terms of cost per calo-rie. A recent study in the *American Journal of Clinical Nutrition* compared the "energy cost" of different foods in the supermarket. The researchers found that a dollar could buy 1,200 calories of potato chips and cookies; spent on a whole food like carrots, the same dollar buys only 250 calories. On the beverage aisle, you can buy 875 calories of soda for a dollar, or 170 calories of fruit juice from concentrate. It makes good economic sense that people with limited money to spend on food would spend it on the cheapest calories they can find, especially when the cheapest calories—fats and sugars—are precisely the ones offering the biggest neurobiological rewards.

Corn is not the only source of cheap energy in the supermarket—much of the fat added to processed foods comes from soybeans—but it is by far the most important. As George Naylor said, growing corn is the most efficient way to get energy—calories—from an acre of Iowa farmland. That corn-made calorie can find its way into our bodies in the form of an animal fat, a sugar, or a starch, such is the protean nature of the carbon in that big kernel. But as productive and protean as the corn plant is, finally it is a set of human choices that have made these molecules quite as cheap as they have become: a quarter century of farm policies designed to encourage the overproduction of this crop and hardly any other. Very simply, we subsidize high-fructose corn syrup in this country, but not carrots. While the surgeon general is raising alarms over the epidemic of obesity, the president is signing farm bills designed to keep the river of cheap corn flowing, guaranteeing that the cheapest calories in the supermarket will continue to be the unhealthiest.

SEVEN

THE MEAL

Fast Food

The meal at the end of the industrial food chain that begins in an Iowa cornfield is prepared by McDonald's and eaten in a moving car. Or at least this was the version of the industrial meal I chose to eat; it could easily have been another. The myriad streams of commodity corn, after being variously processed and turned into meat, converge in all sorts of different meals I might have eaten, at KFC or Pizza Hut or Applebee's, or prepared myself from ingredients bought at the supermarket. Industrial meals are all around us, after all; they make up the food chain from which most of us eat most of the time.

My eleven-year-old son, Isaac, was more than happy to join me at McDonald's; he doesn't get there often, so it's a treat. (For most American children today, it is no longer such a treat: One in three of them eat fast food every single day.) Judith, my wife, was less enthusiastic. She's careful about what she eats, and having a fast-food lunch meant giving up a "real meal," which seemed a shame. Isaac pointed out that she could order one of McDonald's new "premium salads" with the Paul Newman dressing. I read in the business pages that these salads are a big hit, but even if they weren't, they'd probably stay on the menu strictly for their rhetorical usefulness. The marketers have a term for what a salad or veggie burger does for a fast-food chain: "denying the denier." These healthier menu items hand the child who wants to eat fast food a sharp tool with which to chip away at his parents' objections. "But Mom, you can get the salad . . ."

Which is exactly what Judith did: order the Cobb salad with Caesar dressing. At $3.99, it was the most expensive item on the menu. I ordered a classic cheeseburger, large fries, and a large Coke. Large turns out to be a full 32 ounces (a quart of soda!) but, thanks to the magical economics of supersizing, it cost only 30 cents more than the 16-ounce "small." Isaac went with the new white-meat Chicken McNuggets, a double-thick vanilla shake, and a large order of fries, followed by a new dessert treat consisting of freeze-dried pellets of ice cream. That each of us ordered something different is a hallmark of the industrial food chain, which breaks the family down into its various demographics and markets separately to each one: Together we would be eating alone together, and therefore probably eating more. The total for the three of us came to fourteen dollars, and was packed up and ready to go in four minutes. Before I left the register I picked up a densely printed handout called "A Full Serving of Nutrition Facts: Choose the Best Meal for You."

We could have slipped into a booth, but it was such a nice day we decided to put the top down on the convertible and eat our lunch in the car, something the food and the car have both been engineered to accommodate. These days 19 percent of American meals are eaten in the car. The car has cup holders, front seat and rear, and, except for the salad, all the food (which we could have ordered, paid for, and picked up without opening the car door) can be readily eaten with one hand. Indeed, this is the genius of the chicken nugget: It liberated chicken from the fork and plate, making it as convenient, waste-free, and automobile-friendly as the precondimented hamburger. No doubt the food scientists at McDonald's corporate headquarters in Oak Brook, Illinois, are right now hard at work on the one-handed salad.

But though Judith's Cobb salad did present a challenge to front-seat dining, eating it at fifty-five miles per hour seemed like the thing to do, since corn was the theme of this meal: The car was eating corn too, being fueled in part by ethanol. Even though the additive promises to diminish air quality in California, new federal mandates pushed by the corn processors require refineries in the state to help eat the corn surplus by diluting their gasoline with 10 percent ethanol.

I ate a lot of McDonald's as a kid. This was in the pre-Wallerstein era, when you still had to order a second little burger or sack of fries if you wanted more, and the chicken nugget had not yet been invented. (One memorable childhood McDonald's meal ended when our station wagon got rear-ended at a light, propelling my milk shake across the car in creamy white lariats.) I loved everything about fast food: the individual portions all wrapped up like presents (not having to share with my three sisters was a big part of the appeal; fast food was private property at its best); the familiar meaty perfume of the French fries filling the car; and the pleasingly sequenced bite into a burger—the soft, sweet roll, the crunchy pickle, the savory moistness of the meat.

Well-designed fast food has a fragrance and flavor all its own, a fragrance and flavor only nominally connected to hamburgers or French fries or for that matter to *any* particular food. Certainly the hamburgers and fries you make at home don't have it. And yet Chicken McNuggets do, even though they're ostensibly an entirely different food made from a different species. Whatever it is (surely the food scientists know), for countless millions of people living now, this generic fast-food flavor is one of the unerasable smells and tastes of childhood—which makes it a kind of comfort food. Like other comfort foods, it supplies (besides nostalgia) a jolt of carbohydrates and fat, which, some scientists now believe, relieve stress and bathe the brain in chemicals that make it feel good.

Isaac announced that his white-meat McNuggets were tasty, a definite improvement over the old recipe. McNuggets have come in for a lot of criticism recently, which might explain the reformulation. Ruling in 2003 in a lawsuit brought against McDonald's by a group of obese teenagers, a federal judge in New York had defamed the McNugget even as he dismissed the suit. "Rather than being merely chicken fried in a pan," he wrote in his decision, McNuggets "are a McFrankensteinian creation of various elements not utilized by the home cook." After cataloging the thirty-eight ingredients in a McNugget, Judge Sweet suggested that McDonald's marketing bordered on deceptive, since the dish is not what it purports to be—that is, a piece of chicken simply fried—and, contrary to what a consumer might reasonably expect, actually contains more fat and total calories than a cheeseburger. Since the lawsuit, McDonald's has reformulated the nugget with white meat, and begun handing out "A Full Serving of Nutrition Facts."* According to the flyer, a serving of six nuggets now has precisely ten fewer calories than a cheeseburger. Chalk up another achievement for food science.

When I asked Isaac if the new nuggets tasted more like chicken than the old ones, he seemed baffled by the question. "No, they taste like what they are, which is nuggets," and then dropped on his dad a withering two-syllable "duh." In this consumer's mind at least, the link between a nugget and the chicken in it was never more than notional, and probably irrelevant. By now the nugget constitutes its own genre of food for American children, many of whom eat nuggets every day. For Isaac, the nugget is a distinct taste of childhood, quite apart from chicken, and no doubt a future vehicle of nostalgia—a madeleine in the making.

Isaac passed one up to the front for Judith and me to sample. It looked and smelled pretty good, with a nice crust and bright white interior reminiscent of chicken breast meat. In appearance and texture a nugget certainly alludes to fried chicken, yet all I could really taste was salt, that all-purpose fast-food flavor, and, okay, maybe a note of chicken bouillon informing the salt. Overall the nugget seemed more like an abstraction than a full-fledged food, an idea of chicken waiting to be fleshed out.

The ingredients listed in the flyer suggest a lot of thought goes into a nugget, that and a lot of corn. Of the thirty-eight ingredients it takes

---

*In 2005 McDonald's announced it would begin printing nutrition information on its packaging.

to make a McNugget, I counted thirteen that can be derived from corn: the corn-fed chicken itself; modified cornstarch (to bind the pulverized chicken meat); mono-, tri-, and diglycerides (emulsifiers, which keep the fats and water from separating); dextrose; lecithin (another emulsifier); chicken broth (to restore some of the flavor that processing leaches out); yellow corn flour and more modified cornstarch (for the batter); cornstarch (a filler); vegetable shortening; partially hydrogenated corn oil; and citric acid as a preservative. A couple of other plants take part in the nugget: There's some wheat in the batter, and on any given day the hydrogenated oil could come from soybeans, canola, or cotton rather than corn, depending on market price and availability.

According to the handout, McNuggets also contain several completely synthetic ingredients, quasi-edible substances that ultimately come not from a corn or soybean field but from a petroleum refinery or chemical plant. These chemicals are what make modern processed foods possible, by keeping the organic materials in them from going bad or looking strange after months in the freezer or on the road. Listed first are the "leavening agents": sodium aluminum phosphate, monocalcium phosphate, sodium acid pyrophosphate, and calcium lactate. These are antioxidants added to keep the various animal and vegetable fats involved in a nugget from turning rancid. Then there are "antifoaming agents" like dimethylpolysiloxene, added to the cooking oil to keep the starches from binding to air molecules, so as to produce foam during the fry. The problem is evidently grave enough to warrant adding a toxic chemical to the food: According to the *Handbook of Food Additives*, dimethylpolysiloxene is a suspected carcinogen and an established mutagen, tumorigen, and reproductive effector; it's also flammable. But perhaps the most alarming ingredient in a Chicken McNugget is tertiary butylhydroquinone, or TBHQ, an antioxidant derived from petroleum that is either sprayed directly on the nugget or the inside of the box it comes in to "help preserve freshness." According to *A Consumer's Dictionary of Food Additives*, TBHQ is a form of butane (i.e., lighter fluid) the FDA allows processors to use sparingly in our food: It can comprise no more than 0.02 percent of the oil in a nugget. Which is probably just as well, considering that ingesting a single gram of TBHQ can cause "nausea, vomiting, ringing in the ears, delirium, a sense of suffocation, and collapse." Ingesting five grams of TBHQ can kill.

With so many exotic molecules organized into a food of such complexity, you would almost expect a chicken nugget to do something more spectacular than taste okay to a child and fill him up inexpensively. What it has done, of course, is to sell an awful lot of chicken for companies like Tyson, which invented the nugget—at McDonald's behest—in 1983. The nugget is the reason chicken has supplanted beef as the most popular meat in America.

Compared to Isaac's nuggets, my cheeseburger is a fairly simple construct. According to "A Full Serving of Nutrition Facts," the cheeseburger contains a mere six ingredients, all but one of them familiar: a 100 percent beef patty, a bun, two American cheese slices, ketchup, mustard, pickles, onions, and "grill seasoning," whatever that is. It tasted pretty good, too, though on reflection what I mainly tasted were the condiments: Sampled by itself, the gray patty had hardly any flavor. And yet the whole package, especially on first bite, did manage to give off a fairly convincing burgerish aura. I suspect, however, that owes more to the olfactory brilliance of the "grill seasoning" than to the 100 percent beef patty.

In truth, my cheeseburger's relationship to beef seemed nearly as metaphorical as the nugget's relationship to a chicken. Eating it, I had to remind myself that there was an actual cow involved in this meal—most likely a burned-out old dairy cow (the source of most fast-food beef) but possibly bits and pieces of a steer like 534 as well. Part of the appeal of hamburgers and nuggets is that their boneless abstractions allow us to forget we're eating animals. I'd been on the feedlot in Garden City only a few months earlier, yet this experience of cattle was so far removed from that one as to be taking place in a different dimension. No, I could not taste the feed corn or the petroleum or the antibiotics or the hormones—or the feedlot manure. Yet while "A Full Serving of Nutrition Facts" did not enumerate *these* facts, they too have gone into the making of this hamburger, are part of its natural history. That perhaps is what the industrial food chain does best: obscure the histories of the foods it produces by processing them to such an extent that they appear as pure products of culture rather than nature—things made from plants and animals. Despite the blizzard of information contained in the helpful McDonald's flyer—the thousands of words and numbers specifying ingredients and portion sizes, calories and nutrients—all this food remains perfectly opaque. Where does it come from? It comes from McDonald's.

But that's not so. It comes from refrigerated trucks and from warehouses, from slaughterhouses, from factory farms in towns like Garden City, Kansas, from ranches in Sturgis, South Dakota, from food science laboratories in Oak Brook, Illinois, from flavor companies on the New Jersey Turnpike, from petroleum refineries, from processing plants owned by ADM and Cargill, from grain elevators in towns like Jefferson, and, at the end of that long and tortuous trail, from a field of corn and soybeans farmed by George Naylor in Churdan, Iowa.

It would not be impossible to calculate exactly how much corn Judith, Isaac, and I consumed in our McDonald's meal. I figure my 4-ounce burger, for instance, represents nearly 2 pounds of corn (based on a cow's feed conversion rate of 7 pounds of corn for every 1 pound of gain, half of which is edible meat). The nuggets are a little harder to translate into corn, since there's no telling how much actual chicken goes into a nugget; but if 6 nuggets contain a quarter pound of meat, that would have taken a chicken half a pound of feed corn to grow. A 32-ounce soda contains 86 grams of high-fructose corn syrup (as does a double-thick shake), which can be refined from a third of a pound of corn; so our 3 drinks used another 1 pound. Subtotal: 6 pounds of corn.

From here the calculations become trickier because, according to the ingredients list in the flyer, corn is everywhere in our meal, but in unspecified amounts. There's more corn sweetener in my cheeseburger, of all places: The bun and the ketchup both contain HFCS. It's in the salad dressing, too, and the sauces for the nuggets, not to mention Isaac's dessert. (Of the sixty menu items listed in the handout, forty-five contain HFCS.) Then there are all the other corn ingredients in the nugget: the binders and emulsifiers and fillers. In addition to corn sweeteners, Isaac's shake contains corn syrup solids, mono- and diglycerides, and milk from corn-fed animals. Judith's Cobb salad is also stuffed with corn, even though there's not a kernel in it: Paul Newman makes his dressing with HFCS, corn syrup, corn starch, dextrin, caramel color, and xanthan gum; the salad itself contains cheese and eggs from corn-fed animals. The salad's grilled chicken breast is injected with a "flavor solution" that contains maltodextrin, dextrose, and monosodium glutamate. Sure, there are a lot of leafy greens in Judith's salad too, but the overwhelming majority of the calories in it (and there are 500 of them, when you count the dressing) ultimately come from corn. And the French fries? You would think those are mostly potatoes. Yet since half of the 540 calories in a large order of fries come from the oil they're fried in, the ultimate source of these calories is not a potato farm but a field of corn or soybeans.

The calculation finally defeated me, but I took it far enough to estimate that, if you include the corn in the gas tank (a whole bushel right there, to make two and a half gallons of ethanol), the amount of corn that went into producing our movable fast-food feast would easily have overflowed the car's trunk, spilling a trail of golden kernels on the blacktop behind us.

Some time later I found another way to calculate just how much corn we had eaten that day. I asked Todd Dawson, a biologist at Berkeley, to run a McDonald's meal through his mass spectrometer and calculate how much of the carbon in it came originally from a corn plant. It is hard to believe that the identity of the atoms in a cheeseburger or a Coke is preserved from farm field to fast-food counter, but the atomic signature of those carbon isotopes is indestructible, and still legible to the mass spectrometer. Dawson and his colleague Stefania Mambelli prepared an analysis showing roughly how much of the carbon in the various McDonald's menu items came from corn, and plotted them on a graph. The sodas came out at the top, not surprising since they consist of little else than corn sweetener, but virtually everything else we ate revealed a high proportion of corn, too. In order of diminishing corniness, this is how the laboratory measured our meal: soda (100 percent corn), milk shake (78 percent), salad dressing (65 percent), chicken nuggets (56 percent), cheeseburger (52 percent), and French fries (23 percent). What in the eyes of the omnivore looks like a meal of impressive variety turns out, when viewed through the eyes of the mass spectrometer, to be the meal of a far more specialized kind of eater. But then, this is what the industrial eater has become: corn's koala.

So WHAT? Why should it matter that we have become a race of corn eaters such as the world has never seen? Is this necessarily a bad thing? The answer all depends on where you stand.

If where you stand is in agribusiness, processing cheap corn into forty-five different McDonald's items is an impressive accomplishment. It represents a solution to the agricultural contradictions of capitalism, the challenge of increasing food industry profits faster than America can increase its population. Supersized portions of cheap corn-fixed carbon solves the problem of the fixed stomach; we may not be expanding the number of eaters in America, but we've figured out how to expand each of their appetites, which is almost as good. Judith, Isaac, and I together consumed a total of 4,510 calories at our lunch—more than half as many as we each should probably consume in a day. We had certainly done our parts in chomping through the corn surplus. (We had also consumed a lot of petroleum, and not just because we were in a car. To grow and process those 4,510 food calories took at least ten times as many calories of fossil energy, the equivalent of 1.3 gallons of oil.)

If where you stand is on one of the lower rungs of America's economic ladder, our cornified food chain offers real advantages: not cheap food exactly (for the consumer ultimately pays the added cost of processing), but cheap calories in a variety of attractive forms. In the long run, however, the eater pays a high price for these cheap calories: obesity, Type II diabetes, heart disease.

If where you stand is at the lower end of the *world's* economic ladder, however, America's corn-fed food chain looks like an unalloyed disaster. I mentioned earlier that all life on earth can be viewed as a competition for the energy captured by plants and stored in carbohydrates, energy we measure in calories. There is a limit to how many of those calories the world's arable land can produce each year, and an industrial meal of meat and processed food consumes—and wastes—an unconscionable amount of that energy. To eat corn directly (as Mexicans and many Africans do) is to consume all the energy in that corn, but when you feed that corn to a steer or a chicken, 90 percent of its energy is lost—to bones or feathers or fur, to living and metabolizing as a steer or chicken. This is why vegetarians advocate eating "low on the food chain"; every step up the chain reduces the amount of food energy by a factor of ten, which is why in any ecosystem there are only a fraction as many predators as there are prey. But processing food also burns energy. What this means is that the amount of food energy lost in the making of something like a Chicken McNugget could feed a great many more children than just mine, and that behind the 4,510 calories the three of us had for lunch stand tens of thousand of corn calories that could have fed a great many hungry people.

And how does this corn-fed food chain look if where you stand is in the middle of a field of corn? Well, it depends on whether you are the corn farmer or the plant. For the corn farmer, you might think the cornification of our food system would have redounded to his benefit, but it has not. Corn's triumph is the direct result of its overproduction, and that has been a disaster for the people who grow it. Growing corn and nothing but corn has also exacted a toll on the farmer's soil, the quality of the local water and the overall health of his community, the biodiversity of his landscape, and the health of all the creatures living on or downstream from it. And not only those creatures, for cheap corn has also changed, and much for the worse, the lives of several billion food animals, animals that would not be living on factory farms if not for the ocean of corn on which these animal cities float.

But return to that Iowa farm field for a moment and look at the matter—at us—from the standpoint of the corn plant itself. Corn, corn,

corn as far as the eye can see, ten-foot stalks soldiering in perfect thirty-inch rows to the far horizon, an 80-million-acre corn lawn rolling across the continent. It's a good thing this plant can't form an impression of us, for how risible that impression would be: the farmers going broke cultivating it; the countless other species routed or emiserated by it; the humans eating and drinking it as fast as they can, some of them—like me and my family—in automobiles engineered to drink it, too. Of all the species that have figured out how to thrive in a world dominated by *Homo sapiens*, surely no other has succeeded more spectacularly—has colonized more acres and bodies—than *Zea mays*, the grass that domesticated its domesticator. You have to wonder why we Americans don't worship this plant as fervently as the Aztecs; like they once did, we make extraordinary sacrifices to it.

These, at least, were my somewhat fevered speculations, as we sped down the highway putting away our fast-food lunch. What is it about fast food? Not only is it served in a flash, but more often than not it's eaten that way too: We finished our meal in under ten minutes. Since we were in the convertible and the sun was shining, I can't blame the McDonald's ambiance. Perhaps the reason you eat this food quickly is because it doesn't bear savoring. The more you concentrate on how it tastes, the less like anything it tastes. I said before that McDonald's serves a kind of comfort food, but after a few bites I'm more inclined to think they're selling something more schematic than that—something more like a signifier of comfort food. So you eat more and eat more quickly, hoping somehow to catch up to the original idea of a cheeseburger or French fry as it retreats over the horizon. And so it goes, bite after bite, until you feel not satisfied exactly, but simply, regrettably, full.

II

PASTORAL
GRASS

EIGHT

ALL FLESH
IS GRASS

1. GREEN ACRES

Early in the afternoon on the first day of summer, I found myself sitting in the middle of an impossibly green pasture, resting. "The longest day of the year" is what I would jot down in my notebook in bed late that night, followed by "literally," which was then struck out and replaced with "figuratively." What can I say? I was tired. I'd spent the afternoon making hay, really just lending a hand to a farmer making hay, and after a few hours in the midday sun hoisting and throwing fifty-pound bales onto a hay wagon, I hurt. We think of grass as soft and hospitable stuff, but once it's been dried in the sun and shredded by machines—once it's become hay—grass is sharp enough to draw blood and dusty enough to thicken lungs. I was covered in chaff, my forearms tattooed red with its pinpricks.

The others—Joel Salatin, whose farm this was; his grown son, Daniel; and two helpers—had gone off to the barn for something, leaving me with a welcome moment in the pasture to gather myself before we cranked up the baler again. We were racing to get this hay in before thunderstorms predicted for the evening. It was Monday, my first of seven days working on the farm, and thus far my principal conclusion was that in the event I survived the labors of the week, I would never again begrudge a farmer any price he cared to name for his produce: one dollar for an egg seemed entirely reasonable; fifty dollars for a steak a steal.

The wail of farm machinery had fallen silent, and in the space it left I could hear the varied sounds of birds: songbirds in the trees, but also the low gossip of hens and the lower throat singing of turkeys. Up on the green, green shoulder of hill rising to the west I could see a small herd of cattle grazing, and, below them on a gentler slope, several dozen portable chicken pens marching in formation down the meadow.

Laid before me was, I realized, a scene of almost classical pastoral beauty—the meadows dotted with contented animals, the backdrop of woods, a twisting brook threading through it all—marred only by the fact that I couldn't just lie here on this springy pasture admiring it for the rest of the afternoon. (Wasn't leisure supposed to be a big part of the pastoral idyll?) Our culture, perhaps even our biology, disposes us to respond to just such a grassy middle landscape, suspended as it is halfway between the wilderness of forest and the artifice of civilization. "The argument of the verdurous vista," Henry James once called it. He had just returned from Europe to tour rural New England, and found himself beguiled by Connecticut's pastoral charms in spite of himself and all he knew—about history, about the inevitable triumph of the machine, about "the bullying railway." A century earlier, of course, Thomas Jefferson had made the argument of the verdurous vista with a force some of us still feel: His agrarian ideal was an attempt to make a literal American reality out of the old world's pastoral dreams, though even he sometimes doubted the middle landscape could survive the advent of industry. But then, the pastoral idyll was already in trouble even in Virgil's time, threatened by the encroaching marshlands on one side, the corruptions of civilization on the other.

The wonder really is that it survives at all. Two centuries and a one-hour drive over the Blue Ridge from Monticello, Joel Salatin, a self-described "Christian-conservative-libertarian-environmentalist-lunatic farmer," is attempting again and against all odds to put real-live grass under the old agrarian-pastoral ideal, to try to make it new long after the triumph of the industrial system Jefferson fretted over has been completed. I'd come here to the Shenandoah Valley to see whether such a farm, and the alternative food chain it is part of, belonged to the past or the future.

Taking in Salatin's verdurous vista that afternoon, it occurred to me that the only thing missing from the scene was a happy shepherd, but then, wasn't that the tall fellow loping toward me in the broad blue suspenders and the floppy hat? Salatin's broad-brimmed straw hat did more than protect his neck and face from the Virginia sun: It declared a political and aesthetic stance, one descended from Virgil through Jefferson with a detour through the sixties counterculture. Whereas a feed company cap emblazoned with the logo of an agribusiness giant would have said labor, would have implied (in more ways than one) a debt to the industrial, Salatin's jaunty chapeau—made of grass, note, rather than plastic—bespoke independence, sufficiency, even ease. "On our farm the animals do most of the work," he had told me the first time we talked. At the moment, too tired to stand, the claim sounded to me like a pretty empty pastoral conceit. But as I would understand by the end of my week on Salatin's farm, the old pastoral idea is alive and, if not well exactly, still useful, perhaps even necessary.

2. THE GENIUS OF THE PLACE

Polyface Farm raises chicken, beef, turkeys, eggs, rabbits, and pigs, plus tomatoes, sweet corn, and berries on one hundred acres of pasture patchworked into another 450 acres of forest, but if you ask Joel Salatin what he does for a living (Is he foremost a cattle rancher? A chicken farmer?) he'll tell you in no uncertain terms, "I'm a grass farmer." The first time I heard this designation I didn't get it at all—hay seemed the least (and least edible) of his many crops, and he brought none of it to market. But undergirding the "farm of many faces," as he calls it, is a single plant—or rather that whole community of plants for which the word "grass" is shorthand.

"Grass," so understood, is the foundation of the intricate food chain Salatin has assembled at Polyface, where a half dozen different animal species are raised together in an intensive rotational dance on the theme of symbiosis. Salatin is the choreographer and the grasses are his verdurous stage; the dance has made Polyface one of the most productive and influential alternative farms in America.

Though it was only the third week of June, the pasture beneath me had already seen several rotational turns. Before being cut earlier in the week for the hay that would feed the farm's animals through the winter, it had been grazed twice by beef cattle, which after each day-long stay had been succeeded by several hundred laying hens. They'd arrived by Eggmobile, a ramshackle portable henhouse designed and built by Salatin. Why chickens? "Because that's how it works in nature," Salatin explained. "Birds follow and clean up after herbivores." And so during their turn in the pasture, the hens had performed several ecological services for the cattle as well as the grass: They'd picked the tasty grubs and fly larvae out of the cowpats, in the process spreading the manure and eliminating parasites. (This is what Joel has in mind when he says the animals do the work around here; the hens are his "sanitation crew," the reason his cattle have no need of chemical parasiticides.) And while they were at it, nibbling on the short cattle-clipped grasses they like best, the chickens applied a few thousand pounds of nitrogen to the pasture—and produced several thousand uncommonly rich and tasty eggs. After a few weeks' rest, the pasture will be grazed again, each steer turning these lush grasses into beef at the rate of two or three pounds a day.

By the end of the season Salatin's grasses will have been transformed by his animals into some 40,000 pounds of beef, 30,000 pounds of pork, 10,000 broilers, 1,200 turkeys, 1,000 rabbits, and 35,000 dozen eggs. This is an astounding cornucopia of food to draw from a hundred acres of pasture, yet what is perhaps still more astonishing is the fact that this pasture will be in no way diminished by the process—in fact, it will be the better for it, lusher, more fertile, even springier underfoot (this thanks to the increased earthworm traffic). Salatin's audacious bet is that feeding ourselves from nature need not be a zero-sum proposition, one in which if there is more for us at the end of the season then there must be less for nature—less topsoil, less fertility, less life. He's betting, in other words, on a very different proposition, one that looks an awful lot like the proverbially unattainable free lunch.

And none of it happens without the grass. In fact, the first time I met Salatin he'd insisted that even before I met any of his animals, I get down on my belly in this very pasture to make the acquaintance of the less charismatic species his farm was nurturing that, in turn, were nurturing his farm. Taking the ant's-eye view, he ticked off the census of a single square foot of pasture: orchard grass, foxtail, a couple of different fescues, bluegrass, and timothy. Then he cataloged the legumes— red clover and white, plus lupines—and finally the forbs, broad-leaved species like plantain, dandelion, and Queen Anne's lace. And those were just the plants, the species occupying the surface along with a handful of itinerant insects; belowdecks and out of sight tunneled earthworms (knowable by their castled mounds of rich castings), woodchucks, moles and burrowing insects, all making their dim way through an unseen wilderness of bacteria, phages, eelish nematodes, shrimpy rotifers, and miles upon miles of mycelium, the underground filaments of fungi. We think of the grasses as the basis of this food chain, yet behind, or beneath, the grassland stands the soil, that inconceivably complex community of the living and the dead. Because a healthy soil digests the dead to nourish the living, Salatin calls it the earth's stomach.

But it is upon the grass, mediator of soil and sun, that the human gaze has always tended to settle, and not just our gaze, either. A great many animals, too, are drawn to grass, which partly accounts for our own deep attraction to it: We come here to eat the animals that ate the grass that we (lacking rumens) can't eat ourselves. "All flesh is grass." The Old Testament's earthy equation reflects a pastoral culture's appreciation of the food chain that sustained it, though the hunter-gatherers living on the African savanna thousands of years earlier would have understood the flesh-grass connection just as well. It's only in our own time, after we began raising our food animals on grain in Concentrated Animal Feeding Operations (following the dubious new equation, All flesh is corn), that our ancient engagement with grass could be overlooked.

Or should I say partly overlooked, for surely our abiding affection for the stuff—reflected in our scrupulously tended lawns and playing fields, as well as in the persistence of so many forms of grassy pastoral, in everything from poetry to supermarket labels—expresses an unconscious recognition of our one-time dependence. Our inclination toward grass, which has the force of a tropism, is frequently cited as a prime example of "biophilia," E. O. Wilson's coinage for what he claims is our inherited genetic attraction for the plants and animals and landscapes with which we coevolved.

Certainly I was feeling the pull of the pastoral that summer afternoon on Joel Salatin's farm; whether or not its wellsprings were in my genes who can really say, but the idea does not strike me as implausible in the least. Our species' coevolutionary alliance with the grasses has deep roots and has probably done more to ensure our success as a species than any other, with the possible exception of our alliance with the trillion or so bacteria that inhabit the human gut. Working together, grass and man have overspread much of the earth, far more of it than would ever have been possible working alone.

This human-grass alliance has, in fact, had two distinct phases, taking us all the way from our time as hunter-gatherers to agriculturists, or, to date this natural history as the grasses might, from the Age of Perennials, like the fescues and bluegrass in these pastures, to the Age of Annuals, such as the corn George Naylor and I had planted in Iowa. In the first phase, which began when our earliest ancestors came down out of the trees to hunt animals on the savanna, the human relationship with grass was mediated by animals that (unlike us) could digest it, in much the same way it still is on Joel Salatin's postmodern savanna. Like Salatin, hunter-gatherers deliberately promoted the welfare of the grasses in order to attract and fatten the animals they depended upon. Hunters would periodically set fire to the savanna to keep it free of trees and nourish the soil. In a sense, they too were "grass farmers," deliberately nurturing grasses so that they might harvest meat.

So at least it appeared to us. Regarded from the grasses' point of view the arrangement appears even cleverer. The existential challenge facing grasses in all but the most arid regions is how to successfully compete against trees for territory and sunlight. The evolutionary strategy they hit upon was to make their leaves nourishing and tasty to animals who in turn are nourishing and tasty to us, the big-brained creature best equipped to vanquish the trees on their behalf. But for this strategy to succeed the grasses needed an anatomy that could withstand the rigors of both grazing and fire. So they developed a deep root system and a ground-hugging crown that in many cases puts out runners, allowing the grasses to recover quickly from fire and to reproduce even when grazers (or lawnmowers) prevent them from ever flowering and going to seed. (I used to think we were dominating the grass whenever we mowed the lawn, but in fact we're playing right into its strategy for world domination, by helping it outcompete the shrubs and trees.)

The second phase of the marriage of grasses and humans is usually called the "invention of agriculture," a self-congratulatory phrase that overlooks the role of the grasses themselves in revising the terms of the relationship. Beginning about ten thousand years ago a handful of particularly opportunistic grass species—the ancestors of wheat, rice, and corn—evolved to produce tremendous, nutritionally dense seeds that could nourish humans directly, thereby cutting out the intermediary animals. The grasses accomplished this feat by becoming annuals, throwing all their energy into making seeds rather than storing some of it underground in roots and rhizomes to get through the winter. These monster annual grasses outcompeted not only the trees, which humans obligingly cut down to expand the annuals' habitats, but bested the perennial grasses, which in most places succumbed to the plow. Their human sponsors ripped up the great perennial-polyculture grasslands to make the earth safe for annuals, which would henceforth be grown in strict monocultures.

3. INDUSTRIAL ORGANIC

Hard to believe, but Joel Salatin and George Naylor are, if regarded from a great enough distance, engaged in much the same pursuit: growing grasses to feed the cattle, chickens, and pigs that feed us. Compared to Salatin, however, Naylor participates in an infinitely more complex industrial system, involving not only corn (and soybeans), but fossil fuels, petrochemicals, heavy machinery, CAFOs, and an elaborate international system of distribution to move all these elements around: the energy from the Persian Gulf, the corn to the CAFOs, the animals to slaughter, and their meat finally to a Wal-Mart or McDonald's near you. Considered as a whole this system comprises a great machine, transforming inputs of seed and fossil energy into outputs of carbohydrate and protein. And, as with any machine, this one generates streams of waste: the nitrogen and pesticides running off the cornfields; the manure pooling in the feedlot lagoons; the heat and exhaust produced by all the machines within the machine—the tractors and trucks and combines.

Polyface Farm stands about as far from this industrialized sort of agriculture as it is possible to get without leaving the planet. Joel's farm stands as a kind of alternative reality to George's: Every term governing a conventional 500–acre corn-and-bean operation in Churdan, Iowa, finds its mirror opposite here on these 550 acres in Swoope, Virginia. To wit:

NAYLOR FARM	POLYFACE FARM
Industrial	Pastoral
Annual species	Perennial species
Monoculture	Polyculture
Fossil energy	Solar energy
Global market	Local market
Specialized	Diversified
Mechanical	Biological

Imported fertility	Local fertility
Myriad inputs	Chicken feed

For half a century now, which is to say for as long as industrial agriculture has held sway in America, the principal alternative to its methods and general approach has gone by the name "organic," a word chosen (by J. I. Rodale, the founding editor of *Organic Gardening and Farming* magazine) to imply that nature rather than the machine should supply the proper model for agriculture. Before my journey through the organic food industry I would have thought that virtually any organic farm would belong on the Polyface side of this ledger. But it turns out that this is not necessarily the case: There are now "industrial organic" farms that belong firmly on the left-hand side. Then there is this further paradox: Polyface Farm is technically *not* an organic farm, though by any standard it is more "sustainable" than virtually any organic farm. Its example forces you to think a lot harder about what these words—sustainable, organic, natural—really mean.

As it happened, the reason I found my way to Polyface Farm in the first place had everything to do with Joel Salatin's unusually strict construction of the word sustainable. As part of my research into the organic food chain, I kept hearing about this organic farmer in Virginia who had no use for the federal government's new organic standards. I also kept hearing about the exceptional food he was producing. So I gave him a call, hoping to get some salty quotes about the organic industry and perhaps get him to ship me a pastured chicken or steak.

The salty quotes I got. Speaking in a rapid-fire delivery that sounded like a cross between Bill Clinton and a hopped-up TV evangelist, Salatin delivered a scathing indictment of the "organic empire." I struggled to keep up with a spirited diatribe that bounced from the "Western conquistador mentality" and the "clash of paradigms" to the "innate distinctive desires of a chicken" and the impossibility of taking a "decidedly Eastern, connected, holistic product, and selling it through a decidedly Western, disconnected, reductionist Wall Streetified marketing system."

"You know what the best kind of organic certification would be?

Make an unannounced visit to a farm and take a good long look at the farmer's bookshelf. Because what you're feeding your emotions and thoughts is what this is really all about. The way I produce a chicken is an extension of my worldview. You can learn more about that by seeing what's sitting on my bookshelf than having me fill out a whole bunch of forms."

I asked him what was on his bookshelf. J. I. Rodale. Sir Albert Howard. Aldo Leopold. Wes Jackson. Wendell Berry. Louis Bromfield. The classic texts of organic agriculture and American agrarianism.

"We never called ourselves organic—we call ourselves 'beyond organic.' Why dumb down to a lesser level than we are? If I said I was organic, people would fuss at me for getting feed corn from a neighbor who might be using atrazine. Well, I would much rather use my money to keep my neighborhood productive and healthy than export my dollars five hundred miles away to get 'pure product' that's really coated in diesel fuel. There are a whole lot more variables in making the right decision than does the chicken feed have chemicals or not. Like what sort of habitat is going to allow that chicken to express its physiological distinctiveness? A ten-thousand-bird shed that stinks to high heaven or a new paddock of fresh green grass every day? Now which chicken shall we call 'organic'? I'm afraid you'll have to ask the government, because now they own the word.

"Me and the folks who buy my food are like the Indians—we just want to opt out. That's all the Indians ever wanted—to keep their tepees, to give their kids herbs instead of patent medicines and leeches. They didn't care if there was a Washington, D.C., or a Custer or a USDA; just leave us alone. But the Western mind can't bear an opt-out option. We're going to have to refight the Battle of the Little Bighorn to preserve the right to opt out, or your grandchildren and mine will have no choice but to eat amalgamated, irradiated, genetically prostituted, barcoded, adulterated fecal spam from the centralized processing conglomerate."

Whew . . .

As I indicated earlier, I got my quotes, but in the end I didn't get my food. Before we got off the phone, I asked Salatin if he could ship me one of his chickens and maybe a steak, too. He said that he couldn't do that. I figured he meant he wasn't set up for shipping, so offered him my FedEx account number.

"No, I don't think you understand. I don't believe it's sustainable— or 'organic,' if you will—to FedEx meat all around the country. I'm sorry, but I can't do it."

This man was serious.

"Just because we *can* ship organic lettuce from the Salinas Valley, or organic cut flowers from Peru, doesn't mean we *should* do it, not if we're really serious about energy and seasonality and bioregionalism. I'm afraid if you want to try one of our chickens, you're going to have to drive down here to Swoope to pick it up."

Which is eventually what I did. But before I traveled to Virginia for my week on the farm ("your Paris Hilton adventure," as my wife called it), I spent several weeks touring the organic empire to see if Salatin's criticisms, which had taken me by surprise, were just. A new, alternative food chain was taking shape in this country, and this seemed to me an unalloyed good: What had been a fringe movement in the 1960s was now a thriving business—the fastest growing corner of the food industry, in fact. Salatin was suggesting that the organic food chain couldn't expand into America's supermarkets and fast-food outlets without sacrificing its ideals. I wondered if this wasn't a case of making the ideal an enemy of the good, but Salatin was convinced that industrial organic was finally a contradiction in terms. I decided I had to find out if he was right.

NINE

BIG ORGANIC

1. SUPERMARKET PASTORAL

I enjoy shopping at Whole Foods nearly as much as I enjoy browsing a good bookstore, which, come to think of it, is probably no accident: Shopping at Whole Foods is a literary experience, too. That's not to take anything away from the food, which is generally of high quality, much of it "certified organic" or "humanely raised" or "free range." But right there, that's the point: It's the evocative prose as much as anything else that makes this food really special, elevating an egg or chicken breast or bag of arugula from the realm of ordinary protein and carbohydrates into a much headier experience, one with complex aesthetic, emotional, and even political dimensions. Take the "range-fed" sirloin steak I recently eyed in the meat case. According to the brochure on the counter, it was formerly part of a steer that spent its days "living in beautiful places" ranging from "plant-diverse, high-mountain meadows to thick aspen groves and miles of sagebrush-filled flats." Now a steak like that has got to taste better than one from Safeway, where the only accompanying information comes in the form of a number: the price, I mean, which you can bet will be considerably less. But I'm evidently not the only shopper willing to pay more for a good story.

With the growth of organics and mounting concerns about the wholesomeness of industrial food, storied food is showing up in supermarkets everywhere these days, but it is Whole Foods that consistently offers the most cutting-edge grocery lit. On a recent visit I filled my shopping cart with eggs "from cage-free vegetarian hens," milk from cows that live "free from unnecessary fear and distress," wild salmon caught by Native Americans in Yakutat, Alaska (population 833), and heirloom tomatoes from Capay Farm ($4.99 a pound), "one of the early pioneers of the organic movement." The organic broiler I picked up even had a name: Rosie, who turned out to be a "sustainably farmed" "free-range chicken" from Petaluma Poultry, a company whose "farming methods strive to create harmonious relationships in nature, sustaining the health of all creatures and the natural world." Okay, not the most mellifluous or even meaningful sentence, but at least their heart's in the right place.

In several corners of the store I was actually forced to choose between subtly competing stories. For example, some of the organic milk in the milk case was "ultrapasteurized," an extra processing step that was presented as a boon to the consumer, since it extends shelf life. But then another, more local dairy boasted about the fact they had said no to ultrapasteurization, implying that their product was fresher, less processed, and therefore more organic. This was the dairy that talked about cows living free from distress, something I was beginning to feel a bit of myself by this point.

This particular dairy's label had a lot to say about the bovine lifestyle: Its Holsteins are provided with "an appropriate environment, including shelter and a comfortable resting area, . . . sufficient space, proper facilities and the company of their own kind." All this sounded pretty great, until I read the story of another dairy selling raw milk—completely unprocessed—whose "cows graze green pastures all year long." Which made me wonder whether the first dairy's idea of an appropriate environment for a cow included, as I had simply presumed, a pasture. All of a sudden the absence from their story of that word seemed weirdly conspicuous. As the literary critics would say, the writer seemed to be eliding the whole notion of cows and grass. Indeed, the longer I shopped in Whole Foods, the more I thought that this is a place where the skills of a literary critic might come in handy—those, and perhaps also a journalist's.

WORDY LABELS, point-of-purchase brochures, and certification schemes are supposed to make an obscure and complicated food chain more legible to the consumer. In the industrial food economy, virtually the only information that travels along the food chain linking producer and consumer is price. Just look at the typical newspaper ad for a supermarket. The sole quality on display here is actually a quantity: tomatoes $0.69 a pound; ground chuck $1.09 a pound; eggs $0.99 a dozen—special this week. Is there any other category of product sold on such a reductive basis? The bare-bones information travels in both directions, of course, and farmers who get the message that consumers care only about price will themselves care only about yield. This is how a cheap food economy reinforces itself.

One of the key innovations of organic food was to allow some more information to pass along the food chain between the producer and the consumer—an implicit snatch of narrative along with the number. A certified organic label tells a little story about how a particular food was produced, giving the consumer a way to send a message back to the farmer that she values tomatoes produced without harmful pesticides or prefers to feed her children milk from cows that haven't been injected with growth hormones. The word "organic" has proved to be one of the most powerful words in the supermarket: Without any help from government, farmers and consumers working together in this way have built an $11 billion industry that is now the fastest growing sector of the food economy.

Yet the organic label itself—like every other such label in the supermarket—is really just an imperfect substitute for direct observation of how a food is produced, a concession to the reality that most people in an industrial society haven't the time or the inclination to follow their food back to the farm, a farm which today is apt to be, on average, fifteen hundred miles away. So to bridge that space we rely on certifiers and label writers and, to a considerable extent, our imagination of what the farms that are producing our food really look like. The organic label may conjure an image of a simpler agriculture, but its very existence is an industrial artifact. The question is, what about the farms themselves? How well do they match the stories told about them?

Taken as a whole, the story on offer in Whole Foods is a pastoral narrative in which farm animals live much as they did in the books we read as children, and our fruits and vegetables grow in well-composted soils on small farms much like Joel Salatin's. "Organic" on the label conjures up a rich narrative, even if it is the consumer who fills in most of the details, supplying the hero (American Family Farmer), the villain (Agribusinessman), and the literary genre, which I've come to think of as Supermarket Pastoral. By now we may know better than to believe this too simple story, but not much better, and the grocery store poets do everything they can to encourage us in our willing suspension of disbelief.

Supermarket Pastoral is a most seductive literary form, beguiling enough to survive in the face of a great many discomfiting facts. I suspect that's because it gratifies some of our deepest, oldest longings, not merely for safe food, but for a connection to the earth and to the handful of domesticated creatures we've long depended on. Whole Foods understands all this better than we do. One of the company's marketing consultants explained to me that the Whole Foods shopper feels that by buying organic he is "engaging in authentic experiences" and imaginatively enacting a "return to a utopian past with the positive aspects of modernity in tact." This sounds a lot like Virgilian pastoral, which also tried to have it both ways. In *The Machine in the Garden* Leo Marx writes that Virgil's shepherd Tityrus, no primitive, "Enjoys the best of both worlds—the sophisticated order of art and the simple spontaneity of nature." In keeping with the pastoral tradition, Whole Foods offers what Marx terms "a landscape of reconciliation" between the realms of nature and culture, a place where, as the marketing consultant put it, "people will come together through organic foods to get back to the origin of things"—perhaps by sitting down to enjoy one of the microwaveable organic TV dinners (four words I never expected to see conjoined) stacked in the frozen food case. How's that for having it both ways?

Of course the trickiest contradiction Whole Foods attempts to reconcile is the one between the industrialization of the organic food industry of which it is a part and the pastoral ideals on which that industry has been built. The organic movement, as it was once called, has come a remarkably long way in the last thirty years, to the point where it now looks considerably less like a movement than a big business. Lining the walls above the sumptuously stocked produce section in my Whole Foods are full-color photographs of local organic farmers accompanied by text blocks setting forth their farming philosophies. A handful of these farms—Capay is one example—still sell their produce to Whole Foods, but most are long gone from the produce bins, if not yet the walls. That's because Whole Foods in recent years has adopted the grocery industry's standard regional distribution system, which makes supporting small farms impractical. Tremendous warehouses buy produce for dozens of stores at a time, which forces them to deal exclusively with tremendous farms. So while the posters still depict family farmers and their philosophies, the produce on sale below them comes primarily from the two big corporate organic growers in California, Earthbound Farm and Grimmway Farms,* which together dominate the market for organic fresh produce in America. (Earthbound alone grows 80 percent of the organic lettuce sold in America.)

As I tossed a plastic box of Earthbound prewashed spring mix salad into my Whole Foods cart, I realized that I was venturing deep into the belly of the industrial beast Joel Salatin had called "the organic empire."

---

*Grimmway Farms owns Cal-Organic, one of the most ubiquitous organic brands in the supermarket.

(Speaking of my salad mix, another small, beyond organic farmer, a friend of Joel's, had told me he "wouldn't use that stuff to make compost"—the organic purist's stock insult.) But I'm not prepared to accept the premise that industrial organic is necessarily a bad thing, not if the goal is to reform a half-trillion-dollar food system based on chain supermarkets and the consumer's expectations that food be convenient and cheap.

And yet to the extent that the organic movement was conceived as a critique of industrial values, surely there comes a point when the process of industrialization will cost organic its soul (to use a word still uttered by organic types without irony), when Supermarket Pastoral becomes more fiction than fact: another lie told by marketers.

The question is, has that point been reached, as Joel Salatin suggests? Just how well does Supermarket Pastoral hold up under close reading and journalistic scrutiny?

ABOUT AS WELL as you would expect anything genuinely pastoral to hold up in the belly of an $11 billion industry, which is to say not very well at all. At least that's what I discovered when I traced a few of the items in my Whole Foods cart back to the farms where they were grown. I learned, for example, that some (certainly not all) organic milk comes from factory farms, where thousands of Holsteins that never encounter a blade of grass spend their days confined to a fenced "dry lot," eating (certified organic) grain and tethered to milking machines three times a day. The reason much of this milk is ultrapasteurized (a high-heat process that damages its nutritional quality) is so that big companies like Horizon and Aurora can sell it over long distances. I discovered organic beef being raised in "organic feedlots" and organic high-fructose corn syrup—more words I never expected to see combined. And I learned about the making of the aforementioned organic TV dinner, a microwaveable bowl of "rice, vegetables, and grilled chicken breast with a savory herb sauce." Country Herb, as the entrée is called, turns out to be a highly industrialized organic product, involving a choreography of thirty-one ingredients assembled from far-flung farms, laboratories, and processing plants scattered over a half-dozen states and two countries, and containing such mysteries of modern food technology as high-oleic safflower oil, guar and xanthan gum, soy lecithin, carrageenan, and "natural grill flavor." Several of these ingredients are synthetic additives permitted under federal organic rules. So much for "whole" foods. The manufacturer of Country Herb is Cascadian Farm, a pioneering organic farm turned processor in Washington State that is now a wholly owned subsidiary of General Mills. (The Country Herb chicken entrée has since been discontinued.)

I also visited Rosie the organic chicken at her farm in Petaluma, which turns out to be more animal factory than farm. She lives in a shed with twenty thousand other Rosies, who, aside from their certified organic feed, live lives little different from that of any other industrial chicken. Ah, but what about the "free-range" lifestyle promised on the label? True, there's a little door in the shed leading out to a narrow grassy yard. But the free-range story seems a bit of a stretch when you discover that the door remains firmly shut until the birds are at least five or six weeks old—for fear they'll catch something outside—and the chickens are slaughtered only two weeks later.

2. FROM PEOPLE'S PARK TO PETALUMA POULTRY

If you walk five blocks north from the Whole Foods in Berkeley along Telegraph Avenue and then turn right at Dwight Street, you'll soon come to a trash-strewn patch of grass and trees dotted with the tattered camps of a few dozen homeless people. Mostly in their fifties and sixties, some still affecting hippie styles of hair and dress, these men and women pass much of their days sleeping and drinking, like so many of the destitute everywhere. Here, though, they also spend time tending scruffy little patches of flowers and vegetables—a few stalks of corn, some broccoli plants gone to seed. People's Park today is the saddest of places, a blasted monument to sixties' hopes that curdled a long time ago. And yet, while the economic and social distances separating the well-heeled shoppers cruising the aisles at Whole Foods from the unheeled homeless in People's Park could not be much greater, the two neighborhood institutions are branches of the same unlikely tree.

Indeed, were there any poetic justice in the world, the executives at Whole Foods would have long ago erected a commemorative plaque at People's Park and a booth to give away organic fruits and vegetables. The organic movement, much like environmentalism and feminism, has deep roots in the sixties' radicalism that briefly flourished on this site; organic is one of several tributaries of the counterculture that ended up disappearing into the American mainstream, but not before significantly altering its course. And if you trace that particular tributary all the way back to its spring, your journey will eventually pass through this park.

People's Park was born on April 20, 1969, when a group calling itself the Robin Hood Commission seized a vacant lot owned by the University of California and set to work rolling out sod, planting trees, and, perhaps most auspiciously, putting in a vegetable garden. Calling themselves "agrarian reformers," the radicals announced that they wanted to establish on the site the model of a new cooperative society built from the ground up; that included growing their own "uncontaminated" food. One of the inspirations for the commission's act of civil disobedience was the example of the Diggers in seventeenth-century England, who had also seized public land with the aim of growing food to give away to the poor. In People's Park that food would be organic, a word that at the time brimmed with meanings that went far beyond any particular agricultural method.

In *Appetite for Change*, his definitive account of how the sixties' counterculture changed the way we eat, historian Warren J. Belasco writes that the events in People's Park marked the "greening" of the counterculture, the pastoral turn that would lead to the commune movement in the countryside, to food co-ops and "guerilla capitalism," and, eventually, to the rise of organic agriculture and businesses like Whole Foods. The moment for such a turn to nature was ripe in 1969: DDT was in the news, an oil spill off Santa Barbara had blackened California's coastline, and Cleveland's Cuyahoga River had caught fire. Overnight, it seemed, "ecology" was on everybody's lips, and "organic" close behind.

As Belasco points out, the word "organic" had enjoyed a currency among nineteenth-century English social critics, who contrasted the social fragmentation and atomism wrought by the Industrial Revolution with the ideal of a lost organic society, one where the bonds of affection and cooperation still held. Organic stood for everything industrial was not. But applying the word "organic" to food and farming occurred much more recently: In the 1940s in the pages of *Organic Gardening and Farming*. Founded in 1940 by J. I. Rodale, a health-food fanatic from New York City's Lower East Side, the magazine devoted its pages to the agricultural methods and health benefits of growing food without synthetic chemicals—"organically." Joel Salatin's grandfather was a charter subscriber.

Organic Gardening and Farming struggled along in obscurity until 1969, when an ecstatic review in the *Whole Earth Catalog* brought it to the attention of hippies trying to figure out how to grow vegetables without patronizing the military-industrial complex. "If I were a dictator determined to control the national press," the *Whole Earth* correspondent wrote,

> *Organic Gardening* would be the first publication I'd squash, because it's the most subversive. I believe that organic gardeners are in the forefront of a serious effort to save the world by changing man's orientation to it, to move away from the collective, centrist, superindustrial state, toward a simpler, realer one-to-one relationship with the earth itself.

Within two years *Organic Gardening and Farming*'s circulation climbed from 400,000 to 700,000.

As the *Whole Earth* encomium suggests, the counterculture had married the broader and narrower definitions of the word "organic". The organic garden planted in People's Park (soon imitated in urban lots across the country) was itself conceived of as a kind of scale model of a more cooperative society, a landscape of reconciliation that proposed to replace industrialism's attitude of conquest toward nature with a softer, more harmonious approach. A pastoral utopia in miniature, such a garden embraced not only the humans who tended and ate from it but "as many life kingdoms as possible," in the words of an early account of Berkeley's People's Gardens in an underground paper called *Good Times*. The vegetables harvested from these plots, which were sometimes called "conspiracies of soil," would supply, in addition to wholesome calories, an "edible dynamic"—a "new medium through which people can relate to one another and their nourishment." For example, organic's rejection of agricultural chemicals was also a rejection of the war machine, since the same corporations—Dow, Monsanto—that manufactured pesticides also made napalm and Agent Orange, the herbicide with which the U.S. military was waging war against nature in Southeast Asia. Eating organic thus married the personal to the political.

Which was why much more was at stake than a method of farming. Acting on the ecological premise that everything's connected to everything else, the early organic movement sought to establish not just an alternative mode of production (the chemical-free farms), but an alternative system of distribution (the anticapitalist food co-ops), and even an alternative mode of consumption (the "countercuisine"). These were the three struts on which organic's revolutionary program stood; since ecology taught "you can never do only one thing," what you ate was inseparable from how it was grown and how it reached your table.

A countercuisine based on whole grains and unprocessed organic ingredients rose up to challenge conventional industrial "white bread food." ("Plastic food" was an epithet thrown around a lot.) For a host of reasons that seem ridiculous in retrospect, brown foods of all kinds—rice, bread, wheat, eggs, sugar, soy sauce, tamari—were deemed morally superior to white foods. Brown foods were less adulterated by industry, of course, but just as important, eating them allowed you to express your solidarity with the world's brown peoples. (Only later would the health benefits of these whole foods be recognized, not the first or last time an organic conceit would find scientific backing.) But perhaps best of all, brown foods were also precisely what your parents *didn't* eat.

How to grow this stuff without chemicals was a challenge, especially to city kids coming to the farm or garden with a head full of pastoral ideals and precisely no horticultural experience. The rural communes served as organic agriculture's ramshackle research stations, places where neophyte farmers could experiment with making compost and devising alternative methods of pest control. The steepness of their learning curve was on display in the food co-ops, where sorry-looking organic produce was the rule for many years. But the freak farmers stuck with it, following Rodale's step-by-step advice, and some of them went on to become excellent farmers.

ONE SUCH NOTABLE success was Gene Kahn, the founder of Cascadian Farm, the company responsible for the organic TV dinner in my Whole Foods cart. Today Cascadian Farm is foremost a General Mills brand, but it began as a quasi-communal hippie farm, located on a narrow, gorgeous shelf of land wedged between the Skagit River and the North Cascades about seventy-five miles northeast of Seattle. (The idyllic little farmstead depicted on the package turns out to be a real place.) Originally called the New Cascadian Survival and Reclamation Project, the farm was started in 1971 by Gene Kahn with the idea of growing food for the collective of environmentally minded hippies he had hooked up with in nearby Bellingham. At the time Kahn was a twenty-four-year-old grad school dropout from the South Side of Chicago, who had been inspired by *Silent Spring* and *Diet for a Small Planet* to go back to the land—and from there to change the American food system. This particular dream was not so outrageous in 1971, but Kahn's success in actually realizing it surely is: He went on to become a pioneer of the organic movement and probably has done as much as anyone to move organic food into the mainstream, getting it out of the food co-op and into the supermarket. Today, the eponymous Cascadian Farm is a General Mills showcase—"a PR farm," as its founder freely acknowledges—and Kahn, erstwhile hippie farmer, is a General Mills vice president. Cascadian Farm is precisely what Joel Salatin has in mind when he talks about an organic empire.

Like most of the early organic farmers, Kahn had no idea what he was doing at first, and he suffered his share of crop failures. In 1971 organic agriculture was in its infancy—a few hundred scattered amateurs learning by trial and error how to grow food without chemicals, an ad hoc grassroots R&D effort for which there was no institutional support. (In fact, the USDA was actively hostile to organic agriculture until recently, viewing it—quite rightly—as a critique of the industrialized agriculture the USDA was promoting.) What the pioneer organic farmers had instead of the USDA's agricultural extension service was *Organic Gardening and Farming* (to which Kahn subscribed) and the model of various premodern agricultural systems, as described in books like *Farmers of Forty Centuries* by F. H. King and Sir Albert Howard's *The Soil and Health* and *An Agricultural Testament*. This last book may fairly be called the movement's bible.

PERHAPS MORE THAN any other single writer, Sir Albert Howard (1873–1947), an English agronomist knighted after his thirty years of research in India, provided the philosophical foundations for organic agricultural. Even those who never read his 1940 *Testament* nevertheless absorbed his thinking through the pages of Rodale's *Organic Gardening and Farming*, where he was lionized, and in the essays of Wendell Berry, who wrote an influential piece about Howard in the *The Last Whole Earth Catalog* in 1971. Berry seized particularly on Howard's arresting—and prescient—idea that we needed to treat "the whole problem of health in soil, plant, animal and man as one great subject."

For a book that devotes so many of its pages to the proper making of compost, *An Agricultural Testament* turns out to be an important work of philosophy as well as of agricultural science. Indeed, Howard's drawing of lines of connection between so many seemingly discrete realms—from soil fertility to "the national health"; from the supreme importance of animal urine to the limitations of the scientific method—is his signal contribution, his method as well as his message. Even though Howard never uses the term "organic", it is possible to tease out all the many meanings of the word—as a program for not just agricultural but social renovation—from his writings. To measure the current definition of organic against his genuinely holistic conception is to appreciate just how much it has shrunk.

Like many works of social and environmental criticism, *An Agricultural Testament* is in broad outline the story of a Fall. In Howard's case, the serpent in question is a nineteenth-century German chemist by the name of Baron Justus von Liebig, his tempting fruit a set of initials: NPK. It was Liebig, in his 1840 monograph *Chemistry in its Application to Agriculture*, who set agriculture on its industrial path when he broke down the quasi-mystical concept of fertility in soil into a straightforward inventory of the chemical elements plants require for growth. At a stroke, soil biology gave way to soil chemistry, and specifically to the three chemical nutrients Liebig highlighted as crucial to plant growth: nitrogen, phosphorus, and potassium, or to use these elements' initials from the periodic table, N-P-K. (The three letters correspond to the three-digit designation printed on every bag of fertilizer.) Much of Howard's work is an attempt to demolish what he called the "NPK mentality."

The NPK mentality embraces a good deal more than fertilizer, however. Indeed, to read Howard is to begin to wonder if it might not be one of the keys to everything wrong with modern civilization. In Howard's thinking, the NPK mentality serves as a shorthand for both the powers and limitations of reductionist science. For as followers of Liebig discovered, NPK "works": If you give plants these three elements, they will grow. From this success it was a short step to drawing the conclusion that the entire mystery of soil fertility had been solved. It fostered the wholesale reimagining of soil (and with it agriculture) from a living system to a kind of machine: Apply inputs of NPK at this end and you will get yields of wheat or corn on the other end. Since treating the soil as a machine seemed to work well enough, at least in the short term, there no longer seemed any need to worry about such quaint things as earthworms and humus.

Humus is the stuff in a handful of soil that gives it its blackish cast and characteristic smell. It's hard to say exactly what humus is because it is so many things. Humus is what's left of organic matter after it has been broken down by the billions of big and small organisms that inhabit a spoonful of earth—the bacteria, phages, fungi, and earthworms responsible for decomposition. (The psalmist who described life as a transit from "dust to dust" would have been more accurate to say "humus to humus.") But humus is not a final product of decomposition so much as a stage, since a whole other group of organisms slowly breaks humus down into the chemical elements plants need to grow, elements including, but not limited to, nitrogen, phosphorus, and potassium. This process is as much biological as chemical, involving the symbiosis of plants and the mycorrhizal fungi that live in and among their roots; the fungi offer soluble nutrients to the roots, receiving a drop of sucrose in return. Another critical symbiotic relationship links plants to the bacteria in a humus-rich soil that fix atmospheric nitrogen, putting it into a form the plants can use. But providing a buffet of nutrients to plants is not the only thing humus does: It also serves as the glue that binds the minute mineral particles in soil together into airy crumbs and holds water in suspension so that rainfall remains available to plant roots instead of instantly seeping away.

To reduce such a vast biological complexity to NPK represented the scientific method at its reductionist worst. Complex qualities are reduced to simple quantities; biology gives way to chemistry. As Howard was not the first to point out, that method can only deal with one or two variables at a time. The problem is that once science has reduced a complex phenomenon to a couple of variables, however important they may be, the natural tendency is to overlook everything else, to assume that what you can measure is all there is, or at least all that really matters. When we mistake what we can know for all there is to know, a healthy appreciation of one's ignorance in the face of a mystery like soil fertility gives way to the hubris that we can treat nature as a machine. Once that leap has been made, one input follows another, so that when the synthetic nitrogen fed to plants makes them more attractive to insects and vulnerable to disease, as we have discovered, the farmer turns to chemical pesticides to fix his broken machine.

In the case of artificial manures—the original term for synthetic fertilizers—Howard contended that our hubris threatened to damage the health not only of the soil (since the harsh chemicals kill off biological activity in humus) but of "the national health" as well. He linked the health of the soil to the health of all the creatures that depended on it, an idea that, once upon a time before the advent of industrial agriculture, was in fact a commonplace, discussed by Plato and Thomas Jefferson, among many others. Howard put it this way: "Artificial manures lead inevitably to artificial nutrition, artificial food, artificial animals and finally to artificial men and women."

Howard's flight of rhetoric might strike our ears as a bit over the top (we are talking about fertilizer, after all), but it was written in the heat of the pitched battle that accompanied the introduction of chemical agriculture to England in the 1930s and 1940s. "The great humus controversy," as it was called, actually reached the floor of the House of Lords in 1943, a year when one might have thought there were more pressing matters on the agenda. But England's agriculture ministry was promoting the new fertilizers, and many farmers complained their pastures and animals had become less robust as a result. Howard and his allies were convinced that "history will condemn [chemical fertilizer] as one of the greatest misfortunes to have befallen agriculture and mankind." He claimed that the wholesale adoption of artificial manures would destroy the fertility of the soil, leave plants vulnerable to pests and disease, and damage the health of the animals and peoples eating those plants, for how could such plants be any more nutritious than the soil in which they grew? Moreover, the short-term boosts in yield that fertilizers delivered could not be sustained; since the chemicals would eventually destroy the soil's fertility, today's high yields were robbing the future.

Needless to say, the great humus controversy of the 1940s was settled in favor of the NPK mentality.

HOWARD POINTED DOWN another path. "We now have to retrace our steps," he wrote, which meant jettisoning the legacy of Liebig and industrial agriculture. "We have to go back to nature and to copy the methods to be seen in the forest and prairie." Howard's call to redesign the farm as an imitation of nature wasn't merely rhetorical; he had specific practices and processes in mind, which he outlined in a paragraph at the beginning of *An Agricultural Testament* that stands as a fair summary of the whole organic ideal:

> Mother earth never attempts to farm without live stock; she always raises mixed crops; great pains are taken to preserve the soil and to prevent erosion; the mixed vegetable and animal wastes are converted into humus; there is no waste; the processes of growth and the processes of decay balance one another; the greatest care is taken to store the rainfall; both plants and animals are left to protect themselves against disease.

Each of the biological processes at work in a forest or prairie could have its analog on a farm: Animals could feed on plant wastes as they do in the wild; in turn their wastes could feed the soil; mulches could protect bare soil in the same way leaf litter in a forest does; the compost pile, acting like the lively layer of decomposition beneath the leaf litter, could create humus. Even the diseases and insects would perform the salutary function they do in nature: to eliminate the weakest plants and animals, which he predicted would be far fewer in number once the system was operating properly. For Howard, insects and diseases—the bane of industrial agriculture—are simply "nature's censors," useful to the farmer for "pointing out unsuitable varieties and methods of farming inappropriate to the locality." On a healthy farm pests would be no more prevalent than in a healthy wood or pasture, which should be agriculture's standard. Howard was thus bidding farmers to regard their farms less like machines than living organisms.

The notion of imitating whole natural systems stands in stark opposition to reductionist science, which works by breaking such systems down into their component parts in order to understand how they work and then manipulating them—one variable at a time. In this sense, Howard's concept of organic agriculture is premodern, arguably even antiscientific: He's telling us we don't need to understand how humus works or what compost does in order to make good use of it. Our ignorance of the teeming wilderness that is the soil (even the act of regarding it *as* a wilderness) is no impediment to nurturing it. To the contrary, a healthy sense of all we don't know—even a sense of mystery— keeps us from reaching for oversimplifications and technological silver bullets.

A charge often leveled against organic agriculture is that it is more philosophy than science. There's some truth to this indictment, if that is what it is, though why organic farmers should feel defensive about it is itself a mystery, a relic, perhaps, of our fetishism of science as the only credible tool with which to approach nature. In Howard's conception, the philosophy of mimicking natural processes precedes the science of understanding them. The peasant rice farmer who introduces ducks and fish to his paddy may not understand all the symbiotic relationships he's put in play—that the ducks and fish are feeding nitrogen to the rice and at the same time eating the pests. But the high yields of food from this ingenious polyculture are his to harvest even so.

The philosophy underlying Howard's conception of organic agriculture is a variety of pragmatism, of course, the school of thought that is willing to call "true" whatever works. Charles Darwin taught us that a kind of pragmatism—he called it natural selection—is at the very heart of nature, guiding evolution: What works is what survives. This is why Howard spent so much time studying peasant agricultural systems in India and elsewhere: The best ones survived as long as they did because they brought food forth from the same ground year after year without depleting the soil.

In Howard's agronomy, science is mostly a tool for describing what works and explaining why it does. As it happens, in the years since Howard wrote, science has provided support for a great many of his unscientific claims: Plants grown in synthetically fertilized soils *are* less nourishing than ones grown in composted soils;[1] such plants *are* more vulnerable to diseases and insect pests;[2] polycultures *are* more productive and less prone to disease than monocultures;[3] and that in fact the health of the soil, plant, animal, human, and even nation are, as Howard claimed, connected along lines we can now begin to draw with empirical confidence. We may not be prepared to act on this knowledge, but we know that civilizations that abuse their soil eventually collapse.[4]

If farms modeled on natural systems work as well as Howard suggests, then why don't we see more of them? The sad fact is that the organic ideal as set forth by Howard and others has been honored mainly in the breach. Especially as organic agriculture has grown more successful, finding its way into the supermarket and the embrace of agribusiness, organic farming has increasingly come to resemble the industrial system it originally set out to replace. The logic of that system has so far proven more ineluctable than the logic of natural systems.

THE JOURNEY OF Cascadian Farm from the New Cascadian Survival and Reclamation Project to a General Mills subsidiary stands as a parable of this process. On an overcast morning a few winters ago, Kahn drove me out to see the original farm, following the twists of the Skagit River east

---

1. Asami, et al (2003); Benbrook (2005); Carbonaro (2001); Davis, et al (2004).
2. Altieri (1995); Tilman (1998).
3. Altieri (1995, 1999); Tilman (1998); Wolfe (2000).
4. Diamond (2005).

in a new forest-green Lexus with vanity plates that say ORGANIC. Kahn is a strikingly boyish-looking man in his midfifties, and after you factor in a shave and twenty pounds, it's not hard to pick his face out from the beards-beads-and-tractors photos on display in his office. Walking me through the history of his company as we drove out to the farm, Gene Kahn spoke candidly and without defensiveness about the compromises made along his path from organic farmer to agribusinessman, and about "how everything eventually morphs into the way the world is."

By the late seventies, Kahn had become a pretty good organic farmer and an even better businessman. He had discovered the economic virtues of adding value to his produce by processing it (freezing blueberries and strawberries, making jam), and once Cascadian Farm started processing food, Kahn discovered he could make more money buying produce from other farmers than by growing it himself—the same discovery conventional agribusiness companies had made a long time before.

"The whole notion of a 'cooperative community' we started with gradually began to mimic the system," Kahn told me. "We were shipping food around the country, using diesel fuel—we were industrial organic farmers. I was bit by bit becoming more of this world, and there was a lot of pressure on the business to become more privatized."

That pressure became irresistible in 1990, when in the aftermath of the "Alar scare" Kahn nearly lost everything—and control of Cascadian Farm wound up in corporate hands. In the history of the organic movement the Alar episode is a watershed, marking the birth pangs of the modern organic industry. Throughout its history, the sharpest growth of organic has closely followed spikes in public concern over the industrial food supply. Some critics condemn organic for profiting time and again from "food scares," and while there is certainly some truth to this charge, whether it represents a more serious indictment of organic or industrial food is open to question. Organic farmers reply that episodes focusing public attention on pesticides, food poisoning, genetically modified crops, and mad cow disease serve as "teachable moments" about the industrial food system and its alternatives. Alar was one of the first.

After a somewhat overheated *60 Minutes* exposé on apple growers' use of Alar, a growth-regulating chemical widely used in conventional orchards that the Environmental Protection Agency had declared a carcinogen, Middle America suddenly discovered organic. "Panic for Organic" was the cover line on one newsweekly, and overnight, demand from the supermarket chains soared. The ragtag industry was not quite ready for prime time, however. Like a lot of organic producers, Gene Kahn borrowed heavily to finance an ambitious expansion, contracted with farmers to grow an awful lot of organic produce—and then watched in horror as the bubble of demand subsided along with the headlines about Alar. Badly overextended, Kahn was forced to sell a majority stake in his company—to Welch's—and the onetime hippie farmer set out on what he calls his "corporate adventure."

"We were part of the food industry now," he told me. "But I wanted to leverage that position to redefine the way we grow food—not what people want to eat or how we distribute it. That sure as hell isn't going to change." Becoming part of the food industry meant jettisoning two of the three original legs on which the organic movement had stood: the countercuisine—what people want to eat—and the food co-ops and other alternative modes of distribution. Kahn's bet was that agribusiness could accommodate itself most easily to the first leg—the new way to grow food—by treating organic essentially as a niche product that could be distributed and marketed through the existing channels. The original organic ideal held that you could not divorce these three elements, since (as ecology taught) everything was connected. But Gene Kahn, for one (and he was by no means the only one), was a realist, a businessman with a payroll to meet. And he wasn't looking back.

"You have a choice of getting sad about all that or moving on. We tried hard to build a cooperative community and a local food system, but at the end of the day it wasn't successful. This is just lunch for most people. *Just lunch*. We can call it sacred, we can talk about communion, but it's just lunch."

———

IN THE YEARS after the Alar bubble burst in 1990, the organic industry recovered, embarking on a period of double-digit annual growth and rapid consolidation, as mainstream food companies began to take organic (or at least the organic market) seriously. Gerber's, Heinz, Dole, ConAgra, and ADM all created or acquired organic brands. Cascadian Farm itself became a miniconglomerate, acquiring Muir Glen, a California organic tomato processor, and the combined company changed its name to Small Planet Foods. Nineteen ninety also marked the beginning of federal recognition for organic agriculture: That year, Congress passed the Organic Food and Production Act (OFPA). The legislation instructed the Department of Agriculture—which historically had treated organic farming with undisguised contempt—to establish uniform national standards for organic food and farming, fixing the definition of a word that had always meant different things to different people.

Settling on that definition turned out to be a grueling decade-long process, as various forces both within and outside the movement battled for control of a word that had developed a certain magic in the marketplace. Agribusiness fought to define the word as loosely as possible, in part to make it easier for mainstream companies to get into organic, but also out of fear that anything deemed not organic—such as genetically modified food—would henceforth carry an official stigma. At first, the USDA, acting out of long-standing habit, obliged its agribusiness clients, issuing a watery set of standards in 1997 that—astoundingly—allowed for the use of genetically modified crops and irradiation and sewage sludge in organic food production. Some saw the dark hand of companies like Monsanto or ADM at work, but it seems more likely the USDA was simply acting on the reasonable assumption that the organic industry, like any other industry, would want as light a regulatory burden as possible. But it turned out organic wasn't like other industries: It still had a lot of the old movement values in its genetic makeup, and it reacted to the weak standards with fury. An unprecedented flood of public comment from outraged organic farmers and consumers forced the USDA back to the drawing board, in what was widely viewed as a victory for the movement's principles.

Yet while the struggle with the government over the meaning of "organic" was making headlines in 1997, another equally important struggle was underway within the USDA between Big and Little Organic—or, put another way, between the organic industry and the organic movement—and here the outcome was decidedly more ambiguous. Could a factory farm be organic? Was an organic dairy cow entitled to graze on pasture? Did food additives and synthetic chemicals have a place in processed organic food? If the answers to these questions seem like no-brainers, then you too are stuck in an outdated pastoral view of organic. Big Organic won all three arguments. The final standards do a good job of setting the bar for a more environmentally responsible kind of farming but, as perhaps was inevitable as soon as bureaucratic and industrial thinking was brought to bear, many of the philosophical values embodied in the word "organic"—the sorts of values expressed by Albert Howard—did not survive the federal rulemaking process.

From 1992 to 1997 Gene Kahn served on the USDA's National Organic Standards Board, where he played a key role in making the standards safe for the organic TV dinner and a great many other organic processed foods. This was no small feat, for Kahn and his allies had to work around the original 1990 legislation, which had prohibited synthetic food additives and manufacturing agents outright. Kahn argued that you couldn't have organic processed food without synthetics, which are necessary to both the manufacture and preservation of such supermarket products. Several of the consumer representatives on the standards board contended that this was precisely the point, and if no synthetics meant no organic TV dinners, then TV dinners were something organic simply shouldn't do. At stake was the very idea of a countercuisine.

Joan Dye Gussow, a nutritionist and an outspoken standards-board member, made the case against synthetics in a 1996 article that was much debated at the time: "Can an Organic Twinkie Be Certified?" Demonstrating that under the proposed rules such a thing was entirely possible, Gussow questioned whether organic should simply mirror the existing food supply, with its highly processed, salted, and sugary junk food, or whether it should aspire to something better—a counter-cuisine based on whole foods. Kahn responded with an argument rooted in the populism of the market: If the consumer wants an organic Twinkie, then we should give it to him. As he put it to me on the drive back from Cascadian Farm, "Organic is not your mother." In the end it came down to an argument between the old movement and the new industry and the new industry won: The final standards simply ignored the 1990 law, drawing up a list of permissible additives and synthetics, from ascorbic acid to xanthan gum.*

"If we had lost on synthetics," Kahn told me, "we'd be out of business."

The same might be said for the biggest organic meat and dairy producers, who fought to make the new standards safe for the organic factory farm. Horizon Organic's Mark Retzloff labored mightily to preserve the ability of his company—which is the Microsoft of organic milk, controlling more than half of the market—to operate its large-scale industrial dairy in southern Idaho. Here in the western desert, where precious little grass can grow, the company was milking several thousand cows that, rather than graze on pasture (as most consumers presume their organic cows are doing), spend their days milling around a dry lot—a grassless fenced enclosure. It's doubtful a dairy could pasture that many cows even if it wanted to—you would need at least an acre of grass per animal and more hours than there are in a day to move that many cows all the way out to their distant acre and then back again to the milking parlor every morning and evening. So in-

---

* After Arthur Harvey, a Maine blueberry farmer, won a 2003 lawsuit forcing the USDA to obey the language of the 1990 law, lobbyists working for the Organic Trade Association managed in 2005 to slip language into a USDA appropriations bill restoring—and possibly expanding—the industry's right to use synthetics in organic foods.

stead, as in the typical industrial dairy, these organic cows stood around eating grain and silage when they weren't being milked three times a day. Their organic feed was shipped in from all over the West, and their waste accumulated in manure ponds. Retzloff argued that keeping cows in confinement meant that his farmhands, who all carried stethoscopes, could keep a closer eye on their health. Of course, cows need this sort of surveillance only when they're living in such close quarters—and can't be given antibiotics.

Such a factory farm didn't sound terribly organic to the smaller dairy farmers on the board, not to mention to the consumer representatives. Also, the OFPA had spelled out that the welfare of organic animals should take into account, and accommodate, their "natural behavior," which in the case of cows—ruminants who have evolved to eat grass—surely meant grazing on pasture. You might say the whole pastoral idea was hardwired into these animals and stood squarely in the way of industrializing them. So how could the logic of industry ever hope to prevail?

The USDA listened to the arguments on both sides and finally ruled that dairy cows must have "access to pasture," which sounds like more of a victory for the pastoral ideal than it turned out to be in practice. By itself "access to pasture" is an extremely vague standard (What constitutes "access"? How much pasture per animal? How often could it graze?), and it was weakened further by a provision stating that even access could be dispensed with at certain stages of the animal's life. Some big organic dairies have decided that lactation constitutes one such stage, and thus far the USDA has not objected. Some of its organic certifiers have complained that "access to pasture" is so vague as to be meaningless—and therefore unenforceable. It's hard to argue with them.

Along with the national list of permissible synthetics, "access to pasture," and, for other organic animals, "access to the outdoors" indicate how the word "organic" has been stretched and twisted to admit the very sort of industrial practices for which it once offered a critique and an alternative. The final standards also demonstrate how, in Gene Kahn's words, "everything eventually morphs into the way the world is." And yet the pastoral values and imagery embodied in that word survive in the minds of many people, as the marketers of organic food well understand: Just look at a container of organic milk, with its happy cows and verdant pastures. Thus is a venerable ideal hollowed out, reduced to a sentimental conceit printed on the side of a milk carton: Supermarket Pastoral.

3. DOWN ON THE INDUSTRIAL ORGANIC FARM

Get over it, Gene Kahn would say. The important thing, the real value of putting organic on an industrial scale, is the sheer amount of acreage it puts under organic management. Behind every organic TV dinner or chicken or carton of industrial organic milk stands a certain quantity of land that will no longer be doused with chemicals, an undeniable gain for the environment and the public health. I could see his point. So I decided to travel around California to see these farms for myself. Why California? Because the state's industrial agriculture grows most of America's produce, and organic has in large part become a subset, or brand, of that agriculture.

No farms I had ever visited before prepared me for the industrial organic farms I saw in California. When I think about organic farming, I think family farm, I think small scale, I think hedgerows and compost piles and battered pickups—the old agrarian idea (which in fact has never had much purchase in California). I don't think migrant labor crews, combines the size of houses, mobile lettuce-packing factories marching across fields of romaine, twenty-thousand-broiler-chicken houses, or hundreds of acres of corn or broccoli or lettuce reaching clear to the horizon. To the eye, these farms look exactly like any other industrial farm in California—and in fact some of the biggest organic operations in the state are owned and operated by conventional mega-farms. The same farmer who is applying toxic fumigants to sterilize the soil in one field is in the next field applying compost to nurture the soil's natural fertility.

Is there anything wrong with this picture? I'm not sure, frankly. Gene Kahn makes the case that the scale of a farm has no bearing on its fidelity to organic principles, and that unless organic "scales up [it will] never be anything more than yuppie food." To prove his point Kahn sent me to visit several of the large-scale farms that supply Small Planet Foods. These included Greenways, the Central Valley operation that grows vegetables for his frozen dinners (and tomatoes for Muir Glen), and Petaluma Poultry, which grows the chicken in his frozen dinner as well as Rosie, the organic chicken I made the acquaintance of in Whole Foods. I also paid a visit to the Salinas Valley, where Earthbound Farm, the largest organic grower in the world, has most of its lettuce fields.

My first stop was Greenways Organic, a successful two-thousand-acre organic produce operation tucked into a twenty-four-thousand-acre conventional farm in the Central Valley outside Fresno; the crops, the machines, the crews, the rotations, and the fields were virtually indistinguishable, and yet two different kinds of industrial agriculture are being practiced here side by side.

In many respects the same factory model is at work in both fields, but for every chemical input used in the farm's conventional fields, a more benign organic input has been substituted in the organic ones. So in place of petrochemical fertilizers, Greenways' organic acres are nourished by compost made by the ton at a horse farm nearby, and by poultry manure. Instead of toxic pesticides, insects are controlled by spraying-approved organic agents (most of them derived from plants) such as rotenone, pyrethrum, and nicotine sulfate, and by introducing beneficial insects like lacewings. Inputs and outputs: a much greener machine, but a machine nevertheless.

Perhaps the greatest challenge to farming organically on an industrial scale is controlling weeds without the use of chemical herbicides. Greenways tackles its weeds with frequent and carefully timed tilling. Even before the crops are planted, the fields are irrigated to germinate the weed seeds present in the soil; a tractor then tills the field to kill them, the first of several passes it will make over the course of the growing season. When the crops stand too high to drive a tractor over, farm workers wielding propane torches will spot kill the biggest weeds by hand. The result is fields that look just as clean as the most herbicide-soaked farmland. But this approach, which I discovered is typical of large-scale organic operations, represents a compromise at best. The heavy tillage—heavier than in a conventional field—destroys the tilth of the soil and reduces its biological activity as surely as chemicals would; frequent tilling also releases so much nitrogen into the air that these weed-free organic fields require a lot more nitrogen fertilizer than they otherwise might. In a less disturbed, healthier soil, nitrogen-fixing bacteria would create much of the fertility that industrial organic growers must add in the form of compost, manures, fish emulsion, or Chilean nitrate—all inputs permitted under federal rules. (International organic rules, however, forbid the use of Chilean nitrate, a mineral form of nitrogen mined in Chile, often using child labor.) Not surprisingly the manufacturers of these inputs lobbied hard to shape the federal organic rules; in the end it proved easier to agree on a simple list of approved and prohibited materials rather than to try to legislate a genuinely more ecological model of farming.

Yet the best organic farmers deplore this sort of input substitution as a fall from the organic ideal, which envisions farms that provide for as much of their own fertility as possible, and control pests by means of crop diversification and rotation. It is too simple to say that smaller organic farms automatically hew closer to the organic ideals set forth by Albert Howard: Many small organic farms practice input substitution as well. The organic ideal is so exacting—a sustainable system modeled on nature that requires not only no synthetic chemicals but also no purchased inputs of any kind, and that returns as much to the soil as it removes—that it is mostly honored in the breach. Still, standing in a 160-acre block of organic broccoli in the Central Valley makes you appreciate why the farmers who come closest to achieving this ideal tend to be smaller in scale. These are the farmers who can plant literally dozens of different crops in fields that resemble quilts and practice long and elaborate rotations, thereby achieving the rich biodiversity in space and time that is the key to making a farm sustainable in something of the way a natural ecosystem is.

For better or worse, these are not the kinds of farms a big company like Small Planet Foods, or Whole Foods, does business with today. It's simply more cost-efficient to buy from one thousand-acre farm than ten hundred-acre farms. That's not because those big farms are necessarily any more productive, however. In fact, study after study has demonstrated that, measured in terms of the amount of food produced per acre, small farms are actually *more* productive than big farms; it is the higher transaction costs involved that makes dealing with them impractical for a company like Kahn's—that and the fact that they don't grow tremendous quantities of any one thing. As soon as your business involves stocking the frozen food case or produce section at a national chain, whether it be Wal-Mart or Whole Foods, the sheer quantities of organic produce you need makes it imperative to buy from farms operating on the same industrial scale you are. *Everything's connected.* The industrial values of specialization, economies of scale, and mechanization wind up crowding out ecological values such as diversity, complexity, and symbiosis. Or, to frame the matter in less abstract terms, as one of Kahn's employees did for me, "The combine just can't make the turn in a five-acre corn field"—and Small Planet Foods now consumes combine quantities of organic corn.

The big question is whether the logic of an industrial food chain can be reconciled to the logic of the natural systems on which organic agriculture has tried to model itself. Put another way, is industrial organic ultimately a contradiction in terms?

Kahn is convinced it is not, but others both inside and outside his company see an inescapable tension. Sarah Huntington is one of Cascadian Farm's oldest employees. She worked alongside Kahn on the original farm and at one time or another has held just about every job in the company. "The maw of that processing beast eats ten acres of cornfield in an hour," she told me. "And you're locked into planting a particular variety like Jubilee that ripens all at once and holds up in processing. So you see how the system is constantly pushing you back toward mono-culture, which is anathema in organic. But that's the challenge—to change the system more than it changes you."

One of the most striking ways companies like Small Planet Foods is changing the system is by helping conventional farms convert a portion of their acreage to organic. Several thousand acres of American farm-land are now organic as a result of the company's efforts, which go well beyond offering contracts to providing instruction and even manage-ment. Kahn has helped to prove to the skeptical that organic farming—dismissed as "hippie farming" only a few short years ago—can work on a large scale. The environmental benefits of this process cannot be overestimated. And yet the industrialization of organic comes at a price. The most obvious is consolidation down on the farm: Today two giant growers sell most of the fresh organic produce from California.

ONE OF THEM is Earthbound Farm, a company that arguably represents industrial organic farming at its best. If Cascadian Farm is a first-generation organic farm, Earthbound is second generation. It was started in the early eighties by Drew and Myra Goodman, two entirely improbable farmers who came to the land from the city with exactly no farming experience. The two had grown up within a few blocks of each other on the Upper East Side of Manhattan, where they attended the same progressive private high school. They didn't get together until af-ter both had gone off to college in California, Drew to Santa Cruz, Myra to Berkeley. While living near Carmel, killing time before heading to graduate school, Drew and Myra started a roadside organic farm on a few rented acres, growing raspberries and the sort of baby greens that chefs were making trendy in the eighties. Every Sunday Myra would wash and bag a bunch of lettuce for their own use, a salad for each night of the week. They discovered that the whole-leaf lettuces held up remarkably well right through to dinner the following Saturday.

One day in 1986 the Goodmans learned that the Carmel chef who

bought the bulk of their lettuce crop had moved on, and that his re-
placement wanted to use his own supplier. Suddenly they were faced
with a field of baby greens to get rid of, greens that wouldn't stay baby
for very long. So they decided to wash and bag them, and try to sell a
prewashed salad mix at retail. Produce managers greeted the novel
product with skepticism, so the Goodmans offered to take back any un-
sold bags at the end of the week. None of them returned. The "spring
mix" business had been born.

So at least goes the Earthbound creation story, as recounted to me by
Myra Goodman, now a tanned, leggy, and loquacious forty-two-year-
old, over lunch at the company's roadside stand in the Carmel Valley.
Like Cascadian Farm, Earthbound still maintains a showplace farm and
roadside stand, a tangible reminder of its roots. Unlike Cascadian, however, Earthbound is still very much in the farming business, though
most of its production land is an hour and a half north of Carmel, in
the Salinas Valley. Opening onto the Pacific near Monterey, the fertile,
sea breeze–conditioned valley offers ideal conditions for growing lettuces nine months of the year. In winter, the company picks up and
moves its operation, and many of its employees, south to Yuma, Arizona.

The prewashed salad business became one of the great success stories in American agriculture during the eighties and nineties, a time
when there wasn't much to celebrate, and the Goodmans are directly
responsible for much of that success. They helped dethrone iceberg,
which used to dominate the valley, by introducing dozens of different
salad mixes and innovating the way lettuces were grown, harvested,
cleaned, and packed. Myra's father is an engineer and inveterate tinkerer, and while the business was still headquartered in their Carmel
Valley living room, he designed gentle-cycle washing machines for lettuce; later the company introduced one of the first customized baby lettuce harvesters, and helped pioneer the packing of greens in specially
formulated plastic bags pumped with inert gases to extend shelf life.

Earthbound Farm's growth exploded after Costco placed an order in
1993. "Costco wanted our prewashed spring mix, but they didn't want
organic," Myra told me. "To them, organic sent the wrong message:

high price and low quality." At the time, organic was still recovering from the boom and bust following the Alar episode. But the Goodmans were committed to organic farming practices, so they decided to sell Costco their organically grown lettuce without calling it that.

"Costco was moving two thousand cases a week to start," Myra said, "and the order kept increasing." Wal-Mart, Lucky's, and Albertson's soon followed. The Goodmans quickly learned that in order to feed the maw of this industrial beast, Earthbound would have to industrialize itself. Their days of washing lettuce in the living room and selling at the Monterey farmer's market were over. "We didn't know how to farm on that scale," Drew told me, "and we needed a lot more land—fast." So the Goodmans entered into partnership with two of the most established conventional growers in the Salinas Valley, first Mission Ranches in 1995, and then Tanimura & Antle in 1999. These growers (no one in the valley calls himself a farmer) controlled some of the best land in the valley; they also knew how to grow, harvest, pack, and distribute tremendous quantities of produce. What they didn't know was organic production; in fact, Mission Ranches had tried it once and failed.

Through these partnerships, the Goodmans have helped convert several thousand acres of prime Salinas Valley land to organic; if you include all the farmland growing produce for Earthbound—which has expanded beyond greens to a full line of fruits and vegetables—the company represents a total of 25,000 organic acres. (This includes the acreage of the 135 farms that grow under contract to Earthbound.) The Goodmans estimate that taking all that land out of conventional production has eliminated some 270,000 pounds of pesticide and 8 million pounds of petrochemical fertilizer that would otherwise have been applied, a boon to both the environment and the people who work in those fields. Earthbound also uses biodiesel fuel in its tractors.

I expected a field of spring mix to look a lot like the stuff in the bag: a dozen varieties tossed together in happy profusion. But it turns out the mixing comes later. Each variety, which has its own slightly different cultural requirements and life span, is grown in a monoculture of several acres each, which has the effect of turning this part of the valley into a mosaic of giant color blocks: dark green, burgundy, pale green, blue green. As you get closer you see that the blocks are divided into a series of eighty-inch-wide raised beds thickly planted with a single variety. Each weed-free strip is as smooth and flat as a tabletop, leveled with a laser so that the custom-built harvester can snip each leaf at precisely the same point. Earthbound's tabletop fields exemplify one of the most powerful industrial ideas: the tremendous gains in efficiency to be had when you can conform the irregularity of nature to the precision and control of a machine.

Apart from the much higher level of precision—time as well as space are scrupulously managed on this farm—the organic practices at Earthbound resemble those I saw at Greenways farm. Frequent tilling is used to control weeds, though crews of migrant workers, their heads wrapped in brightly colored cloths against the hot sun, do a last pass through each block before harvest, pulling weeds by hand. To provide fertility—the farm's biggest expense—compost is trucked in; some crops also receive fish emulsion along with their water and a side dressing of pelleted chicken manure. Over the winter a cover crop of legumes is planted to build up nitrogen in the soil.

To control pests, every six or seven strips of lettuce is punctuated with a strip of flowers: sweet alyssum, which attracts the lacewings and syrphid flies that eat the aphids that can molest lettuces. Aside from some insecticidal soap to control insects in the cruciferous crops, pesticides are seldom sprayed. "We prefer to practice resistance and avoidance," Drew Goodman explained. Or, as their farm manager put it, "You have to give up the macho idea that you can grow anything you want anywhere you want to." So they closely track insect or disease outbreaks in their many fields and keep vulnerable crops at a safe distance; they also search out varieties with a strong natural resistance. Occasionally they'll lose a block to a pest, but as a rule growing baby greens is less risky since, by definition, the crop stays in the ground for so short a period of time—usually thirty days or so. Indeed, baby lettuce is one crop that may well be easier to grow organically than conventionally: Harsh chemicals can scorch young leaves, and nitrogen fertilizers render lettuces more vulnerable to insects. It seems the bugs are attracted to the free nitrogen in their leaves, and because of the more rapid growth of chemically nourished plants, insects find their leaves easier to pierce.

From the moment an organic lettuce plant is ready for harvest, the rest of its journey from field to produce aisle follows a swift and often ingenious industrial logic that is only nominally organic. "The only way we can sell organic produce at a reasonable price is by moving it into a conventional supply chain the moment it's picked," Drew Goodman explained. There is nothing particularly sustainable about that chain: It relies on the same crews of contract workers who pick produce throughout the Valley on a per piece basis, and on the same prodigious quantities of energy required to deliver any bag of prewashed salad to supermarkets across the country. (Though Earthbound does work to offset its fossil fuel consumption by planting trees.)

That conventional supply chain begins with the clever machine Earthbound developed to harvest baby greens: a car-size lettuce-shaving machine that moves down the rows, cutting the baby greens at a precise point just above the crown. Spidery arms extended in front of the machine gently rake through the bed in advance of the blade, scaring off any mice that might find their way into the salad. A fan blows the cut leaves over a screen to shake out any pebbles or soil, after which a belt conveys the greens into white plastic totes that workers stack on pallets on a wagon trailing alongside. At the end of each row the pallets are loaded onto a refrigerated tractor trailer, entering a "cold chain" that will continue unbroken all the way to the produce section at your supermarket.

Earthbound's own employees (who receive generous benefits by Valley standards, including health insurance and retirement) operate the baby greens harvester, but on the far side of the field I saw a contract crew of Mexicans, mostly women, slowly moving through the rows pulling weeds. I noticed that some of the workers had blue Band-Aids on their fingers. The Band-Aids are colored so inspectors at the plant can easily pick them out of the greens; each Band-Aid also contains a metal filament so that the metal detector through which every Earthbound leaf passes will pick it up before it winds up in a customer's salad.

Once filled, the trucks deliver their cargo of leaves to the loading dock at the processing plant in San Juan Bautista, essentially a 200,000-square-foot refrigerator designed to maintain the lettuce at exactly thirty-six degrees through the entire process of sorting, mixing, washing, drying, and packaging. These employees, most of them Mexicans, are dressed in full-length down coats; they empty totes of arugula, radicchio, and frisée into stainless steel rivers of lightly chlorinated water, the first of three washes each leaf will undergo. Viewed from overhead, the lettuce-packing operation looks like a hugely intricate Rube Goldberg contraption, a tangle of curving silver watercourses, shaking trays, and centrifuges, blue Band-Aid detectors, scales, and bagging stations that in about a half hour propels a freshly harvested leaf of baby lettuce into a polyethylene bag or box of ready-to-dress spring mix. The plant washes and packs 2.5 million pounds of lettuce a week; when you think just how many baby leaves it takes to make a pound, that represents a truly stupendous amount of lettuce. It also represents a truly stupendous amount of energy: to run the machines and chill the building, not to mention to transport all that salad to supermarkets across the country in refrigerated trucks and to manufacture the plastic containers it's packed in. A one-pound box of prewashed lettuce contains 80 calories of food energy. According to Cornell ecologist David Pimentel, growing, chilling, washing, packaging, and transporting that box of organic salad to a plate on the East Coast takes more than 4,600 calories of fossil fuel energy, or 57 calories of fossil fuel energy for every calorie of food. (These figures would be about 4 percent higher if the salad were grown conventionally.)

I had never before spent quite so much time looking at and thinking about lettuce, which when you do think about it—at least in the confines of the world's biggest refrigerator packed to the rafters with the stuff—is truly peculiar stuff. There are few things humans eat that are quite so elemental—a handful of leaves, after all, consumed raw.

When we're eating salad we're behaving a lot like herbivores, drawing as close as we ever do to all those creatures who bend their heads down to the grass, or reach up into the trees, to nibble on plant leaves. We add only the thinnest veneer of culture to these raw leaves, dressing them in oil and vinegar. Much virtue attaches to this kind of eating, for what do we regard as more wholesome than tucking into a pile of green leaves?

The contrast of the simplicity of this sort of eating, with all its pastoral overtones, and the complexity of the industrial process behind it produced a certain cognitive dissonance in my refrigerated mind. I began to feel that I no longer understood what this word I'd been following across the country and the decades really meant—I mean, of course, the word "organic." It is an unavoidable and in some ways impolite question, and very possibly besides the point if you look at the world the way Gene Kahn or Drew and Myra Goodman do, but in precisely what sense can that box of salad on sale in a Whole Foods three thousand miles and five days away from this place truly be said to be organic? And if that well-traveled plastic box deserves that designation, should we then perhaps be looking for another word to describe the much shorter and much less industrial food chain that the first users of the word "organic" had in mind?

This at least is the thinking of the smaller organic farmers who, not surprisingly, are finding it impossible to compete against the impressive industrial efficiencies achieved by a company like Earthbound Farm. Supermarket chains don't want to deal with dozens of different organic farmers; they want one company to offer them a complete line of fruits and vegetables, every SKU in the produce section. And Earthbound has obliged, consolidating its hold on the organic produce section of the American supermarket, and in the process growing into a $350 million company. "Everything eventually morphs into the way the world is." Drew Goodman told me a day had come several years ago when he suddenly no longer felt comfortable manning his usual stall at the Monterey farmer's market. He looked around and understood "we didn't belong here anymore. We're really in a whole different business now." Goodman makes no apologies for that, and rightly so: His company has done a world of good, for its land, its workers, the growers it works with, and its customers.

Yet his success, like Gene Kahn's, has opened up a gulf between Big and Little Organic and convinced many of the movement's founders, as well as pioneering farmers like Joel Salatin, that the time has come to move beyond organic—to raise the bar on the American food system once again. Some of these innovating farmers are putting their emphasis on quality, others on labor standards, some on local systems of distribution, and still others on achieving a more thoroughgoing sustainability. Michael Ableman, one of the self-described beyond organic farmers I interviewed in California, said, "We may have to give up on the word 'organic,' leave it to the Gene Kahns of the world. To be honest, I'm not sure I want that association, because what I'm doing on my farm is not just substituting inputs."

A few years ago, at a conference on organic agriculture in California, a corporate organic grower suggested to a small farmer struggling to survive in the competitive world of industrial organic agriculture that "you should really try to develop a niche to distinguish yourself in the market." Holding his fury in check, the small farmer replied as levelly as he could manage:

"I believe I developed that niche twenty years ago. It's called 'organic.' And now you, sir, are sitting on it."

4. MEET ROSIE, THE ORGANIC FREE-RANGE CHICKEN

The last stop on my tour of California industrial organic farming took me to Petaluma, where I tried without success to find the picturesque farmstead, with its red barn, cornfield, and farmhouse, depicted on the package in which the organic roasting chicken I bought at Whole Foods had been wrapped; nor could I find Rosie herself, at least not outdoors, ranging freely.

Petaluma Poultry has its headquarters not on a farm but in a sleek modern office building in an industrial park just off Route 101; there's little farmland left in Petaluma, which is now a prosperous San Francisco bedroom community. The survival of Petaluma Poultry in the face of this development (it's one of what were once dozens of chicken farms in the area) is a testament to the company's marketing acumen. When its founder, Allen Shainsky, recognized the threat from integrated national chicken processors like Tyson and Perdue, he decided that the only way to stay in business was through niche marketing. So he started processing, on different days of the week, chickens for the kosher, Asian, natural, and organic markets. Each required a slightly different protocol: to process a kosher bird you needed a rabbi on hand, for example; for an Asian bird you left the head and feet on; for the natural market you sold the same bird minus head and feet, but played up the fact that Rocky, as this product was called, received no antibiotics or animal by-products in its feed, and you provided a little exercise yard outside the shed so Rocky could, at his option, range free. And to call a bird organic, you followed the natural protocol except that you also fed it certified organic feed (corn and soy grown without pesticides and chemical fertilizer) and you processed the bird slightly younger and smaller, so it wouldn't seem quite so expensive. Philosophy didn't really enter into it.

(Petaluma Eggs, a nearby egg producer with corporate ties to Petaluma Poultry, pursues a similar niche strategy, offering natural free-range eggs [no drugs in the chickens' feed, no battery cages]; fertile eggs [all of the above plus the hens have access to a rooster]; enhanced omega-3 natural eggs [all of the above, save the rooster, plus kelp in the feed to boost levels of omega-3 fatty acids]; and certified organic eggs [cage- and drug-free plus certified organic feed]. These last are sold under the label Judy's Family Farm, a brand that until my visit to Petaluma I hadn't connected to Petaluma Eggs. The Judy's label had always made me picture a little family farm, or maybe even a commune of back-to-the-land lesbians up in Sonoma. But it turns out Judy is the name of the wife of Petaluma's principal owner, a marketer who has clearly mastered the conventions of Supermarket Pastoral. Who could begrudge a farmer named Judy $3.59 for a dozen organic eggs she presumably has to get up at dawn each morning to gather? Just how big and sophisticated an operation Petaluma Eggs really is I was never able to ascertain: The company was too concerned about biosecurity to let a visitor get past the office.)

Rosie the organic chicken's life is little different from that of her kosher and Asian cousins, all of whom are conventional Cornish Cross broilers processed according to state-of-the-art industrial practice. (Though Petaluma Poultry sets the bar higher than many of its competitors, who routinely administer antibiotics and use feed made from animal by-products.) The Cornish Cross represents the pinnacle of industrial chicken breeding. It is the most efficient converter of corn into breast meat ever designed, though this efficiency comes at a high physiological price: The birds grow so rapidly (reaching oven-roaster proportions in seven weeks) that their poor legs cannot keep pace, and frequently fail.

After a tour of the fully automated processing facility, which can translate a chicken from a clucking, feathered bird to a shrink-wrapped pack of parts inside of ten minutes, the head of marketing drove me out to meet Rosie—preprocessing. The chicken houses don't resemble a farm so much as a military barracks: a dozen long, low-slung sheds with giant fans at either end. I donned what looked like a hooded white hazmat suit—since the birds receive no antibiotics yet live in close confinement, the company is ever worried about infection, which could doom a whole house overnight—and stepped inside. Twenty thousand birds moved away from me as one, like a ground-hugging white cloud, clucking softly. The air was warm and humid and smelled powerfully of ammonia; the fumes caught in my throat. Twenty thousand is a lot of chickens, and they formed a gently undulating white carpet that stretched nearly the length of a football field. After they adjusted to our presence, the birds resumed sipping from waterers suspended from the ceiling, nibbled organic food from elevated trays connected by tubes to a silo outside, and did pretty much everything chickens do except step outside the little doors located at either end of the shed.

Compared to conventional chickens, I was told, these organic birds have it pretty good: They get a few more square inches of living space per bird (though it was hard to see how they could be packed together much more tightly), and because there are no hormones or antibiotics in their feed to accelerate growth, they get to live a few days longer. Though under the circumstances it's not clear that a longer life is necessarily a boon.

Running along the entire length of each shed was a grassy yard maybe fifteen feet wide, not nearly big enough accommodate all twenty thousand birds inside should the group ever decide to take the air en masse. Which, truth be told, is the last thing the farm managers want to see happen, since these defenseless, crowded, and genetically identical birds are exquisitely vulnerable to infection. This is one of the larger ironies of growing organic food in an industrial system: It is even more precarious than a conventional industrial system. But the federal rules say an organic chicken should have "access to the outdoors," and Supermarket Pastoral imagines it, so Petaluma Poultry provides the doors and the yard and everyone keeps their fingers crossed.

It would appear Petaluma's farm managers have nothing to worry about. Since the food and water and flock remain inside the shed, and since the little doors remain shut until the birds are at least five weeks old and well settled in their habits, the chickens apparently see no reason to venture out into what must seem to them an unfamiliar and terrifying world. Since the birds are slaughtered at seven weeks, free range turns out to be not so much a lifestyle for these chickens as a two-week vacation option.

After I stepped back outside into the fresh air, grateful to escape the humidity and ammonia, I waited by the chicken door to see if any of the birds would exercise that option and stroll down the little ramp to their grassy yard, which had been mowed recently. And waited. I finally had to conclude that Rosie the organic free-range chicken doesn't really grok the whole free-range conceit. The space that has been provided to her for that purpose is, I realized, not unlike the typical American front lawn it resembles—it's a kind of ritual space, intended not so much for the use of the local residents as a symbolic offering to the larger community. Seldom if ever stepped upon, the chicken-house lawn is scrupulously maintained nevertheless, to honor an ideal nobody wants to admit has by now become something of a joke, an empty pastoral conceit.

5. MY ORGANIC INDUSTRIAL MEAL

My shopping foray to Whole Foods yielded all the ingredients for a comforting winter Sunday night dinner: roast chicken (Rosie) with roasted vegetables (yellow potatoes, purple kale, and red winter squash from Cal-Organics), steamed asparagus, and a spring mix salad from Earthbound Farm. Dessert would be even simpler: organic ice cream from Stonyfield Farm topped with organic blackberries from Mexico.

On a hunch it probably wasn't quite ready for prime time (or at least for my wife), I served the Cascadian Farm organic TV dinner I'd bought to myself for lunch, right in its microwaveable plastic bowl. Five minutes on high and it was good to go. Peeling back the polyethylene film covering the dish, I felt a little like a flight attendant serving meals, and indeed the entrée looked and tasted very much like airline food. The chunks of white meat chicken had been striped nicely with grill marks and impregnated with a salty marinade that gave the meat that slightly abstract chicken taste processed chicken often has, no doubt owing to the "natural chicken flavor" mentioned on the box's list of ingredients. The chicken chunks and allied vegetables (soft carrots, peas, green beans, and corn) were "blanketed in a creamy rosemary dill sauce"—a creaminess that had evidently been achieved synthetically, since no dairy products appeared among the ingredients. I'm betting it's the xanthan gum (or maybe the carrageenan?) that bears responsibility for the sauce's unfortunate viscosity. To be fair, one shouldn't compare an organic TV dinner to real food but to a conventional TV dinner, and by that standard (or at least my recollection of it) Cascadian Farm has nothing to be ashamed of, especially considering that an organic food scientist must work with only a tiny fraction of the synthetic preservatives, emulsifiers, and flavor agents available to his colleagues at Swanson or Kraft.

Rosie and her consort of fresh vegetables fared much better at dinner, if I don't mind saying so myself. I roasted the bird in a pan surrounded by the potatoes and chunks of winter squash. After removing the chicken from the oven, I spread the crinkled leaves of kale on a cookie sheet, sprinkled them with olive oil and salt, and slid them into the hot oven to roast. After ten minutes or so, the kale was nicely crisped and the chicken was ready to carve.

All but one of the vegetables I served that night bore the label of Cal-Organic Farms, which, along with Earthbound, dominates the organic produce section in the supermarket. Cal-Organic is a big grower of organic vegetables in the San Joaquin Valley. As part of the consolidation of the organic industry, the company was acquired by Grimmway Farms, which already enjoyed a virtual monopoly in organic carrots. Unlike Earthbound, neither Grimmway nor Cal-Organic has ever been part of the organic movement. Both companies were started by conventional growers looking for a more profitable niche and worried that the state might ban certain key pesticides. "I'm not necessarily a fan of organic," a spokesman for Grimmway recently told an interviewer. "Right now I don't see that conventional farming does harm. Whether we stay with organic for the long haul depends on profitability." Philosophy, in other words, has nothing to do with it.

The combined company now controls seventeen thousand acres across California, enough land that it can, like Earthbound, rotate production up and down the West Coast (and south into Mexico) in order to ensure a twelve-month national supply of fresh organic produce, just as California's conventional growers have done for decades. It wasn't many years ago that organic produce had only a spotty presence in the supermarket, especially during the winter months. Today, thanks in large part to Grimmway and Earthbound, you can find pretty much everything, all year round.

Including asparagus in January, I discovered. This was the one vegetable I prepared that wasn't grown by Cal-Organic or Earthbound; it had been grown in Argentina and imported by a small San Francisco distributor. My plan had been a cozy winter dinner, but I couldn't resist the bundles of fresh asparagus on sale at Whole Foods, even though it set me back six dollars a pound. I had never tasted organic South American asparagus in January, and felt my foray into the organic empire demanded that I do. What better way to test the outer limits of the word "organic" than by dining on a springtime delicacy that had been grown according to organic rules on a farm six thousand miles (and two seasons) away, picked, packed, and chilled on Monday, flown by jet to Los Angeles Tuesday, trucked north to a Whole Foods regional distribution center, then put on sale in Berkeley by Thursday, to be steamed, by me, Sunday night?

The ethical implications of buying such a product are almost too numerous and knotty to sort out: There's the expense, there's the prodigious amounts of energy involved, the defiance of seasonality, and the whole question of whether the best soils in South America should be devoted to growing food for affluent and overfed North Americans. And yet you can also make a good argument that my purchase of organic asparagus from Argentina generates foreign exchange for a country desperately in need of it, and supports a level of care for that country's land—farming without pesticides or chemical fertilizer—it might not otherwise receive. Clearly my bunch of asparagus had delivered me deep into the thicket of trade-offs that a global organic marketplace entails.

Okay, but how did it taste?

My jet-setting Argentine asparagus tasted like damp cardboard. After the first spear or two no one touched it. Perhaps if it had been sweeter and tenderer we would have finished it, but I suspect the fact that asparagus was out of place in a winter supper made it even less appetizing. Asparagus is one of a dwindling number of foods still firmly linked in our minds to the seasonal calendar.

All the other vegetables and greens were much tastier—really good, in fact. Whether they would have been quite so sweet and bright after a cross-country truck ride is doubtful, though the Earthbound greens, in their polyethylene bag, stayed crisp right up to the expiration date, a full eighteen days after leaving the field—no small technological feat. The inert gases, scrupulous cold chain and space-age plastic bag (which allows the leaves to respire just enough) account for much of this longevity, but some of it, as the Goodmans had explained to me, owes to the fact that the greens were grown organically. Since they're not pumped up on synthetic nitrogen, the cells of these slower-growing leaves develop thicker walls and take up less water, making them more durable.

And, I'm convinced, tastier, too. When I visited Greenways Organic, which grows both conventional and organic tomatoes, I learned that the organic ones consistently earn higher Brix scores (a measure of sugars) than the same varieties grown conventionally. More sugars means less water and more flavor. It stands to reason the same would hold true for other organic vegetables: slower growth, thicker cell walls, and less water should produce more concentrated flavors. That at least has always been my impression, though in the end freshness probably affects flavor even more than growing method.

To serve such a scrupulously organic meal begs an unavoidable question: Is organic food better? Is it worth the extra cost? My Whole Foods dinner certainly wasn't cheap, considering I made it from scratch: Rosie cost $15 ($2.99 a pound), the vegetables another $12 (thanks to that six-buck bunch of asparagus), and the dessert $7 (including $3 for a six-ounce box of blackberries). Thirty-four dollars to feed a family of three at home. (Though we did make a second meal from the leftovers.) Whether organic is better and worth it are certainly fair, straightforward questions, but the answers, I've discovered, are anything but simple.

Better for what? is the all-important corollary to that question. If the answer is "taste," then the answer is, as I've suggested, very likely, at least in the case of produce—but not necessarily. Freshly picked conventional produce is bound to taste better than organic produce that's been riding the interstates in a truck for three days. Meat is a harder call. Rosie was a tasty bird, yet, truth be told, not quite as tasty as Rocky, her bigger nonorganic brother. That's probably because Rocky is an older chicken, and older chickens generally have more flavor. The fact that the corn and soybeans in Rosie's diet were grown without chemicals probably doesn't change the taste of her meat. Though it should be said that Rocky and Rosie both taste more like chicken than mass-market birds fed on a diet of antibiotics and animal by-products, which makes for mushier and blander meat. What's in an animal's feed naturally affects how it will taste, though whether that feed is organic or not probably makes no difference.

Better for what? If the answer is "for my health" the answer, again, is probably—but not automatically. I happen to believe the organic dinner I served my family is healthier than a meal of the same foods conventionally produced, but I'd be hard-pressed to prove it scientifically. What I could prove, with the help of a mass spectrometer, is that it contained little or no pesticide residue—the traces of the carcinogens, neurotoxins, and endocrine disruptors now routinely found in conventional produce and meat. What I probably can't prove is that the low levels of these toxins present in these foods will make us sick—give us cancer, say, or interfere with my son's neurological or sexual development. But that does not mean those poisons are *not* making us sick: Remarkably little research has been done to assess the effects of regular exposure to the levels of organophosphate pesticide or growth hormone that the government deems "tolerable" in our foods. (One problem with these official tolerances is that they don't adequately account for children's exposure to pesticides, which, because of children's size and eating habits, is much greater than adults'.) Given what we do know about exposure to endocrine disruptors, the biological impact of which depends less on dose than timing, minimizing a child's exposure to these chemicals seems like a prudent idea. I very much like the fact that the milk in the ice cream I served came from cows that did not receive injections of growth hormone to boost their productivity, or that the corn those cows are fed, like the corn that feeds Rosie, contains no residues of atrazine, the herbicide commonly sprayed on American cornfields. Exposure to vanishingly small amounts (0.1 part per billion) of this herbicide has been shown to turn normal male frogs into hermaphrodites. Frogs are not boys, of course. So I can wait for that science to be done, or for our government to ban atrazine (as European governments have done), or I can act now on the presumption that food from which this chemical is absent is better for my son's health than food that contains it.

Of course, the healthfulness of a food is not simply a question of its toxicity; we have also to consider its nutritional quality. Is there any reason to think my Whole Foods meal is any more nutritious than the same meal prepared with conventionally grown ingredients?

Over the years there have been sporadic efforts to demonstrate the nutritional superiority of organic produce, but most have foundered on the difficulty of isolating the great many variables that can affect the nutritional quality of a carrot or a potato—climate, soils, geography, freshness, farming practices, genetics, and so on. Back in the fifties, when the USDA routinely compared the nutritional quality of produce from region to region, it found striking differences: carrots grown in the deep soils of Michigan, for example, commonly had more vitamins than carrots grown in the thin, sandy soils of Florida. Naturally this information discomfited the carrot growers of Florida, which probably explains why the USDA no longer conducts this sort of research. Nowadays U.S. agricultural policy, like the Declaration of Independence, is founded on the principle that all carrots are created equal, even though there's good reason to believe this isn't really true. But in an agricultural system dedicated to quantity rather than quality, the fiction that all foods are created equal is essential. This is why, in inaugurating the federal organic program in 2000, the secretary of agriculture went out of his way to say that organic food is no better than conventional food.

"The organic label is a marketing tool," Secretary Glickman said. "It is not a statement about food safety. Nor is 'organic' a value judgment about nutrition or quality."

Some intriguing recent research suggests otherwise. A study by University of California–Davis researchers published in the *Journal of Agriculture and Food Chemistry* in 2003 described an experiment in which identical varieties of corn, strawberries, and blackberries grown in neighboring plots using different methods (including organically and conventionally) were compared for levels of vitamins and polyphenols. Polyphenols are a group of secondary metabolites manufactured by plants that we've recently learned play an important role in human health and nutrition. Many are potent antioxidants; some play a role in preventing or fighting cancer; others exhibit antimicrobial properties. The Davis researchers found that organic and otherwise sustainably grown fruits and vegetables contained significantly higher levels of both ascorbic acid (vitamin C) and a wide range of polyphenols.

The recent discovery of these secondary metabolites in plants has brought our understanding of the biological and chemical complexity of foods to a deeper level of refinement; history suggests we haven't gotten anywhere near the bottom of this question, either. The first level was reached early in the nineteenth century with the identification of the macronutrients—protein, carbohydrate, and fat. Having isolated these compounds, chemists thought they'd unlocked the key to human nutrition. Yet some people (such as sailors) living on diets rich in macronutrients nevertheless got sick. The mystery was solved when scientists discovered the major vitamins—a second key to human nutrition. Now it's the polyphenols in plants that we're learning play a critical role in keeping us healthy. (And which might explain why diets heavy in processed food fortified with vitamins still aren't as nutritious as fresh foods.) You wonder what else is going on in these plants, what other undiscovered qualities in them we've evolved to depend on.

In many ways the mysteries of nutrition at the eating end of the food chain closely mirror the mysteries of fertility at the growing end: The two realms are like wildernesses that we keep convincing ourselves our chemistry has mapped, at least until the next level of complexity comes into view. Curiously, Justus von Liebig, the nineteenth-century German chemist with the spectacularly ironic surname, bears responsibility for science's overly reductive understanding of both ends of the food chain. It was Liebig, you'll recall, who thought he had found the chemical key to soil fertility with the discovery of NPK, and it was the same Liebig who thought he had found the key to human nutrition when he identified the macronutrients in food. Liebig wasn't wrong on either count, yet in both instances he made the fatal mistake of thinking that what we knew about nourishing plants and people was all we needed to know to keep them healthy. It's a mistake we'll probably keep repeating until we develop a deeper respect for the complexity of food and soil and, perhaps, the links between the two.

But back to the polyphenols, which may hint at the nature of that link. Why in the world should organically grown blackberries or corn contain significantly more of these compounds? The authors of the Davis study haven't settled the question, but they offer two suggestive theories. The reason plants produce these compounds in the first place is to defend themselves against pests and diseases; the more pressure from pathogens, the more polyphenols a plant will produce. These compounds, then, are the products of natural selection and, more specifically, the coevolutionary relationship between plants and the species that prey on them. Who would have guessed that humans evolved to profit from a diet of these plant pesticides? Or that we would invent an agriculture that then deprived us of them? The Davis authors hypothesize that plants being defended by man-made pesticides don't need to work as hard to make their own polyphenol pesticides. Coddled by us and our chemicals, the plants see no reason to invest their resources in mounting a strong defense. (Sort of like European nations during the cold war.)

A second explanation (one that subsequent research seems to support) may be that the radically simplified soils in which chemically fertilized plants grow don't supply all the raw ingredients needed to synthesize these compounds, leaving the plants more vulnerable to attack, as we know conventionally grown plants tend to be. NPK might be sufficient for plant growth yet still might not give a plant everything it needs to manufacture ascorbic acid or lycopene or resveratrol in quantity. As it happens, many of the polyphenols (and especially a subset called the flavonols) contribute to the characteristic taste of a fruit or vegetable. Qualities we can't yet identify in soil may contribute qualities we've only just begun to identify in our foods and our bodies.

Reading the Davis study I couldn't help thinking about the early proponents of organic agriculture, people like Sir Albert Howard and J. I. Rodale, who would have been cheered, if unsurprised, by the findings. Both men were ridiculed for their unscientific conviction that a reductive approach to soil fertility—the NPK mentality—would diminish the nutritional quality of the food grown in it and, in turn, the health of the people who lived on that food. All carrots are not created equal, they believed; how we grow it, the soil we grow it in, what we feed that soil all contribute qualities to a carrot, qualities that may yet escape the explanatory net of our chemistry. Sooner or later the soil scientists and nutritionists will catch up to Sir Howard, heed his admonition that we begin "treating the whole problem of health in soil, plant, animal and man as one great subject."

So it happens that these organic blackberries perched on this mound of vanilla ice cream, having been grown in a complexly fertile soil and forced to fight their own fights against pests and disease, are in some quantifiable way more nutritious than conventional blackberries. This would probably not come as earthshaking news to Albert Howard or J. I. Rodale or any number of organic farmers, but at least now it is a claim for which we can supply a scientific citation: *J. Agric. Food. Chem.* vol. 51, no. 5, 2003. (Several other such studies have appeared since; see the Sources section at the back of this book.)

Obviously there is much more to be learned about the relationship of soil to plant, animals, and health, and it would be a mistake to lean too heavily on any one study. It would also be a mistake to assume that the word "organic" on a label automatically signifies healthfulness, especially when that label appears on heavily processed and long-distance foods that have probably had much of their nutritional value, not to mention flavor, beaten out of them long before they arrive on our tables.

The better for what? question about my organic meal can of course be answered in a much less selfish way: Is it better for the environment? Better for the farmers who grew it? Better for the public health? For the taxpayer? The answer to all three questions is an (almost) unqualified yes. To grow the plants and animals that made up my meal, no pesticides found their way into any farmworker's bloodstream, no nitrogen runoff or growth hormones seeped into the watershed, no soils were poisoned, no antibiotics were squandered, no subsidy checks were written. If the high price of my all-organic meal is weighed against the comparatively low price it exacted from the larger world, as it should be, it begins to look, at least in karmic terms, like a real bargain.

And yet, and yet . . . an industrial organic meal such as mine does leave deep footprints on our world. The lot of the workers who harvested the vegetables and gathered up Rosie for slaughter is not appreciably different from that of those on nonorganic factory farms. The chickens lived only marginally better lives than their conventional counterparts; in the end a CAFO is a CAFO, whether the food served in it is organic or not. As for the cows that produced the milk in our ice cream, they may well have spent time outdoors in an actual pasture (Stonyfield Farm buys most—though not all—of its milk from small dairy farmers), but the organic label guarantees no such thing. And while the organic farms I visited don't receive direct government payments, they do receive other subsidies from taxpayers, notably subsidized water and electricity in California. The two-hundred-thousand-square-foot refrigerated processing plant where my salad was washed pays half as much for its electricity as it would were Earthbound not classified as a "farm enterprise."

But perhaps most discouraging of all, my industrial organic meal is nearly as drenched in fossil fuel as its conventional counterpart. Asparagus traveling in a 747 from Argentina; blackberries trucked up from Mexico; a salad chilled to thirty-six degrees from the moment it was picked in Arizona (where Earthbound moves its entire operation every winter) to the moment I walk it out the doors of my Whole Foods. The food industry burns nearly a fifth of all the petroleum consumed in the United States (about as much as automobiles do). Today it takes between seven and ten calories of fossil fuel energy to deliver one calorie of food energy to an American plate. And while it is true that organic farmers don't spread fertilizers made from natural gas or spray pesticides made from petroleum, industrial organic farmers often wind up burning more diesel fuel than their conventional counterparts: in trucking bulky loads of compost across the countryside and weeding their fields, a particularly energy-intensive process involving extra irrigation (to germinate the weeds before planting) and extra cultivation. All told, growing food organically uses about a third less fossil fuel than growing it conventionally, according to David Pimentel, though that savings disappears if the compost is not produced on site or nearby.

Yet growing the food is the least of it: only a fifth of the total energy used to feed us is consumed on the farm; the rest is spent processing the food and moving it around. At least in terms of the fuel burned to get it from the farm to my table, there's little reason to think my Cascadian Farm TV dinner or Earthbound Farm spring mix salad is any more sustainable than a conventional TV dinner or salad would have been.

Well, at least we didn't eat it in the car.

So is an industrial organic food chain finally a contradiction in terms? It's hard to escape the conclusion that it is. Of course it is possible to live with contradictions, at least for a time, and sometimes it is necessary or worthwhile. But we ought at least face up to the cost of our compromises. The inspiration for organic was to find a way to feed ourselves more in keeping with the logic of nature, to build a food system that looked more like an ecosystem that would draw its fertility and energy from the sun. To feed ourselves otherwise was "unsustainable," a word that's been so abused we're apt to forget what it very specifically means: *Sooner or later it must collapse*. To a remarkable extent, farmers succeeded in creating the new food chain on their farms; the trouble began when they encountered the expectations of the super-market. As in so many other realms, nature's logic has proven no match for the logic of capitalism, one in which cheap energy has always been a given. And so, today, the organic food industry finds itself in a most unexpected, uncomfortable, and, yes, unsustainable position: floating on a sinking sea of petroleum.

TEN

GRASS

Thirteen Ways of Looking at a Pasture

1. MONDAY

For something people profess to like so much, grass is peculiarly hard for us to see. Oh, you can see it well enough in a general sense, but how much do you *really* see when you look at a patch of grass? The color green, of course, perhaps a transitory recording of the breeze: an abstraction. Grass to us is more ground than figure, a backdrop to more legible things in the landscape—trees, animals, buildings. It's less a subject in its own right than a context. Maybe this has to do with the disparity in scale between us and the uncountable tiny beings that make up a pasture. Maybe we're just too big to see what's going on down there in any detail.

Curiously, we seem to like grass less for what it is than for what it isn't—the forest, I mean—and yet we're much more likely to identify with a tree than a blade of grass. When poets liken people to blades of grass it's usually to humble us, to pull the rug out from under our individuality and remind us of our existential puniness. Composed of so many tiny seemingly indistinguishable parts, a patch of grass—which on closer inspection isn't even composed of grasses half the time but of legumes and broad-leafed plants of many kinds—resolves itself in our perception into an undifferentiated mass, a more or less shaggy field of color. This way of looking at, or not looking at, grass must suit us, or why would we work so hard to keep it mowed? Mowing only adds to the abstractness of grass.

This is not at all how grass looks to a cow or for that matter to a grass farmer like Joel Salatin. When one of his cows moves into a new paddock, she doesn't just see the color green; she doesn't even see grass. She sees, out of the corner of her eye, this nice tuft of white clover, the emerald-green one over there with the heart-shaped leaves, or, up ahead, that grassy spray of bluish fescue tightly cinched at ground level. These two entities are as different in her mind as vanilla ice cream is from cauliflower, two dishes you would never conflate just because they both happen to be white. The cow opens her meaty wet lips, curls her sandpaper tongue around the bunched clover like a fat rope, and with the pleasing sound of tearing foliage, rips the mouthful of tender leaves from its crown. She'll get to the fescue eventually, and the orchard grass, and even to quite a few of the weeds, but not before she's eaten all the clover ice cream she can find.

Joel calls his pastures the "salad bar," and to his cows they contain at least as many different things to eat. As well as a few things not to eat. Though we might fail to notice the handful of Carolina nightshades or thistles lurking in this pasture, when the cows are done grazing it to-morrow, those plants will still be standing, like forlorn florets of cauli-flower languishing on a picky child's plate.

What watching this cow eat her supper tells me is that the scale argument doesn't really hold. The reason we don't see very much when we look at grass has less to do with our relative proportions than with our interests. The cow I'm following in Joel Salatin's pasture this evening is a far sight bigger than I am, and in most matters a good deal less perceptive, yet she can pick a clump of timothy out of this illegible green chaos in less time than it would take me to remember that plant's name. I don't eat timothy, or even clover. But if I did I'd probably perceive the order and beauty and delectability of this salad bar as vividly as she does. Legibility, too, is in the eye of the beholder.

Joel doesn't eat grass either—it's one of the few nutritious things in nature the human omnivore, lacking a rumen to break down its cellulose, can't digest—yet he can see the salad bar almost as vividly as his cows. That first day I spent on his farm, when he insisted that before I met any animals I join him down on his belly in a pasture, he introduced me to orchard grass and fescue, to red and white clover, to millet and bluegrass, plantain and timothy and sweet grass, which he pulled a blade of for me to taste (and a very sweet grass it is). Joel wanted me to understand why he calls himself a grass farmer rather than a rancher or a pig farmer or a chicken farmer or a turkey farmer or a rabbit farmer or an egg farmer. The animals come and go, but the grasses, which directly or indirectly feed all the animals, abide, and the well-being of the farm depends more than anything else on the well-being of its grass.

Grass farming is a relatively new term in American agriculture, imported from New Zealand by Allan Nation, the editor of *Stockman Grass Farmer*, in the 1980s. *Stockman* is a tabloid monthly, chock-full of ads for portable electric fencing, mineral supplements, and bull semen, that has become the bible for the growing band of livestock producers who practice something called "management-intensive grazing," or as abbreviated in the pages of Nation's magazine, MiG. (It's sometimes also called rotational grazing.) Joel writes a column for the *Stockman Grass Farmer* called The Pastoralist, and has become close friends with Nation, whom he regards as something of a mentor.

When Allan Nation went to New Zealand in 1984 and heard sheep ranchers there refer to themselves as grass farmers something clicked, he says, and he began to regard the growing of food in a completely fresh light. Nation promptly changed the name of his little journal from the *Stockman* to *Stockman Grass Farmer* and "got pretty evangelical about grass." He gathered around his magazine a group of like-minded grass evangelists, including Joel, Jim Gerrish, an Idaho rancher and teacher (who coined the phrase "management-intensive grazing"), Gerald Fry, a breeding specialist, Jo Robinson, a health writer who studies the health benefits of grass-fed meat, and an Argentine agronomist named Dr. Anibal Pordomingo. Many of these people first encountered the theory of rotational grazing in the work of André Voisin, a French agronomist whose 1959 treatise, *Grass Productivity*, documented that simply by applying the right number of ruminants at the right time pastures could produce far more grass (and, in turn, meat and milk) than anyone had ever thought possible.

Grass farmers grow animals—for meat, eggs, milk, and wool—but regard them as part of a food chain in which grass is the keystone species, the nexus between the solar energy that powers every food chain and the animals we eat. "To be even more accurate," Joel has said, "we should call ourselves sun farmers. The grass is just the way we capture the solar energy." One of the principles of modern grass farming is that to the greatest extent possible farmers should rely on the contemporary energy of the sun, as captured every day by photosynthesis, instead of the fossilized sun energy contained in petroleum.

For Allan Nation, who grew up on a cattle ranch in Mississippi, doing so is as much a matter of sound economics as environmental virtue. "All agriculture is at its heart a business of capturing free solar energy in a food product that can then be turned into high-value human energy," he recently wrote in his column, Al's Obs; here each month he applies the theories of a decidedly eclectic group of thinkers (ranging from business gurus like Peter Drucker and Michael Porter to writers like Arthur Koestler) to the problems of farming. "There are only two efficient ways to do this," he wrote in his column. "One is for you to walk out in your garden, pull a carrot and eat it. This is a direct transfer of solar energy to human energy. The second most efficient way is for you to send an animal out to gather this free solar food and then you eat the animal.

"All other methods of harvest and transfer require higher capital and petroleum energy inputs and these necessarily lower the return to the farmer/rancher. As Florida rancher Bud Adams once told me,

'Ranching is a very simple business. The really hard part is keeping it simple.'"

The simplest way to capture the sun's energy in a form food animals can use is by growing grass: "These blades are our photovoltaic panels," Joel says. And the most efficient—if not the simplest—way to grow vast quantities of solar panels is by management-intensive grazing, a method that as its name implies relies more heavily on the farmer's brain than on capital—or on energy-intensive inputs. All you need, in fact, is some portable electric fencing, a willingness to move your livestock onto fresh pasture every day, and the kind of intimate knowledge of grass that Joel tried to impart to me that early spring afternoon, down on our bellies in his pasture.

"The important thing to know about any grass is that its growth follows a sigmoid, or S, curve," Joel explained. He grabbed my pen and notebook and began drawing a graph, based on one that appears in Voisin's book. "This vertical axis here is the height of our grass plant, okay? And the horizontal axis is time: the number of days since this paddock was last grazed." He started tracing a big S on the page, beginning in the lower left-hand corner where the two axes met. "See, the growth starts out real slow like this, but then after a few days it begins to zoom. That's called 'the blaze of growth,' when the grass has recovered from the first bite, rebuilt its reserves and root mass, and really taken off. But after a while"—the curve leveled out at around day fourteen or so—"it slows down again, as the grass gets ready to flower and seed. It's entering its period of senescence, when the grass begins to lignify [get woody] and becomes less palatable to the cow.

"What you want to do is graze a pasture right at this point here"—he tapped my pad sharply—"at the very top of the blaze of growth. But what you never, ever want to do is violate the law of the second bite. You can't let your cows take a second bite of a grass before it has had a chance to fully recover."

If the law of the second bite were actually on the books, most of the world's ranchers and dairy farmers would be outlaws, since they allow their stock to graze their pastures continuously. By allowing cattle a second or third bite, the most desirable "ice cream" species—clover, orchard grass, sweet grass, bluegrass, timothy—weaken and gradually disappear from the sward, giving way to bald spots and to weedy and brushy species the cows won't touch. Any plant wants to keep its roots and shoots roughly in balance, so grasses kept short by overgrazing lack the deep roots needed to bring water and minerals up from the subsoil. Over time a closely cropped grassland deteriorates, and in a dry or brittle environment, it will eventually turn into a desert. The reason environmentalists in the western United States take such a dim view of grazing is that most ranchers practice continuous grazing, degrading the land by flouting the law of the second bite.

Joel pulled a single blade of orchard grass, showing me exactly where a cow had sheared it the week before, and pointing out the finger of fresh green growth that had emerged from the cut in the days since. The blade was a kind of timeline, sharply demarcated between the dark growth predating the bite, and the bright green blade coming after it. "That's the blaze of growth, right there. I'd say this paddock will be ready for the cows to come back in three or four more days."

"Management intensive" it is. Joel is constantly updating the spreadsheet he keeps in his head to track the precise stage of growth of the farm's several dozen paddocks, which range in size from one to five acres, depending on the season and the weather. This particular paddock, a flattish five acres directly behind the barn that is bordered to the north by a hedgerow and to the south by the creek and dirt road that links Polyface's various parts and pastures like a crooked tree trunk, now took its place on the mental schedule. The sheer number of local variables involved in making such a determination hurt my head to consider, and help explain the difficulty of fitting intensive grazing into an industrial agriculture founded on standardization and simplicity. The amount of time it takes a paddock to recover is constantly changing, depending on temperature, rainfall, exposure to the sun, and the time of year, as does the amount of forage any given cow requires, depending on its size, age, and stage of life: A lactating cow, for example, eats twice as much grass as a dry one.

The unit in which a grass farmer performs and records all these calculations, deciding exactly when and where to move the herd, is a "cow
day," which is simply the average amount of forage a cow will eat in one
day; for his rotations to work, the farmer needs to know just how many
cow days each paddock will yield. Though it turns out that, as a unit of
measurement a cow day is a good deal more rubbery than, say, the speed
of light, since the number of cow days any given paddock can supply
rises and falls in response to all the aforementioned variables.

As destructive as overgrazing can be to a pasture, undergrazing can
be almost as damaging, since it leads to woody, senescent grasses and a
loss of productivity. But getting it just right—grazing the optimal number of cattle at the optimal moment to exploit the blaze of growth—
yields tremendous amounts of grass, all the while improving the quality
of the land. Joel calls this optimal grazing rhythm "pulsing the pastures" and says that at Polyface it has boosted the number of cow days
to as much as four hundred per acre; the county average is seventy. "In
effect we've bought a whole new farm for the price of some portable
fencing and a lot of management."

Grass farming done well depends almost entirely on a wealth of nu-
anced local knowledge at a time when most of the rest of agriculture
has come to rely on precisely the opposite: on the off-farm brain, and
the one-size-fits-all universal intelligence represented by agrochemicals
and machines. Very much on his own in a very particular place, the
grass farmer must continually juggle the various elements of his farm
in space as well as time, relying on his powers of observation and organization to arrange the appointed daily meeting of animal and grass
in such a way as to ensure maximum benefit for both.

So is this sort of low-tech pastoralism simply a throwback to preindustrial agriculture? Salatin adamantly begged to differ: "It might not
look that way, but this is all information-age stuff we're doing here.
Polyface Farm is a postindustrial enterprise. You'll see."

2. MONDAY EVENING

As I neared the blessed, longed-for end of my first day as a Polyface farmhand, I must say I didn't feel at all the way I normally do after a day spent laboring in the information economy. And there was still one more daunting chore before dinner: moving the cows, an operation that, Joel wanted me to understand, is a whole lot easier than it sounds. I certainly hoped so. Throwing and stacking fifty-pound bales of hay all afternoon had left me bone tired, sore, and itchy all over from pricks of the chaff, so I was mightily relieved when Joel proposed we ride the four-wheeler to the upper pasture where the cows had spent their day. (It's axiomatic that the more weary you feel the more kindly you look on fossil fuel.) We stopped by the toolshed for a freshly charged car battery to power the electrified paddock fence, and sped up the rutted dirt road, Joel behind the wheel, me hanging on behind him, trying to keep my rear end planted on the little wooden deck he'd rigged up for hauling stuff around the farm.

"My neighbors think I'm insane, moving my cows as often as I do. That's because when most people hear the words 'moving the cattle' they picture a long miserable day, featuring a couple of pickup trucks, a bunch of barking dogs, several cans of Skoal, and a whole lot of hollering," Joel said, hollering himself to be heard over the ATV's engine. "But honestly, it's not like that at all."

Like most grass farmers who practice rotational grazing, Joel moves his cattle onto fresh grass every day. The basic principle is "mob and move," he explained, as we bumped to a halt at the gate to the upper pasture. Eighty or so cattle were milling or lying around what looked like relatively tight quarters in a fenced-off section of a much larger pasture that sloped to the south.

"What we're trying to do here is mimic on a domestic scale what herbivore populations do all over the world. Whether it is wildebeests on the Serengeti, caribou in Alaska, or bison on the American plains, multistomached herds are always moving onto fresh ground, following the cycles of the grass. Predators forced the buffalo to move frequently, and stay mobbed-up together for safety."

These intense but brief stays completely change the animals' interaction with the grass and the soil. They eat down just about everything in the paddock, and then they move on, giving the grasses a chance to recover. Native grasses evolved to thrive under precisely such grazing patterns; indeed, they depend on them for their reproductive success. Not only do ruminants spread and fertilize seed with their manure, but their hoofprints create shady little pockets of exposed soil where water collects—ideal conditions for germinating a grass seed. And in brittle lands during the driest summer months, when microbial life in the soil all but stops, the rumen of the animals takes over the soil's nutrient-cycling role, breaking down dry plant matter into basic nutrients and organic matter, which the animals then spread in their urine and manure.

The mob-and-move routine also helps to keep the ruminants healthy. "Short-duration stays allow the animals to follow their instinct to seek fresh ground that hasn't been fouled by their own droppings, which are incubators for parasites."

Joel disconnected the electric fence from its battery and held down the wire with his boot to let me into the paddock. "We achieve the same objective domestically with our portable electric fences. The fence plays the role of predator in our system, keeping the animals mobbed up and making it possible for us to move them every day." The technology for this light, inexpensive electric fencing (elements of which Joel's father invented in the 1960s) was the breakthrough that made management-intensive grazing practical. (Though much earlier, dogs allowed shepherds to practice a rough approximation of rotational grazing.)

Clearly Joel's cattle knew the drill; I could feel their anticipation. Cows that had been lying around roused themselves, and the bolder ones slowly lumbered over in our direction, one of them—"That's Budger"—stepping right up to nuzzle us like a big cat. Joel's herd is an exceptionally amiable if somewhat motley crew of black, brown, and yellowish animals, crosses of Brahman, Angus, and shorthorn blood-lines. He doesn't believe in artificial insemination or put much stock in fancy genetics. Instead he picks a new bull from his crop of calves every couple of years, naming him for a celebrated Lothario: Slick Willie had the job for much of the Clinton administration. You wouldn't mistake Slick's progeny for show cattle, yet their coats were sleek, their tails were clean, and for cows on a steamy afternoon in June, they had re-markably few flies on them.

It took the two of us working together no more than fifteen min-utes to fence a new paddock next to the old one, drag the watering tub into it, and set up the water line. (The farm's irrigation system is gravity-fed from a series of ponds Joel's dug on the hillside.) The grasses in the new paddock were thigh-high and lush, and the cattle plainly couldn't wait to get at them.

The moment arrived. Looking more like a maître d' than a rancher, Joel opened the gate between the two paddocks, removed his straw hat and swept it grandly in the direction of the fresh salad bar, and called his cows to their dinner. After a moment of bovine hesitation, the cows began to move, first singly, then two by two, and then all eighty of them sauntered into the new pasture, brushing past us as they looked about intently for their favorite grasses. The animals fanned out in the new paddock and lowered their great heads, and the evening air filled with the muffled sounds of smacking lips, tearing grass, and the low snuf-fling of contented cows.

The last time I had stood watching a herd of cattle eat their supper I was standing up to my ankles in cow manure in Poky Feeders pen number 43 in Garden City, Kansas. The difference between these two bovine dining scenes could not have been starker. The single most ob-vious difference was that these cows were harvesting their own feed in-stead of waiting for a dump truck to deliver a total mixed ration of corn that had been grown hundreds of miles away and then blended by an-imal nutritionists with urea, antibiotics, minerals, and the fat of other cattle in a feedlot laboratory. Here we'd brought the cattle to the food rather than the other way around, and at the end of their meal there'd be nothing left for us to clean up, since the cattle would spread their waste exactly where it would do the most good.

Cows eating grasses that had themselves eaten the sun: The food chain at work in this pasture could not be any shorter or simpler. Especially when I compared it to the food chain passing through the feedlot, with its transcontinental tentacles reaching all the way back to cornfields in Iowa, from there to the hypoxic zone in the Gulf of Mexico, and farther still, to the oil fields of the Persian Gulf that had supplied much of the energy to grow the corn. The flaked number 2 corn in steer 534's feed bunk linked him to an industrial (not to mention military) complex that reached halfway around the world.

And yet if I could actually see everything that was going on right here in this pasture, could trace all the ecological connections involved, the scene unfolding directly before me was not nearly as simple as it looked. In fact, there was easily as much complexity present in a single square foot of this pasture as there is in the whole industrial complex into which 534 was plugged; what makes this pasture's complexity so much harder for us to comprehend is that it is not a complexity of our making.

But try anyway. Focus in for a moment on just the relationship between Budger and the tuft of fescue she's tearing from its crown. Those blades of grass have spent this long June day turning sunlight into sugars. (The reason Joel moves his cattle at the end of the day is because that's when sugar levels in the grass hit their peak; overnight the plant will gradually use up these reserves.) To feed the photosynthetic process the grass's roots have drawn water and minerals up from deep in the soil (some grasses can sink their roots as much as six feet down), minerals that soon will become part of this cow. Chances are Budger has also chosen exactly which grasses to eat first, depending on whatever minerals her body craves that day; some species supply her more magnesium, others more potassium. (If she's feeling ill she might go for the plantain, a forb whose leaves contain antibiotic compounds; grazing cattle instinctively use the diversity of the salad bar to medicate themselves.) By contrast 534, who never got to pick and choose his dinner, let alone his medications, depends on animal nutritionists to design his total ration—which of course is only as total as the current state of knowledge in animal science permits.

So far the relationship between Budger and this square foot of pasture might seem a little one-sided, since viewed at least from where I stood, Budger's bite appears to have diminished the pasture. But if I could view the same event from underground and over time, I would see that that bite is not a zero-sum transaction between cow and grass plant. The moment Budger shears the clump of grass, she sets into motion a sequence of events that will confer a measurable benefit on this square foot of pasture. The shorn grass plant, endeavoring to restore the rough balance between its roots and leaves, will proceed to shed as much root mass as it's just lost in leaf mass. When the discarded roots die, the soil's resident population of bacteria, fungi, and earthworms will get to work breaking them down into rich brown humus. What had been the grass plant's root runs will become channels through which worms, air, and rainwater will move through the earth, stimulating the process by which new topsoil is formed.

It is in this manner that the grazing of ruminants, when managed properly, actually builds new soil from the bottom up. Organic matter in a pasture also builds from the top down, as leaf litter and animal wastes break down on the surface, much as it does on a forest floor. But in a grassland decaying roots are the biggest source of new organic matter, and in the absence of grazers the soil-building process would be nowhere near as swift or productive.

Back up to the surface now. Over the next few days, Budger's shearing of this grass plant will stimulate new growth, as the crown redirects reserves of carbohydrate energy from the roots upward to form new shoots. This is the critical moment when a second bite would derail the grass's recovery, since the plant has to live on these reserves until it has grown new leaves and resumed photosynthesis. As the plant adds leaves it adds new roots too, reaching deeper into the soil, making good use of the humus the first bite helped sponsor, and bringing nutrients up to the surface. Over the course of the season this one grass plant will convert more sunlight into more biomass, both above and below the surface of the pasture, than it ever would have had it never encountered a cow.

Yet it's misleading to speak about any grass plant in isolation, since many different plant species, performing many different functions, occupy even this one square foot of pasture, and Budger's bite subtly alters the composition of this community. The shearing of the tallest grasses exposes the pasture's shorter plants to sunlight, stimulating their growth. This is why a well-grazed pasture will see its population of ground-hugging clovers increase, a boon to grasses and grazers alike. These legumes fix nitrogen in the soil, fertilizing the neighboring grasses from below while supplying nitrogen to the grazers above; the bacteria living in the animal's rumen will use the nitrogen in these clover leaves to construct new molecules of protein.

Side-by-side comparisons of intensive and continuously grazed pastures have demonstrated that intensive grazing increases the diversity of species in pastures. That's because rotated cattle don't eliminate favored species by overgrazing them and their equal-opportunity shearing ensures that no one species of grass ever dominates by rising to hog all the sunlight. This biodiversity confers a great many benefits on all parties. At the most fundamental level, it allows the farm's land to capture the maximum amount of solar energy, since one kind of photosynthesizer or another is occupying every conceivable niche—niches in space as well as time. For example, when the early season grasses slow down in June, the late season grasses step in, and when drought hits, the deep-rooted species will take over from the shallower ones. A diverse enough polyculture of grasses can withstand virtually any shock and in some places will produce in a year nearly as much total biomass as a forest receiving the same amount of rainfall.

This productivity means Joel's pastures will, like his woodlots, remove thousands of pounds of carbon from the atmosphere each year; instead of sequestering all that carbon in trees, however, grasslands store most of it underground, in the form of soil humus. In fact, grassing over that portion of the world's cropland now being used to grow grain to feed ruminants would offset fossil fuel emissions appreciably. For example, if the sixteen million acres now being used to grow corn to feed cows in the United States became well-managed pasture, that would remove fourteen billion pounds of carbon from the atmosphere each year, the equivalent of taking four million cars off the road. We seldom focus on farming's role in global warming, but as much as a third of all the greenhouse gases that human activity has added to the atmosphere can be attributed to the saw and the plow.

The benefits of a food chain rooted in a perennial polyculture are so many and so great that they've inspired dreams of converting our agriculture of annual grains into something that would look a lot more like Joel Salatin's pastures. That particular vision hatched more than thirty years ago in the mind of a graduate student in plant genetics named Wes Jackson. Today breeders at his Land Institute, in Salina, Kansas, are working on a (very) long-term project to "perennialize" many of our principal grain crops (including corn) and then grow them in polycultures that farmers would seldom if ever have to plow or replant. The basic idea is to allow us to live off the land (and the sun) more like ruminants do, by coaxing perennial grasses (which we can't digest) to yield bigger and more nutritious seeds (which we can). Of course, the same goal would be accomplished by changing us rather than the grasses—giving people rumens, that is, so they could digest grasses. And there are skeptics who believe perennializing the major crops is no less of a pipe dream than outfitting humans with rumens. Jackson claims his group is making slow but steady progress, however, and has already disproved the conventional wisdom, widely held among botanists, that plants must choose, in effect, between devoting their energy to the production of seeds, as annuals do, or using it to survive the winter in the manner of perennials.

For the time being, though, I'll have to eat Budger herself if I want to make use of the food energy contained in the grasses growing in Joel Salatin's pastures. For me, Wes Jackson's audacious vision of an agriculture that might someday feed us without diminishing the earth's sub-stance (its soil), as even the most sustainable annual agriculture must do, only deepens my appreciation for the grass-based food chain we already have—the one, I mean, that links Budger to the soil and sun and, eventually, to me. It's true that prodigious amounts of food energy are wasted every time an animal eats another animal—nine calories for every one we consume. But if all that energy has been drawn from the boundless storehouse of the sun, as in the case of eating meat off this pasture, that meal comes as close to a free lunch as we can hope to get. Instead of mining the soil, such a meal builds more of it. Instead of diminishing the world, it has added to it.

ALL OF WHICH begs a rather large question: Why did we ever turn away from this free lunch in favor of a biologically ruinous meal based on corn? Why in the world did Americans ever take ruminants off the grass? And how could it come to pass that a fast-food burger produced from corn and fossil fuel actually costs less than a burger produced from grass and sunlight?

I asked myself these questions standing there in Joel's pasture that evening, and in the months since I've thought of several answers. The most obvious answer turns out not to be true. I had thought that the victory of corn over grass might owe to the fact that a field of corn simply produces more total food energy than an acre of grass; it certainly looks that way. But researchers at the Land Institute have studied this question and calculated that in fact more nutrients are produced—protein and carbohydrate—in an acre of well-managed pasture than in an acre of field corn. How can this be? Because a polyculture of grass, with its wide diversity of photosynthesizers exploiting every inch of land as well as every moment of growing season, captures more solar energy and therefore produces more biomass than a cornfield; also, only the kernels are harvested from a cornfield, whereas virtually all the grass grown in a pasture finds its way into the rumen.

Even so, the temptations of cheap corn are powerful, as irresistible as the temptations of cheap energy. Even before the advent of the feed-lot, farmers had begun using a little corn to finish their cattle—fatten them for slaughter—whenever they ran out of good grass, especially in the fall and winter. "When you're trying to finish cattle," Allan Nation pointed out, "corn covers a multitude of sins." Cattlemen found that corn, being such a dense source of calories, produced meat more quickly than grass; it also produced a more reliably consistent product, eliminating the seasonal and regional differences you often find in grass-finished beef. Over time, the knowledge that went into growing grass good enough to finish cattle all the year round gradually was lost.

Along the way corn kept getting more plentiful and ever cheaper. When the farmer found that he could buy corn more cheaply than he could ever hope to grow it, it no longer made economic sense to feed animals on the farm, so they moved into CAFOs. The farmer who then plowed up his pastures to grow corn to market found he could take off to Florida in the winter, not work so hard. To help dispose of the rising mountain of cheap corn farmers were now producing, the government did everything it could to help wean cattle off grass and onto corn, by subsidizing the building of feedlots (through tax breaks) and promoting a grading system based on marbling that favored corn-fed over grass-fed beef. (The government also declined to make CAFOs obey clean air and clean water laws.) In time the cattle themselves changed, as the industry selected for animals that did well on corn; these animals, generally much bigger, had trouble getting all the energy they needed from grass. In dairy, farmers moved to superproductive breeds like the Holstein, whose energy requirements were so great they could barely survive on a diet of grass.

So feeding ruminants corn came to make a certain economic sense—I say "certain" because that statement depends on the particular method of accounting our economy applies to such questions, one that tends to hide the high cost of cheap food produced from corn. The ninety-nine-cent price of a fast-food hamburger simply doesn't take account of that meal's true cost—to soil, oil, public health, the public purse, etc., costs which are never charged directly to the consumer but, indirectly and invisibly, to the taxpayer (in the form of subsidies), the health care system (in the form of food-borne illnesses and obesity), and the environment (in the form of pollution), not to mention the welfare of the workers in the feedlot and the slaughterhouse and the welfare of the animals themselves. If not for this sort of blind-man's accounting, grass would make a lot more sense than it now does.

So there are a great many reasons American cattle came off the grass and into the feedlot, and yet all of them finally come down to the same one: Our civilization and, increasingly, our food system are strictly organized on industrial lines. They prize consistency, mechanization, predictability, interchangeability, and economies of scale. Everything about corn meshes smoothly with the gears of this great machine; grass doesn't.

Grain is the closest thing in nature to an industrial commodity: storable, portable, fungible, ever the same today as it was yesterday and will be tomorrow. Since it can be accumulated and traded, grain is a form of wealth. It is a weapon, too, as Earl Butz once had the bad taste to mention in public; the nations with the biggest surpluses of grain have always exerted power over the ones in short supply. Throughout history governments have encouraged their farmers to grow more than enough grain, to protect against famine, to free up labor for other purposes, to improve the trade balance, and generally to augment their own power. George Naylor is not far off when he says the real beneficiary of his crop is not America's eaters but its military-industrial complex. In an industrial economy, the growing of grain supports the larger economy: the chemical and biotech industries, the oil industry, Detroit, pharmaceuticals (without which they couldn't keep animals healthy in CAFOs), agribusiness, and the balance of trade. Growing corn helps drive the very industrial complex that drives it. No wonder the government subsidizes it so lavishly.

You cannot say any of these things about grass. The government writes no subsidy checks to grass farmers. Grass farmers, who buy little in the way of pesticides and fertilizers (none, in the case of Joel Salatin), do little to support agribusiness or the pharmaceutical industry or big oil. A surplus of grass does nothing for a nation's power or its balance of payments. Grass is not a commodity. What grass farmers grow can't easily be accumulated, traded, transported, or stored, at least for very long. Its quality is highly variable, different from region to region, season to season, even farm to farm; there is no number 2 hay. Unlike grain, grass can't be broken down into its constituent molecules and reassembled as value-added processed foods; meat, milk, and fiber is about all you can make out of grass, and the only way to do that is with a living organism, not a machine. Grass farming with skill involves so many variables, and so much local knowledge, that it is difficult to systematize. As faithful to the logic of biology as a carefully grazed pasture is, it meshes poorly with the logic of industry, which has no use for anything it cannot bend to its wheels and bottom line. And, at least for the time being, it is the logic of industry that rules.

3. MONDAY SUPPER

Once the cows were settled in their paddock for the night, Joel showed me how to hook the electric fence to its battery and we rolled down the hill to dinner. We ditched our boots by the back door, washed up in a basin in the mudroom, and sat down to a meal prepared by Joel's wife, Teresa, and Rachel, the Salatins' eighteen-year-old daughter. The farm's two young interns, Galen and Peter, joined us at the big pine table, and focused so intently on eating they uttered not a word. The Salatins' son Daniel, twenty-two, is a full partner in the farm, but most nights he has dinner with his wife and baby son in the new house they recently built themselves, up the hill. Joel's mother, Lucille, also lives on the property, in a trailer home next to the house. It was in Lucille's guest room that I was sleeping.

The Salatins' brick colonial dates to the eighteenth century, and my first impression of the big, cozy kitchen was that it looked strangely familiar. Then it dawned on me: This is exactly the sort of farmhouse kitchen—wood-paneled and decorated with all things quaint and hearth-like, up to and including the neatly framed needlepoints—that countless kitchens in American suburbs and sitcoms have been striving to simulate at least since World War II. This was what all that nostalgia pointed to, the real McCoy.

Indeed, much about dining with the Salatins had, at least for me, the flavor of a long-ago time and faraway place in America. Joel began the meal by closing his eyes and saying a rambling and strikingly non-generic version of grace, offering a fairly detailed summary of the day's doings to a Lord who, to judge by Joel's tone of easy familiarity, was present and keenly interested. Everything we ate had been grown on the farm, with the exception of the cream of mushroom soup that tied together Teresa's tasty casserole of Polyface chicken and broccoli from the garden. Rachel passed a big platter of delicious deviled eggs, eggs that in this form or some other would appear at every meal that week. Though it wasn't even the end of June, we tasted the first sweet corn of the season, which had been grown in the hoop house where the laying hens spend the winter. There was plenty of everything, and the interns endured many jokes about their stupendous appetites. To drink there was only a pitcher of ice water. Caffeine and alcohol, both of which I sorely felt the need of at the end of my first day, were nowhere in evidence. It was going to be a long week.

At dinner I mentioned that this was probably the all-time most local meal I'd ever eaten. Teresa joked that if Joel and Daniel could just figure out how to mill paper towels and toilet paper from the trees on the farm, she'd never have to go to the supermarket. It was true: We were eating almost completely off the grid. I realized that the sort of agriculture practiced at Polyface was very much of a piece with the sort of life the Salatins led. They had largely detached their household from industrial civilization, and not just by eating from land that had virtually no economic or ecological ties to what Joel variously called "the empire," "the establishment," and "Wall Street." Joel, who had described his politics as Christian libertarian environmentalist, wanted nothing to do with "institutional anything," but especially the institutions of government. Daniel and Rachel had both been homeschooled.

There were plenty of books in the house, but, aside from the Staunton daily newspaper, which devoted more space to local car crashes than the war in Iraq, little media (and no television) penetrated the Salatin household.

The farm and the family comprised a remarkably self-contained world, in the way I imagined all American farm life once did. But the agrarian self-sufficiency that Thomas Jefferson celebrated used to be a matter of course and a product of necessity; nowadays that sort of independence constitutes a politics and economics and way of life both deliberate and hard-won—an achievement. Were Jefferson to return today he would no doubt be gratified to learn that there were still farmers down the road from Monticello as Jeffersonian as Joel Salatin. Until, that is, Jefferson got around a bit more and discovered there weren't many others like him.

At dinner I got Joel and Teresa talking about the history of Polyface, a history in which the roots of Salatin's politics and agriculture become fairly easy to trace. "I'm actually a third-generation alternative farmer," Joel said. "My grandfather was a charter subscriber to Rodale's *Organic Gardening and Farming.*" Fred Salatin had farmed a half-acre in-town lot in Anderson, Indiana, supplying the local markets with fruit, honey, and eggs sold in boxes that bore the Salatin name. Fred Salatin, who was as much an inventor and tinkerer as he was a farmer, held the patent for the very first walking garden sprinkler.

To hear Joel and Teresa describe him, Joel's father William was an ingenious and somewhat eccentric farmer, a man who wore bow ties and sandals and drove a '58 Plymouth sedan that he'd converted into a pickup by removing all the seats and the lid of the trunk. ("He would drive it into town sitting on a bucket," Joel explained. "It embarrassed us kids terribly.") From the time he was a young boy, William had wanted to farm; after flying planes in World War II and earning an economics degree from the University of Indiana, he bought a farm in the highlands of Venezuela, where he and Lucille began raising chickens. Why Venezuela? "Dad felt he could farm the way he wanted there, get out from under both convention and regulations."

The chicken farm thrived until 1959, when a leftist coup toppled the government and "we got caught as ugly Americans in the middle of this political mess." Joel's father refused on principle to buy protection from the local authorities, who proceeded to look the other way when guerillas came after the family's property. "We fled out the back door as the guerillas were coming in the front. We stayed in the country nine or ten months after that, living with a missionary friend while my dad tried to get the government to return our land. We had a deed, but not a single official would look at us without a bribe. And the whole time the American ambassador was dutifully reporting that everything was under control."

In 1961 the Salatins were forced to flee the country, leaving behind everything they'd built and saved. "Now that I'm hitting the age he was then, I just can't imagine what it must have felt like to walk away from it all." The episode clearly left its mark on Joel, undermining his faith that a government, right or left, could protect its citizens and their property, much less do the morally right thing.

Determined to start over again, William Salatin went shopping for farmland within a day's drive of Washington, D.C., so that he might continue petitioning the Venezuelan embassy for compensation. He ended up buying 550 acres of badly eroded and hilly farmland on the western edge of the Shenandoah Valley, in the tiny town of Swoope. (It's pronounced Swope.) After Drew Pearson, the muckraking journalist, publicized his case against the Venezuelans, Salatin won a small settlement that he used to buy a small herd of Hereford cattle.

"The farm had been abused by tenant farmers for 150 years," Joel said. On land that was really too steep for row crops, several generations of tenant farmers had grown corn and other grains until most of the soil had been either exhausted or lost to erosion. "We measured gullies fourteen feet deep. This farm couldn't stand any more plowing. In many places there was no topsoil left whatsoever—just outcroppings of granite and clay. Some spots you couldn't even dig a posthole, so Dad would fill tires with concrete and sink fence posts in that. We've been working to heal this land ever since."

William Salatin quickly discovered the farm couldn't support both a mortgage and a family, so he took on work in town as an accountant. "He turned the farm into an R&D project instead of a salary project." William was now free to experiment, to turn his back on conventional thinking about how to farm.

His instinct to go against conventional agricultural wisdom was confirmed by his accounting clients, many of whom were struggling farmers. "One look at their books convinced him that all the advice he'd been hearing from consultants and extension agents—to build silos, graze the forest, plant corn, and sell commodities—was a recipe for financial ruin."

"So instead of building bankruptcy tubes"—farmer lingo for silos—"he started down a whole other path." William read André Voisin's treatise on grass and began practicing rotational grazing. He stopped buying fertilizer and started composting. He also let the steeper, north-facing hillsides return to forest.

"Dad was very much a visionary and an inventor. He figured out the key to success on a farm like this was first, grass, and second, mobility." This last guiding principle, which Joel claims goes all the way back to Frederick Salatin's patented walking sprinkler ("moving things must be in our genes"), inspired his father to invent a movable electric fence, a portable veal calf barn, and a portable chicken coop for the laying hens Joel raised as a boy. (Until he went off to college, Joel sold eggs every Saturday at a farmer's market in Staunton.) When William noticed that on hot days the cattle gathered under the trees, concentrating their manure in one place, he built a portable "shademobile"—basically a big section of canvas stretched over a steel frame on wheels. Now he could induce the cattle to spread their manure evenly over his pastures, simply by towing the shademobile to a new spot every few days.

Innovations like these helped rebuild the fertility of the soil, and gradually the farm began to recover. Grasses colonized the gullies, the thin soils deepened, and the rock outcrops disappeared under a fresh mantle of sod. And though William Salatin was never quite able to support his family from the farm, he did live to see Joel make a success of the place by building on his example, especially the devotion to grass and mobility—and a determination to go his own way. Joel had returned to the farm in 1982 after four years at Bob Jones University and a stint as a newspaper reporter. Six years later, when Joel was thirty-one, William Salatin died of prostate cancer.

"I still miss him every day," Joel said. "Dad was definitely a little odd, but in a good way. How many other Christian conservatives were reading *Mother Earth News*? He lived out his beliefs. I can remember when the Arab oil embargo hit in 1974, Dad rode his bicycle thirty-five miles back and forth to work every day because he refused to buy another drop of imported oil. He would have been a wonderful tent dweller, always living on less than you have and more lightly than you need to." I felt a tiny flush of embarrassment at ever having asked Joel to FedEx me a steak; I also better understood why he had refused.

"But you want to know when I miss him the most? When I see thick hay and earthworm castings and slick cows, all the progress we've made since he left us. Oh, how proud he would be to see this place now!"

ELEVEN

THE ANIMALS

Practicing Complexity

1. TUESDAY MORNING

It's not often I wake up at six in the morning to discover I've overslept, but by the time I had hauled my six-foot self out of the five-foot bed in Lucille's microscopic guest room, everyone was already gone and morning chores were nearly done. Shockingly, chores at Polyface commence as soon as the sun comes up (five-ish this time of year) and always before breakfast. Before coffee, that is, not that there was a drop of it to be had on this farm. I couldn't recall the last time I'd even attempted to do anything consequential before breakfast, or before caffeine at the very least.

When I stepped out of the trailer into the warm early morning mist, I could make out two figures—the interns, probably—moving around up on the broad shoulder of hill to the east, where a phalanx of portable chicken pens formed a checkerboard pattern on the grass. Among other things, morning chores consist of feeding and watering the broilers and moving their pens one length down the hillside. I was supposed to be helping Galen and Peter do this, so I started up the path, somewhat groggily, hoping to get there before they finished.

As I stumbled up the hill, I was struck by how very beautiful the farm looked in the hazy early light. The thick June grass was silvered with dew, the sequence of bright pastures stepping up the hillside dramatically set off by broad expanses of blackish woods. Birdsong stitched the thick blanket of summer air, pierced now and again by the wood clap of chicken pen doors slamming shut. It was hard to believe this hillside had ever been the gullied wreck Joel had described at dinner, and even harder to believe that farming such a damaged landscape so intensively, rather than just letting it be, could restore it to health and yield this beauty. This is not the environmentalist's standard prescription. But Polyface is proof that people can sometimes do more for the health of a place by cultivating it rather than by leaving it alone.

By the time I reached the pasture Galen and Peter had finished moving the pens. Fortunately they were either too kind or too timid to give me a hard time for oversleeping. I grabbed a pair of water buckets, filled them from the big tub in the center of the pasture, and lugged them to the nearest pen. Fifty of these pens were spread out across the damp grass in a serrated formation that had been calibrated to cover every square foot of this meadow in the course of the fifty-six days it takes a broiler to reach slaughter weight; the pens moved ten feet each day, the length of one pen. Each ten-by-twelve, two-foot-tall floorless pen houses seventy birds. A section of the roof is hinged to allow access, and a five-gallon bucket perched atop each unit feeds a watering device suspended inside.

Directly behind each pen was a perfectly square patch of closely cropped grass resembling a really awful Jackson Pollock painting, thickly spattered with chicken crap in pigments of white, brown, and green. It was amazing what a mess seventy chickens could make in a day. But that was the idea: Give them twenty-four hours to eat the grass and fertilize it with their manure, and then move them onto fresh ground.

Joel developed this novel method for raising broiler chickens in the 1980s and popularized it in his 1993 book, *Pastured Poultry Profit$*, something of a cult classic among grass farmers. (Joel has self-published four other how-to books on farming, and all but one of them has a $ stepping in for an S somewhere in its title.) Left to their own devices, a confined flock of chickens will eventually destroy any patch of land, by pecking the grass down to its roots and poisoning the soil with its extremely "hot," or nitrogenous, manure. This is why the typical free-range chicken yard quickly winds up bereft of plant life and hard as brick. Moving the birds daily keeps both the land and the birds healthy; the broilers escape their pathogens and the varied diet of greens supplies most of their vitamins and minerals. The birds also get a ration of corn, toasted soybeans, and kelp, which we scooped into long troughs in their pens, but Joel claims the fresh grass, along with the worms, grasshoppers, and crickets they peck out of the grass, provides as much as 20 percent of their diet—a significant savings to the farmer and a boon to the birds. Meanwhile, their manure fertilizes the grass, supplying all the nitrogen it needs. The chief reason Polyface Farm is completely self-sufficient in nitrogen is that a chicken, defecating copiously, pays a visit to virtually every square foot of it at several points during the season. Apart from some greensand (a mineral supplement to replace calcium lost in the meadows), chicken feed is the only important input Joel buys, and the sole off-farm source of fertility. ("The way I look at it, I'm just returning some of the grain that's been extracted from this land over the last 150 years.") The chicken feed not only feeds the broilers but, transformed into chicken crap, feeds the grass that feeds the cows that, as I was about to see, feed the pigs and the laying hens.

After we had finished watering and feeding the broilers, I headed up to the next pasture, where I could hear a tractor idling. Galen had told me Joel was moving the Eggmobile, an operation I'd been eager to watch. The Eggmobile, one of Joel's proudest innovations, is a ramshackle cross between a henhouse and a prairie schooner. Housing four hundred laying hens, this rickety old covered wagon has hinged nesting boxes lined up like saddlebags on either side, allowing someone to retrieve eggs from the outside. I'd first laid eyes on the Eggmobile the night before, parked a couple of paddocks away from the cattle herd.

The hens had already climbed the little ramp into the safety of the coop for the night, and before we went down to dinner Joel had latched the trapdoor behind them. Now it was time to move them into a fresh paddock, and Joel was bolting the Eggmobile to the hitch of his tractor. It wasn't quite 7:00 A.M. yet, but Joel seemed delighted to have someone to talk to, holding forth being one of his greatest pleasures.

"In nature you'll always find birds following herbivores," Joel explained, when I asked him for the theory behind the Eggmobile. "The egret perched on the rhino's nose, the pheasants and turkeys trailing after the bison—that's a symbiotic relationship we're trying to imitate." In each case the birds dine on the insects that would otherwise bother the herbivore; they also pick insect larvae and parasites out of the animal's droppings, breaking the cycle of infestation and disease. "To mimic this symbiosis on a domestic scale, we follow the cattle in their rotation with the Eggmobile. I call these gals our sanitation crew."

Joel climbed onto the tractor, threw it into gear, and slowly towed the rickety contraption fifty yards or so across the meadow to a paddock the cattle had vacated three days earlier. It seems the chickens eschew fresh manure, so he waits three or four days before bringing them in—but not a day longer. That's because the fly larvae in the manure are on a four-day cycle, he explained. "Three days is ideal. That gives the grubs a chance to fatten up nicely, the way the hens like them, but not quite long enough to hatch into flies." The result is prodigious amounts of protein for the hens, the insects supplying as much as a third of their total diet—and making their eggs unusually rich and tasty. By means of this simple little management trick, Joel is able to use his cattle's waste to "grow" large quantities of high-protein chicken feed for free; he says this trims his cost of producing eggs by twenty-five cents per dozen. (Very much his accountant father's son, Joel can tell you the exact economic implication of every synergy on the farm.) The cows further oblige the chickens by shearing the grass; chickens can't navigate in grass more than about six inches tall.

After Joel had maneuvered the Eggmobile into position, he opened the trapdoor, and an eager, gossipy procession of Barred Rocks, Rhode

Island Reds, and New Hampshire Whites filed down the little ramp, fanning out across the pasture. The hens picked at the grasses, especially the clover, but mainly they were all over the cowpats, doing this frantic backward-stepping break-dance with their claws to scratch apart the caked manure and expose the meaty morsels within. Unfolding here before us, I realized, was a most impressive form of alchemy: cowpatties in the process of being transformed into exceptionally tasty eggs.

"I'm convinced an Eggmobile would be worth it even if the chickens never laid a single egg. These birds do a more effective job of sanitizing a pasture than anything human, mechanical, or chemical, and the chickens love doing it." Because of the Eggmobile, Joel doesn't have to run his cattle through a headgate to slather Ivomectrin, a systemic paraciticide, on their hides or worm them with toxic chemicals. This is what Joel means when he says the animals do the real work around here. "I'm just the orchestra conductor, making sure everybody's in the right place at the right time."

THAT DAY, my second on the farm, as Joel introduced me to each of his intricately layered enterprises, I began to understand just how radically different this sort of farming is from the industrial models I'd observed before, whether in an Iowa cornfield or an organic chicken farm in California. Indeed, it is so different that I found Polyface's system difficult to describe to myself in an orderly way. Industrial processes follow a clear, linear, hierarchical logic that is fairly easy to put into words, probably because words follow a similar logic: First this, then that; put this in here, and then out comes that. But the relationship between cows and chickens on this farm (leaving aside for the moment the other creatures and relationships present here) takes the form of a loop rather than a line, and that makes it hard to know where to start, or how to distinguish between causes and effects, subjects and objects.

Is what I'm looking at in this pasture a system for producing exceptionally tasty eggs? If so, then the cattle and their manure are a means to an end. Or is it a system for producing grass-fed beef without the use of any chemicals, in which case the chickens, by fertilizing and sanitizing the cow pastures, comprise the means to that end. So does that make their eggs a product or a by-product? And is manure—theirs or the cattle's—a waste product or a raw material? (And what should we call the fly larvae?) Depending on the point of view you take—that of the chicken, the cow, or even the grass—the relationship between subject and object, cause and effect, flips.

Joel would say this is precisely the point, and precisely the distinction between a biological and an industrial system. "In an ecological system like this everything's connected to everything else, so you can't change one thing without changing ten other things.

"Take the issue of scale. I could sell a whole lot more chickens and eggs than I do. They're my most profitable items, and the market is telling me to produce more of them. Operating under the industrial paradigm, I could boost production however much I wanted—just buy more chicks and more feed, crank up that machine. But in a biological system you can never do just one thing, and I couldn't add many more chickens without messing up something else.

"Here's an example: This pasture can absorb four hundred units of nitrogen a year. That translates into four visits from the Eggmobile or two passes of a broiler pen. If I ran any more Eggmobiles or broiler pens over it, the chickens would put down more nitrogen than the grass could metabolize. Whatever the grass couldn't absorb would run off, and suddenly I have a pollution problem." Quality would suffer, too: Unless he added more cattle, to produce more grubs for the chickens and to keep the grass short enough for them to eat it, those chickens and eggs would not taste nearly as good as they do.

"It's all connected. This farm is more like an organism than a machine, and like any organism it has its proper scale. A mouse is the size of a mouse for a good reason, and a mouse that was the size of an elephant wouldn't do very well."

Joel likes to quote from an old agricultural textbook he dug out of the stacks at Virginia Tech many years ago. The book, which was published in 1941 by a Cornell Ag professor, offers a stark conclusion that, depending on your point of view, will sound either hopelessly quaint or arresting in its gnomic wisdom: "Farming is not adapted to large-scale operations because of the following reasons: Farming is concerned with plants and animals that live, grow, and die."

"EFFICIENCY" is the term usually invoked to defend large-scale industrial farms, and it usually refers to the economies of scale that can be achieved by the application of technology and standardization. Yet Joel Salatin's farm makes the case for a very different sort of efficiency—the one found in natural systems, with their coevolutionary relationships and reciprocal loops. For example, in nature there is no such thing as a waste problem, since one creature's waste becomes another creature's lunch. What could be more efficient than turning cow pies into eggs? Or running a half-dozen different production systems—cows, broilers, layers, pigs, turkeys—over the same piece of ground every year?

Most of the efficiencies in an industrial system are achieved through simplification: doing lots of the same thing over and over. In agriculture, this usually means a monoculture of a single animal or crop. In fact, the whole history of agriculture is a progressive history of simplification, as humans reduced the biodiversity of their landscapes to a small handful of chosen species. (Wes Jackson calls our species "homo the homogenizer.") With the industrialization of agriculture, the simplifying process reached its logical extreme—in monoculture. This radical specialization permitted standardization and mechanization, leading to the leaps in efficiency claimed by industrial agriculture. Of course, how you choose to measure efficiency makes all the difference, and industrial agriculture measures it, simply, by the yield of one chosen species per acre of land or farmer.

By contrast, the efficiencies of natural systems flow from complexity and interdependence—by definition the very opposite of simplification. To achieve the efficiency represented by turning cow manure into chicken eggs and producing beef without chemicals you need at least two species (cows and chickens), but actually several more as well, including the larvae in the manure and the grasses in the pasture and the bacteria in the cows' rumens. To measure the efficiency of such a complex system you need to count not only all the products it produces (meat, chicken, eggs) but also all the costs it eliminates: antibiotics, wormers, paraciticides, and fertilizers.

Polyface Farm is built on the efficiencies that come from mimicking relationships found in nature and layering one farm enterprise over another on the same base of land. In effect, Joel is farming in time as well as in space—in four dimensions rather than three. He calls this intricate layering "stacking" and points out that "it is exactly the model God used in building nature." The idea is not to slavishly imitate nature, but to model a natural ecosystem in all its diversity and interdependence, one where all the species "fully express their physiological distinctiveness." He takes advantage of each species' natural proclivities in a way that benefits not only that animal but other species as well. So instead of treating the chicken as a simple egg or protein machine, Polyface honors—and exploits—"the innate distinctive desires of a chicken," which include pecking in the grass and cleaning up after herbivores. The chickens get to do, and eat, what they evolved to do and eat, and in the process the farmer and his cattle both profit. What is the opposite of zero-sum? I'm not sure, but this is it.

Joel calls each of his stacked farm enterprises a "holon," a word I'd never encountered before. He told me he picked it up from Allan Nation; when I asked Nation about it, he pointed me to Arthur Koestler, who coined the term in *The Ghost in the Machine*. Koestler felt English lacked a word to express the complex relationship of parts and wholes in a biological or social system. A holon (from the Greek *holos*, or whole, and the suffix *on*, as in proton, suggesting a particle) is an entity that from one perspective appears a self-contained whole, and from another a dependent part. A body organ like the liver is a holon; so is an Eggmobile.

At any given time, Polyface has a dozen or more holons up and running, and on my second day Joel and Daniel introduced me to a handful of them. I visited the Raken House, the former toolshed where

Daniel has been raising rabbits for the restaurant trade since he was ten. ("Raken?" "Half rabbit, half chicken," Daniel explained.) When the rabbits aren't out on the pasture in portable hutches, they live in cages suspended over a deep bedding of woodchips, in which I watched several dozen hens avidly pecking away in search of earthworms. Daniel explained that the big problem in raising rabbits indoors is their powerful urine, which produces so much ammonia that it scars their lungs and leaves them vulnerable to infection. To cope with the problem most rabbit farmers add antibiotics to their feed. But the scratching of the hens turns the nitrogenous rabbit pee into the carbonaceous bedding, creating a rich compost teeming with earthworms that feed the hens. Drugs become unnecessary and, considering how many rabbits and chickens lived in it, the air in the Raken was, well, tolerable. "Believe me," Daniel said, "if it weren't for these chickens, you'd be gagging right about now, and your eyes would sting something awful."

Before lunch I helped Galen and Peter move the turkeys, another holon. Moving the turkeys, which happens every three days, means setting up a new "feathernet"—a paddock outlined by portable electric fencing so lightweight I could carry and lay out the entire thing by myself—and then wheeling into it the shademobile, called the Gobbledy-Go. The turkeys rest under the Gobbledy-Go by day and roost on top of it at night. They happily follow the contraption into the fresh pasture to feast on the grass, which they seemed to enjoy even more than the chickens do. A turkey consumes a long blade of grass by neatly folding it over and over again with its beak, as if making origami. Joel likes to run his turkeys in the orchard, where they eat the bugs, mow the grass, and fertilize the trees and vines. (Turkeys will eat much more grass than chickens, and they don't damage crops the way chickens can.) "If you run turkeys in a grape orchard," Joel explained, "you can afford to stock the birds at only seventy percent of normal density, and space the vines at seventy percent of what's standard, because you're getting two crops off the same land. And at seventy percent you get much healthier birds and grapevines than you would at one hundred percent. That's the beauty of stacking." By industry standards, the turkey and grape holons are each less than 100 percent efficient; together, however, they produce more than either enterprise would yield if fully stocked, and they do so without fertilizer, weeding, or pesticide.

I had witnessed one of the most winning examples of stacking in the cattle barn during my first visit to Polyface back in March. The barn is an unfancy open-sided structure where the cattle spend three months during the winter, each day consuming twenty-five pounds of hay and producing fifty pounds of manure. (Water makes up the difference.) But instead of regularly mucking out the barn, Joel leaves the manure in place, every few days covering it with another layer of woodchips or straw. As this layer cake of manure, woodchips, and straw gradually rises beneath the cattle, Joel simply raises the adjustable feed gate from which they get their ration of hay; by winter's end the bedding, and the cattle, can be as much as three feet off the ground. There's one more secret ingredient Joel adds to each layer of this cake: a few bucketfuls of corn. All winter long the layered bedding composts, in the process generating heat to warm the barn (thus reducing the animals' feed requirements), and fermenting the corn. Joel calls it his cattle's electric blanket.

Why the corn? Because there's nothing a pig enjoys more than forty-proof corn, and there's nothing he's better equipped to do than root it out with his powerful snout and exquisite sense of smell. "I call them my pigaerators," Salatin said proudly as he showed me into the barn. As soon as the cows head out to pasture in the spring, several dozen pigs come in, proceeding systematically to turn and aerate the compost in their quest for kernels of alcoholic corn. What had been an anaerobic decomposition suddenly turns aerobic, which dramatically heats and speeds up the process, killing any pathogens. The result, after a few weeks of pigaerating, is a rich, cakey compost ready to use.

"This is the sort of farm machinery I like: never needs its oil changed, appreciates over time, and when you're done with it you eat it." We were sitting on the rail of a wooden paddock, watching the pigs do their thing—a thing, of course, we weren't having to do ourselves. The line about the pigaerators was obviously well-worn. But the cliché that kept banging around in my head was "happy as a pig in shit."

Buried clear to their butts in composting manure, a bobbing sea of wriggling hams and corkscrew tails, these were the happiest pigs I'd ever seen.

I couldn't look at their spiraled tails, which cruised above the earthy mass like conning towers on submarines, without thinking about the fate of pigtails in industrial hog production. Simply put, there *are* no pigtails in industrial hog production. Farmers "dock," or snip off, the tails at birth, a practice that makes a certain twisted sense if you follow the logic of industrial efficiency on a hog farm. Piglets in these CAFOs are weaned from their mothers ten days after birth (compared with thirteen weeks in nature) because they gain weight faster on their drug-fortified feed than on sow's milk. But this premature weaning leaves the pigs with a lifelong craving to suck and chew, a need they gratify in confinement by biting the tail of the animal in front of them. A normal pig would fight off his molester, but a demoralized pig has stopped caring. "Learned helplessness" is the psychological term, and it's not uncommon in CAFOs, where tens of thousands of hogs spend their entire lives ignorant of earth or straw or sunshine, crowded together beneath a metal roof standing on metal slats suspended over a septic tank. It's not surprising that an animal as intelligent as a pig would get depressed under these circumstances, and a depressed pig will allow his tail to be chewed on to the point of infection. Since treating sick pigs is not economically efficient, these underperforming production units are typically clubbed to death on the spot.

Tail docking is the USDA's recommended solution to the porcine "vice" of tail chewing. Using a pair of pliers and no anesthetic, most— but not quite all—of the tail is snipped off. Why leave the little stump? Because the whole point of the exercise is not to remove the object of tail biting so much as to render it even more sensitive. Now a bite to the tail is so painful that even the most demoralized pig will struggle to resist it. Horrible as it is to contemplate, it's not hard to see how the road to such a hog hell is smoothly paved with the logic of industrial efficiency.

A very different concept of efficiency sponsors the hog heaven on display here in Salatin's barn, one predicated on what he calls "the pigness of the pig." These pigs too were being exploited—in this case, tricked into making compost as well as pork. What distinguishes Salatin's system is that it is designed around the natural predilections of the pig rather than around the requirements of a production system to which the pigs are then conformed. Pig happiness is simply the by-product of treating a pig as a pig rather than as "a protein machine with flaws"—flaws such as pigtails and a tendency, when emiserated, to get stressed.

Salatin reached down deep where his pigs were happily rooting and brought a handful of fresh compost right up to my nose. What had been cow manure and woodchips just a few weeks before now smelled as sweet and warm as the forest floor in summertime, a miracle of transubstantiation. As soon as the pigs complete their alchemy, Joel will spread the compost on his pastures. There it will feed the grasses, so the grasses might again feed the cows, the cows the chickens, and so on until the snow falls, in one long, beautiful, and utterly convincing proof that in a world where grass can eat sunlight and food animals can eat grass, there is indeed a free lunch.

2. TUESDAY AFTERNOON

After our own quick lunch (ham salad and deviled eggs), Joel and I drove to town in his pickup to make a delivery and take care of a few errands. It felt sweet to be sitting down for a while, especially after a morning taken up with loading the hay we'd baled the day before into the hayloft. For me this rather harrowing operation involved attempting to catch fifty-pound bales that Galen tossed in my general direction from the top of the hay wagon. The ones that didn't completely knock me over I hoisted onto a conveyor belt that carried them to Daniel and Peter, stationed up in the hayloft. It was an assembly line, more or less, and as soon as I fell behind (or just fell, literally) the hay bales piled up fast at my station; I felt like Lucille Ball at the candy factory. I joked to Joel that, contrary to his claims that the animals did most of the real work on this farm, it seemed to me they'd left plenty of it for us.

On a farm, complexity sounds an awful lot like hard work, Joel's claims to the contrary notwithstanding. As much work as the animals do, that's still us humans out there moving the cattle every evening, dragging the broiler pens across the field before breakfast (something I'd pledged I'd wake up in time for the next day), and towing chicken coops hither and yon according to a schedule tied to the life cycle of fly larvae and the nitrogen load of chicken manure. My guess is that there aren't too many farmers today who are up for either the physical or mental challenge of this sort of farming, not when industrializing promises to simplify the job. Indeed, a large part of the appeal of industrial farming is its panoply of labor- and thought-saving devices: machines of every description to do the physical work, and chemicals to keep crops and animals free from pests with scarcely a thought from the farmer. George Naylor works his fields maybe fifty days out of the year; Joel and Daniel and two interns are out there every day sunrise to sunset for a good chunk of the year.

Yet Joel and Daniel plainly relish their work, partly because it is so varied from day to day and even hour to hour, and partly because they find it endlessly interesting. Wendell Berry has written eloquently about the intellectual work that goes into farming well, especially into solving the novel problems that inevitably crop up in a natural system as complex as a farm. You don't see much of this sort of problem-solving in agriculture today, not when so many solutions come ready-made in plastic bottles. So much of the intelligence and local knowledge in agriculture has been removed from the farm to the laboratory, and then returned to the farm in the form of a chemical or machine. "Whose head is the farmer using?" Berry asks in one of his essays. "Whose head is using the farmer?"

"Part of the problem is, you've got a lot of D students left on the farm today," Joel said, as we drove around Staunton running errands. "The guidance counselors encouraged all the A students to leave home and go to college. There's been a tremendous brain drain in rural America. Of course that suits Wall Street just fine; Wall Street is always trying to extract brainpower and capital from the countryside. First they take the brightest bulbs off the farm and put them to work in Dilbert's cubicle, and then they go after the capital of the dimmer ones who stayed behind, by selling them a bunch of gee-whiz solutions to their problems." This isn't just the farmer's problem, either. "It's a foolish culture that entrusts its food supply to simpletons."

It isn't hard to see why there isn't much institutional support for the sort of low-capital, thought-intensive farming Joel Salatin practices: He buys next to nothing. When a livestock farmer is willing to "practice complexity"—to choreograph the symbiosis of several different animals, each of which has been allowed to behave and eat as it evolved to—he will find he has little need for machinery, fertilizer, and, most strikingly, chemicals. He finds he has no sanitation problem or any of the diseases that result from raising a single animal in a crowded monoculture and then feeding it things it wasn't designed to eat. This is perhaps the greatest efficiency of a farm treated as a biological system: health.

I was struck by the fact that for Joel abjuring agrochemicals and pharmaceuticals is not so much a goal of his farming, as it so often is in organic agriculture, as it is an indication that his farm is functioning well. "In nature health is the default," he pointed out. "Most of the time pests and disease are just nature's way of telling the farmer he's doing something wrong."

At Polyface no one ever told me not to touch the animals, or asked me to put on a biohazard suit before going into the brooder house. The reason I had to wear one at Petaluma Poultry is because that system—a monoculture of chickens raised in close confinement—is inherently precarious, and the organic rules' prohibition on antibiotics puts it at a serious disadvantage. Maintaining a single-species animal farm on an industrial scale isn't easy without pharmaceuticals and pesticides. Indeed, that's why these chemicals were invented in the first place, to keep shaky monocultures from collapsing. Sometimes the large-scale organic farmer looks like someone trying to practice industrial agriculture with one hand tied behind his back.

By the same token, a reliance on agrochemicals destroys the information feedback loop on which an attentive farmer depends to improve his farming. "Meds just mask genetic weaknesses," Joel explained one afternoon when we were moving the cattle. "My goal is always to improve the herd, adapt it to the local conditions by careful culling. To do this I need to know: Who has a propensity for pinkeye? For worms? You simply have no clue if you're giving meds all the time."

"So you tell me, who's really in this so-called information economy? Those who learn from what they observe on their farm, or those who rely on concoctions from the devil's pantry?"

OF COURSE the simplest, most traditional measure of a farm's efficiency is how much food it produces per unit of land; by this yardstick too Polyface is impressively efficient. I asked Joel how much food Polyface produces in a season, and he rattled off the following figures:

    30,000 dozen eggs
    10,000 broilers
    800 stewing hens
    50 beeves (representing 25,000 pounds of beef)
    250 hogs (25,000 pounds of pork)
    1,000 turkeys
    500 rabbits.

This seemed to me a truly astonishing amount of food from one hundred acres of grass. But when I put it that way to Joel that afternoon—we were riding the ATV up to the very top of the hill to visit the hogs in their summer quarters—he questioned my accounting method. It was far too simple.

"Sure, you can write that we produced all that food from a hundred open acres, but if you really want to be accurate about it, then you've got to count the four hundred and fifty acres of woodlot too." I didn't get that at all. I knew the woodlot was an important source of farm income in the winter—Joel and Daniel operate a small sawmill from which they sell lumber and mill whatever wood they need to build sheds and barns (and Daniel's new house). But what in the world did the forest have to do with producing food?

Joel proceeded to count the ways. Most obviously, the farm's water supply depended on its forests to hold moisture and prevent erosion. Many of the farm's streams and ponds would simply dry up if not for the cover of trees. Nearly all of the farm's 550 acres had been deforested when the Salatins arrived; one of the first things Bill Salatin did was plant trees on all the north-facing slopes.

"Feel how cool it is in here." We were passing through a dense stand of oak and hickory. "Those deciduous trees work like an air conditioner. That reduces the stress on the animals in summer."

Suddenly we arrived at a patch of woodland that looked more like a savanna than a forest: The trees had been thinned and all around them grew thick grasses. This was one of the pig paddocks that Joel had carved out of the woods with the help of the pigs themselves. "All we do to make a new pig paddock is fence off a quarter acre of forest, thin out the saplings to let in some light, and then let the pigs do their thing." Their thing includes eating down the brush and rooting around in the stony ground, disturbing the soil in a way that induces the grass seed already present to germinate. Within several weeks, a lush stand of wild rye and foxtail emerges among the trees, and a savanna is born. Shady and cool, this looked like ideal habitat for the sunburn-prone pigs, who were avidly nosing through the tall grass and scratching their backs against the trees. There is something viscerally appealing about a savanna, with its pleasing balance of open grass and trees, and something profoundly heartening about the idea that, together, farmer and pigs could create such beauty here in the middle of a brushy second-growth forest.

But Joel wasn't through counting the benefits of woodland to a farm; idyllic pig habitat was the least of it.

"There's not a spreadsheet in the world that can measure the value of maintaining forest on the northern slopes of a farm. Start with those trees easing the swirling of the air in the pastures. That might not seem like a big deal, but it reduces evaporation in the fields—which means more water for the grass. Plus, a grass plant burns up fifteen percent of its calories just defying gravity, so if you can stop it from being wind whipped, you greatly reduce the energy it uses keeping its photovoltaic array pointed toward the sun. More grass for the cows. That's the efficiency of a hedgerow surrounding a small field, something every farmer used to understand before 'fencerow to fencerow' became USDA mantra."

Then there is the water-holding capacity of trees, he explained, which on a north slope literally pumps water uphill. Next was all the ways a forest multiples a farm's biodiversity. More birds on a farm mean fewer insects, but most birds won't venture more than a couple hundred yards from the safety of cover. Like many species, their preferred habitat is the edge between forest and field. The biodiversity of the forest edge also helps control predators. As long as the weasels and coyotes have plenty of chipmunks and voles to eat, they're less likely to venture out and prey on the chickens.

There was more. On a steep northern slope trees will produce much more biomass than will grass. "We're growing carbon in the woods for the rest of the farm—not just the firewood to keep us warm in the winter, but also the woodchips that go into making our compost." Making good compost depends on the proper ratio of carbon to nitrogen; the carbon is needed to lock down the more volatile nitrogen. It takes a lot of woodchips to compost chicken or rabbit waste. So the carbon from the woodlots feeds the fields, finding its way into the grass and, from there, into the beef. Which it turns out is not only grass fed but tree fed as well.

These woods represented a whole other order of complexity that I had failed to take into account. I realized that Joel didn't look at this land the same way I did, or had before this afternoon: as a hundred acres of productive grassland patchworked into four hundred and fifty acres of unproductive forest. It was all of a biological piece, the trees and the grasses and the animals, the wild and the domestic, all part of a single ecological system. By any conventional accounting, the forests here represented a waste of land that could be put to productive use.

But if Joel were to cut down the trees to graze more cattle, as any conventional accounting would recommend, the system would no longer be quite as whole or as healthy as it is. *You can't just do one thing.*

For some reason the image that stuck with me from that day was that slender blade of grass in a too-big, wind-whipped pasture, burning all those calories just to stand up straight and keep its chloroplasts aimed at the sun. I'd always thought of the trees and grasses as antagonists—another zero-sum deal in which the gain of the one entails the loss of the other. To a point, this is true: More grass means less forest; more forest less grass. But either-or is a construction more deeply woven into our culture than into nature, where even antagonists depend on one another and the liveliest places are the edges, the in-betweens or both-ands. So it is with the blade of grass and the adjacent forest as, indeed, with all the species sharing this most complicated farm. Relations are what matter most, and the health of the cultivated turns on the health of the wild. Before I came to Polyface I'd read a sentence of Joel's that in its diction had struck me as an awkward hybrid of the economic and the spiritual. I could see now how characteristic that mixing is, and that perhaps the sentence isn't so awkward after all: "One of the greatest assets of a farm is the sheer ecstasy of life."

TWELVE

SLAUGHTER

In a Glass Abattoir

1. WEDNESDAY

Today promised not to be about the ecstasy of life on a farm. Today was the day we were "processing" broilers or, to abandon euphemism, killing chickens.

For all the considerable beauty I'd witnessed following a food chain in which the sun fed the grass, the grass the cattle, the cattle the chickens, and the chickens us, there was one unavoidable link in that chain few would consider beautiful: the open-air processing shed out behind the Salatins' house where, six times a month in the course of a long morning, several hundred chickens are killed, scalded, plucked, and eviscerated.

I said this link was "unavoidable," but of course most of us, including most of the farmers who raise food animals, do our very best to avoid thinking about, let alone having anything directly to do with, their slaughter. "You have just dined," Emerson once wrote, "and however scrupulously the slaughterhouse is concealed in the graceful distance of miles, there is complicity."

The killing of the animals we eat generally takes place behind high walls, well beyond our gaze or ken. Not here. Joel insists on slaughtering chickens on the farm, and would slaughter his beeves and hogs here too if only the government would let him. (Under an old federal exemption, farmers are still permitted to process a few thousand birds on farms, but most other food animals must be processed in a state or federally inspected facility.) Joel's reasons for wanting to do this work here himself are economic, ecological, political, ethical, and even spiritual. "The way I produce a chicken is an extension of my worldview," he'd told me the first time we'd talked; by the end of the morning I had a much better idea of what he meant.

WEDNESDAY MORNING I managed to get up right on time—5:30 A.M., to be exact—and to make my way to the broilers' pasture before the interns had finished chores. Which today, in addition to watering, feeding, and moving the chickens, included catching and crating the three hundred we planned to process immediately after breakfast. While we waited for Daniel to show up with the chicken crates, I helped Peter move pens, a two-man operation in which one man slides a customized, extrawide hand truck beneath the pen's back edge (thereby raising it up on wheels), while the other grabs a broad loop of cable attached in front and slowly drags the pen forward onto fresh grass. The chickens, familiar with the daily drill, scooted along in step with their slowly moving mobile home. The pens were much heavier than they appeared, though, and it took every ounce of my strength to drag one a few feet across the uneven ground; "moving the broilers" was not as easy as Joel had made it sound or the interns had made it look, but then, I wasn't nineteen, either.

After a while Daniel drove up on the tractor, towing a wagon piled high with plastic chicken crates. We stacked four of them in front of each of the pens housing the doomed birds and then he and I got to work catching chickens. After lifting the top off the pen, Daniel used a big plywood paddle to crowd the birds into one corner, so they'd be easier to catch. He reached in and grabbed a flapping bird by one leg and flipped it upside down, which seemed to settle it. Then, in a deft, practiced move, he switched the dangling bird from his right hand to his left, freeing his right hand to grab another. When he had five birds in one hand, I held open the crate door and he stuffed them in. He could fill a crate with ten birds in less than a minute.

"Your turn," Daniel said, nodding toward the cornered mass of feathers remaining in the pen. To me, the way he'd grabbed and flipped the chickens seemed unduly rough, their pencil legs so fragile-looking, yet when I tried to coddle the birds as I grabbed them, they flapped around even more violently, until I was forced to let go. This clearly wasn't going to work. So finally I just reached into the flapping mass and blindly clutched at a leg with one hand and flipped it over. When I saw the chicken was none the worse for it, I switched it to my right hand (I'm a lefty), and went for a second and a third, until I had five chicken legs and a giant white pom-pom of feathers in my right hand. Daniel flipped open the lid on a crate and I pushed the pom-pom in. I don't know if there is a more humane way to catch three hundred chickens, but I could see why doing it as fast and as surely as possible was best for all concerned.

Before we sat down to breakfast (scrambled Polyface eggs and Polyface bacon), Daniel turned on the gas under the scalding tank; the water had to reach 140 degrees before we could start. At breakfast Joel talked a little about the importance of on-farm processing, not only to Polyface but to the prospects for rebuilding a viable local food chain. To hear him describe it, what we were about to do—kill a bunch of chickens in the backyard—was nothing less than a political act.

"When the USDA sees what we're doing here they get weak in the knees," Joel said with a chuckle. "The inspectors take one look at our processing shed, and they don't know what to do with us. They'll tell me the regulations stipulate a processing facility must have impermeable white walls so they can be washed down between shifts. They'll quote me a rule that says all doors and windows must have screens. I point out we don't have *any* walls at all, not to mention doors and windows, because the best disinfectant in the world is fresh air and sunshine. Well, that really gets them scratching their heads!"

The problem with current food-safety regulations, in Joel's view, is that they are one-size-fits-all rules designed to regulate giant slaughterhouses that are mindlessly applied to small farmers in such a way that "before I can sell my neighbor a T-bone steak I've got to wrap it up in a million dollars' worth of quintuple-permitted processing plant." For example, federal rules stipulate that every processing facility have a bathroom for the exclusive use of the USDA inspector. Such regulations favor the biggest industrial meatpackers, who can spread the costs of compliance over the millions of animals they process every year, at the expense of artisanal enterprises like Polyface.

The fact that Polyface can prove its chickens have much lower bacteria counts than supermarket chickens (Salatin's had them both tested by an independent lab) doesn't cut any mustard with the inspectors, either. USDA regulations spell out precisely what sort of facility and system is permissible, but they don't set thresholds for food-borne pathogens. (That would require the USDA to recall meat from packers who failed to meet the standards, something the USDA, incredibly, lacks the authority to do.) "I'd be happy to swab-test my chickens for salmonella, listeria, campylobacter, you name it, but the USDA refuses to set any levels!" As breakfast-time conversation, the topic left a lot to be desired, but once Joel gets started on the government, there's no stopping him. "Just tell me where the finish line is, and I'll figure out the best way to get there."

The processing shed in question resembles a sort of outdoor kitchen on a concrete slab, protected from (some of) the elements by a sheet-metal roof perched on locust posts. Arranged in an orderly horseshoe along the edge are stainless steel sinks and counters, a scalding tank, a feather-plucking machine, and a brace of metal cones to hold the birds upside down while they're being killed and bled out. It's not hard to see how a plein-air abattoir like this might give a USDA inspector conniptions.

"Make no mistake, we're in a war with the bureaucrats, who would like nothing better than to put us out of business." I couldn't tell whether Joel wasn't perhaps being a tad paranoid on this point; the pastoral idyll has always felt itself besieged by malign outside forces, and on this farm that role is played by the government and the big processing companies whose interests they serve. Joel said state inspectors have tried to close down his chicken-processing operation more than once, but so far he's managed to stave them off.

It was a little early in the day for a full-blown prairie populist stemwinder, but clearly I was going to get one anyway. "The USDA is being used by the global corporate complex to impede the clean-food movement. They aim to close down all but the biggest meat processors, and to do it in the name of biosecurity. Every government study to date has shown that the reasons we're having an epidemic of food-borne illness in this country is centralized production, centralized processing, and long-distance transportation of food. You would think therefore that they'd want to decentralize the food system, especially after 9/11. But no! They'd much rather just irradiate everything instead."

By the time we finished breakfast, a couple of cars had pulled into the driveway—two women from downstate, who had read *Pastured Poultry Profit$* and wanted to learn how to process the chickens they'd started, and a neighbor or two Joel sometimes hires when he needs extra hands on processing day. Joel had once told me he regarded the willingness of neighbors to work for a business as the true mark of its sustainability, that it operated on the proper scale socially and economically, as well as environmentally.

"That's another reason we don't raise a hundred thousand chickens. It's not just the land that couldn't take it, but the community, too. We'd be processing six days a week, so we'd have to do what the industrial folks do: bring in a bunch of migrant workers because no one around here would want to gut chickens every day. Scale makes all the difference."

After a few minutes of neighborly chitchat, everybody drifted toward their stations in the processing shed. I volunteered to join Daniel, the designated executioner, at the first station on the line. Why? Because I'd been dreading this event all week and wanted to get it over with. Nobody was insisting I personally slaughter a chicken, but I was curious to learn how it was done and to see if I could bring myself to do it. The more I'd learned about the food chain, the more obligated I felt to take a good hard look at all of its parts. It seemed to me not too much to ask of a meat eater, which I was then and still am, that at least once in his life he take some direct responsibility for the killing on which his meat-eating depends.

I stacked several chicken crates in the corner by the killing cones and, while Daniel sharpened his knives, began lifting chickens from the crates and placing them, head first, into the killing cones, which have an opening at the bottom for the chicken's head. Taking the squawking birds out of the crate was actually the hard part; as soon as they were snug in the cones, which kept their wings from flapping, the chickens fell silent. Once all eight cones were loaded, Daniel reached underneath and took a chicken head between his first finger and thumb, holding it still. Gently, he gave the head a quarter turn and then quickly drew his knife across the artery running alongside the bird's windpipe. A stream of blood erupted from the cut, pulsing slightly as it poured down into a metal gutter that funneled it into a bucket. Daniel explained that you wanted to sever only the artery, not the head, so that the heart would continue to beat and pump out the blood. The bird shuddered in its cone, its yellow feet dancing spastically.

It was hard to watch. I told myself the spasms were involuntary, and they probably were. I told myself that the birds waiting their turn appeared to have no idea what was going on in the cone next to them. I told myself that their suffering, once their throats were slit, was brief. Yet it took several long minutes for the spasms to subside. Could they smell the blood on Daniel's hands? Recognize the knife? I have no idea, but the waiting birds did not seem panicked, and I took solace in their seeming obliviousness. Yet, honestly, there wasn't much time for these reflections, because you're working on an assembly (or, really, disassembly) line, and it has a rhythm of its own that soon overpowers your mind as well as your body. Within minutes the first eight chickens had been bled out and transferred to the scalding tank. Daniel was calling for eight more, and I had to hustle so as not to fall behind.

After I had loaded and he had slaughtered several batches, Daniel offered me his knife. He showed me how to hold the chicken's little head in a V between my thumb and forefinger, how to turn it to expose the artery and avoid the windpipe, and how to slice down toward you at a spot just beneath the skull. Since I am left-handed, every step had to be reverse engineered, which tangled us in an excruciating moment of delay. I looked into the black eye of the chicken and, thankfully, saw nothing, not a flicker of fear. Holding his head in my right hand, I drew the knife down the left side of the chicken's neck. I worried about not cutting hard enough, which would have prolonged the bird's suffering, but needn't have: The blade was sharp and sliced easily through the white feathers covering the bird's neck, which promptly blossomed a brilliant red. Before I could let go of the bird's suddenly limp head my hand was painted in a gush of warm blood. Somehow, an errant droplet spattered the lens of my glasses, leaving a tiny, fogged red blot in my field of vision for the rest of the morning. Daniel voiced his approval of my technique and, noticing the drop of blood on my glasses, offered one last bit of advice: "The first rule of chicken killing is that if you ever feel anything on your lip, you don't want to lick it off." Daniel smiled. He's been killing chickens since he was ten years old and doesn't seem to mind it.

Daniel gestured toward the next cone; I guess I wasn't done. In the end I personally killed a dozen or so chickens before moving on to try another station. I got fairly good at it, though once or twice I sliced too deeply, nearly severing a whole head. After a while the rhythm of the work took over from my misgivings, and I could kill without a thought to anything but my technique. I wasn't at it long enough for slaughtering chickens to become routine, but the work did begin to feel mechanical, and that feeling, perhaps more than any other, was disconcerting: how quickly you can get used to anything, especially when the people around you think nothing of it. In a way, the most morally troubling thing about killing chickens is that after a while it is no longer morally troubling.

When Daniel and I got ahead of the scalder, which could accommodate only a few birds at a time, I stepped away from the killing area for a break. Joel clapped me on the back for having taken my turn at the killing cones. I told him killing chickens wasn't something I would want to do every day.

"Nobody should," Joel said. "That's why in the Bible the priests drew lots to determine who would conduct the ritual slaughter, and they rotated the job every month. Slaughter is dehumanizing work if you have to do it every day." Temple Grandin, the animal-handling expert who's helped design many slaughterhouses, has written that it is not uncommon for full-time slaughterhouse workers to become sadistic. "Processing but a few days a month means we can actually think about what we're doing," Joel said, "and be as careful and humane as possible."

I'd had enough of the killing station, so after my break I moved down the line. Once the birds were bled out and dead, Daniel handed them, by their feet, to Galen, who dropped them into the scalder, a tub outfitted with moving shelves that plunged the birds up and down in the hot water to loosen their feathers. They came out of the scalder looking very dead and soaked—floppy wet rags with beaks and feet. Next they went into the plucker, a stainless steel cylinder that resembles a top-loading washing machine with dozens of black rubber fingers projecting from the sides. As the chickens spin at high speed, they flop and jostle against the stiff fingers, which pull their feathers off. After a few minutes they emerge as naked as supermarket broilers. This is the moment the chickens passed over from looking like dead animals to looking like food.

Peter pulled the birds from the plucker, yanked off the heads, and cut off the feet before passing the birds to Galen for gutting. I joined him at his station, and he showed me what to do—where to make the incision with your knife, how to reach your hand into the cavity without tearing too much skin, and how to keep the digestive tract intact as you pull the handful of warm viscera from the belly. As the viscera spilled out onto the stainless steel counter he named the parts: gullet, gizzard, gallbladder (which you must be careful not to pierce), liver, heart, lungs, and intestines (have to be careful here again); then he showed me which organs to keep for sale, and which ones to drop in the gutbucket at our feet. The viscera were unexpectedly beautiful, glistening in a whole palette of slightly electric colors, from the steely blue striations of the heart muscle to the sleek milk chocolate liver to the dull mustard of the gallbladder. I was curious to see the gizzard, the stomachlike organ where a chicken uses bits of ingested grit to crush its food after it's passed down the gullet. I slit open the tight, hard nut of gizzard and there inside found tiny pieces of stone and a blade of bright green grass folded like an accordion. I couldn't make out any insects in the gizzard, but its contents recapitulated the Polyface food chain: pasture on its way to becoming meat.

I didn't get very good at evisceration; my clumsy hands tore unacceptably large openings in the skin, giving my chickens a ragged appearance, and I accidentally broke a gallbladder, spilling a thin yellow bile that I then had to painstakingly rinse off the carcass. "After you gut a few thousand chickens," Galen said dryly after I'd torn another chicken, "you'll either get really good at it, or you'll stop gutting chickens." Galen had clearly gotten really good at it, and he seemed to enjoy the work.

Everybody was making desultory conversation as they went about their jobs, and the morning had something of the flavor that I imagine a barn raising or a November session of corn shucking once had: people who ordinarily work alone having a chance to visit with one another while getting something useful done. Much of the work was messy and unpleasant, but it did allow for conversation, and you weren't going to be at it long enough to get bored or sore. And by the end of the morning you had something to show for it—and a great deal more than you would have had had you been working alone. We hadn't been at it much more than three hours before there were three hundred or so chickens floating in the big steel tank of iced water. Each of them had made the transition from clucking animal to oven-ready roaster, from killing cone to holding tank, in ten minutes, give or take.

While we were cleaning up, scrubbing the blood off the tables and hosing down the floor, customers began arriving to pick up their chickens. This was when I began to appreciate what a morally powerful idea an open-air abattoir is. Polyface's customers know to come after noon on a chicken day, but there's nothing to prevent them from showing up earlier and watching their dinner being killed—indeed, customers are welcome to watch, and occasionally one does. More than any USDA rule or regulation, this transparency is their best assurance that the meat they're buying has been humanely and cleanly processed.

"You can't regulate integrity," Joel is fond of saying; the only genuine accountability comes from a producer's relationship with his or her customers, and their freedom "to come out to the farm, poke around, sniff around. If after seeing how we do things they want to buy food from us, that should be none of the government's business." Like fresh air and sunshine, Joel believes transparency is a more powerful disinfectant than any regulation or technology. It is a compelling idea. Imagine if the walls of every slaughterhouse and animal factory were as transparent as Polyface's—if not open to the air then at least made of glass. So much of what happens behind those walls—the cruelty, the carelessness, the filth—would simply have to stop.

The customers pick their chicken out of the tank and bag it themselves before putting it on the scale in the shop next door to the processing shed. (Having customers bag their own chickens preserves the fiction that they're not buying a processed food product, which is illegal in an area zoned for agriculture. Rather, they're buying the live bird, which Polyface has slaughtered and cleaned as a courtesy.) If you buy one at the farm, a Polyface chicken costs $2.05 a pound, compared to $1.29 at the local supermarket. To keep that premium as low as possible is yet another reason for processing on the farm. Having to take beeves and hogs to the packing plant in Harrisonburg adds a dollar to every pound of beef or pork Polyface sells, and two dollars to every pound of ham or bacon, which regulations prohibit Joel from smoking himself. Curing meat is considered manufacturing, he explained, smoking slightly now himself, and manufacturing is prohibited in an area zoned for agriculture. Joel is convinced "clean food" could compete with supermarket food if the government would exempt farmers from the thicket of regulations that prohibit them from processing and selling meat from the farm. For him, regulation is the single biggest impediment to building a viable local food chain, and what's at stake is our liberty, nothing less. "We do not allow the government to dictate what religion you can observe, so why should we allow them to dictate what kind of food you can buy?" He believes "freedom of food"—the freedom to buy a pork chop from the farmer who raised the hog— should be a constitutional right.

While Theresa chatted with customers as she checked them out, occasionally dispatching Daniel or Rachel to fetch a dozen eggs from the fridge or a roast from the walk-in freezer, Galen and I helped Joel compost chicken waste. This just may be the grossest job on the farm—or anywhere else for that matter. Yet I came to see that even the way Polyface handles its chicken guts is, as Joel would say, an extension of his worldview.

Joel went off on the tractor to get a load of woodchips from the big pile he keeps across the road, while Galen and I hauled five-gallon buckets of blood and guts and feathers from the processing shed to the compost pile, which is only a stone's throw from the house. The day was getting steamy, and the heaping mound of woodchips, beneath which simmered earlier installments of chicken waste, exhaled a truly evil stink. I've encountered some funky compost piles, but this one smelled like, well, exactly what it was: rotting flesh. I realized that this was what I had caught the occasional waft of during my first sleepless night in the trailer.

Beside the old pile Joel dumped a few yards of fresh woodchips, which Galen and I raked into a broad rectangular mound about the size of a double bed, leaving a slight depression in the middle. Into this dip we spilled the buckets of guts, forming a glistening, parti-colored stew. On top of this we added the pillowy piles of feathers, and finally the blood, which now had the consistency of house paint. By now Joel was back with another load of chips, which he proceeded to dump onto the top of the pile. Galen climbed up onto the mass of woodchips with his rake, and I followed him with mine. The top layer of woodchips was dry, but you could feel the viscera sliding around underfoot; it felt like walking on a mattress filled with Jell-O. We raked the pile level and got out of there.

The compost pile repulsed me, but what did that say? Beyond the stench in my nostrils (which, believe me, was not so easy to get beyond), the pile offered an inescapable reminder of all that eating chicken involves—the killing, the bleeding, the evisceration. And no matter how well it is masked or how far it is hidden away, this death smell—and the reality that gives rise to it—shadows the eating of any meat, industrial, organic, or whatever, is part and parcel of even this grassy pastoral food chain whose beauty had so impressed me. I wondered whether my disgust didn't cover a certain shame I was feeling about the morning's work. Just at the moment, I wasn't sure I could imagine eating chicken again any time soon.

I certainly couldn't imagine keeping this rotting heap of chicken guts an errant summer's breeze away from my dinner table. But Joel probably saw that pile in a very different light than I did; who knows, by now it might not even smell all that bad to him. For Joel, yet another of the advantages of processing chickens here is that it allows him to keep the whole cycle of birth, growth, death, and decay on the land. Otherwise, the waste would end up in a rendering plant, there to be superheated, dried, and pelleted, turned into "protein meal," and fed to factory-farmed pigs and cattle and even other chickens, a dubious practice that mad cow disease has rendered even more dubious. This is not a system he wants any part of.

It could be that Joel even finds a certain beauty in that compost pile, or at least in its redemptive promise. He certainly hasn't hidden it away. Like every other bit of "waste" on this farm, he regards chicken guts as a form of biological wealth—nitrogen he can return to the land by locking it down with carbon he's harvested from the woodlot. Having seen what happened to last year's pile, and all the piles before that, Joel can see the future of this one in a way I can't, its promise to transubstantiate this mass of blood and guts and feathers into a particularly rich, cakey black compost, improbably sweet-smelling stuff that, by spring, will be ready for him to spread onto the pastures and turn back into grass.

THIRTEEN

THE MARKET

"Greetings from the
Non-Barcode People"

1. WEDNESDAY AFTERNOON

Following the corn-based industrial food chain had taken me on a journey of several thousand miles, from George Naylor's Iowa fields to the feedlots and packing plants of Kansas, through any number of far-flung food processors before ending up in a Marin County McDonald's. After that, it didn't surprise me to read that the typical item of food on an American's plate travels some fifteen hundred miles to get there, and is frequently better traveled and more worldly than its eater. By comparison, the grass-based food chain rooted in these Virginia pastures is, for all its complexity, remarkably short; I had been able to follow it for most of its length without leaving the Salatins' property. The farm work in Virginia may have been more taxing than in Iowa—killing chickens as compared to planting corn—but the detective work here was a relative cinch. And all that remained to do now was to trace the grass-based food chain along the various marketing paths linking Joel's pastures to his customers' plates.

What had brought me to Polyface in the first place, you'll recall, was Joel's refusal to FedEx me a steak. I was given to understand that his concept of sustainability was not limited to agricultural technique or processing method, but extended the entire length of the food chain. Joel is no more likely to sell his grass-finished beef to Whole Foods (let alone Wal-Mart) than he would be to feed his cows grain, chicken manure, or Rumensin; as far as he's concerned, it is all of the same industrial piece. So Polyface does not ship long distance, does not sell into supermarkets, and does not wholesale its food. All three hundred chickens we'd processed Wednesday morning would be eaten within a few dozen miles or, at the most, half a day's drive, of the farm. Originally I assumed Joel's motive for keeping his food chain so short was strictly environmental—to save on the prodigious quantities of fossil fuel Americans burn moving their food around the country and, increasingly today, the world. But it turns out Joel aims to save a whole lot more than energy.

A chicken—or steak, or ham, or carton of eggs—can find its way from Polyface Farm to an eater's plate by five possible routes: direct sales at the farm store, farmer's markets, metropolitan buying clubs, a handful of small shops in Staunton, and Joel's brother Art's panel truck, which makes deliveries to area restaurants every Thursday. Each of these outlets seems quite modest in itself, yet taken together they comprise the arteries of a burgeoning local food economy that Joel believes is indispensable to the survival of his kind of agriculture (and community), not to mention to the reform of the entire global food system.

In Joel's view, that reformation begins with people going to the trouble and expense of buying directly from farmers they know— "relationship marketing," as he calls it. He believes the only meaningful guarantee of integrity is when buyers and sellers can look one another in the eye, something few of us ever take the trouble to do. "Don't you find it odd that people will put more work into choosing their mechanic or house contractor than they will into choosing the person who grows their food?"

Joel often speaks of his farming as his ministry, and certainly his four hundred or so regular customers hear plenty of preaching. Each spring he sends out a long, feisty, single-spaced letter that could convince even a fast-food junkie that buying a pastured broiler from Polyface Farm qualifies as an act of social, environmental, nutritional, and political redemption.

"Greetings from the non-Barcode people," began one recent missive, before it launched into a high-flying jeremiad against our "disconnected multi-national global corporate techno-glitzy food system" with its "industrial fecal factory concentration camp farms." (The dangerous pileup of modifiers is a hallmark of Joel's rhetorical style.) He darkly warns that the government "and their big-food-system fraternity-mates" are exploiting worries about bioterrorism to regulate small food producers out of business, and beseeches his customers "to stand with Polyface during these paranoid, hysterical days." Like any good jeremiad, this one eventually transits from despair to hope, noting that the "yearning in the human soul to smell a flower, pet a pig and enjoy food with a face has never been stronger," before moving into a matter-of-fact discussion of this year's prices and the paramount importance of sending in your order blanks and showing up to collect your chickens on time.

I'd met several of Polyface's four hundred parishioners on Wednesday afternoon, and then again on Friday, as they came to collect the fresh chickens they'd reserved. It was a remarkably diverse group of people: a schoolteacher, several retirees, a young mom with her towheaded twins, a mechanic, an opera singer, a furniture maker, a woman who worked in a metal fabrication plant in Staunton. They were paying a premium over supermarket prices for Polyface food, and in many cases driving more than an hour over a daunting (though gorgeous) tangle of county roads to come get it. But no one would ever mistake these people for the well-heeled urban foodies generally thought to be the market for organic or artisanal food. There was plenty of polyester in this crowd and a lot more Chevrolets than Volvos in the parking lot.

So what exactly had they come all the way out here to the farm to buy? Here are some of the comments I jotted down:

"This is the chicken I remember from my childhood. It actually tastes like chicken."

"I just don't trust the meat in the supermarket anymore."

"These eggs just jump up and slap you in the face!"

"You're not going to find fresher chickens anywhere."

"All this meat comes from happy animals—I know because I've seen them."

"I drive 150 miles one way in order to get clean meat for my family."

"It's very simple: I trust the Salatins more than I trust the Wal-Mart. And I like the idea of keeping my money right here in town."

I was hearing, in other words, the same stew of food fears and food pleasures (and memories) that has driven the growth of the organic food industry over the past twenty years—that and the satisfaction many Polyface customers clearly take in spending a little time on a farm, porch chatting with the Salatins, and taking a beautiful drive in the country to get here. For some people, reconnecting with the source of their food is a powerful idea. For the farmer, these on-farm sales allow him to recapture the ninety-two cents of a consumer's food dollar that now typically winds up in the pockets of processors, middlemen, and retailers.

LATER THAT AFTERNOON, Joel and I took a long drive down to Moneta, at the southern end of the Shenandoah Valley. He was eager for me to meet Bev Eggleston, whose one-man marketing company, EcoFriendly Foods, is a second route along which Polyface food finds its way to eaters. Eggleston, a former herb and livestock farmer who discovered he had a greater gift for marketing food than producing it, sells Polyface meat and eggs from his booths at farmer's markets in the Washington, D.C., area. On the drive, Joel and I talk about the growing local-food movement, the challenges it faces, and the whole sticky issue of price. I asked Joel how he answers the charge that because food like his is more expensive it is inherently elitist.

"I don't accept the premise. First off, those weren't any elitists you met on the farm this morning. We sell to all kinds of people. Second, whenever I hear people say clean food is expensive, I tell them it's actually the cheapest food you can buy. That always gets their attention. Then I explain that with our food all of the costs are figured into the price. Society is not bearing the cost of water pollution, of antibiotic resistance, of food-borne illnesses, of crop subsidies, of subsidized oil and water—of all the hidden costs to the environment and the taxpayer that make cheap food *seem* cheap. No thinking person will tell you they don't care about all that. I tell them the choice is simple: You can buy honestly priced food or you can buy irresponsibly priced food."

He reminded me that his meat would be considerably cheaper than it is if not for government regulations and the resulting high cost of processing—at least a dollar cheaper per pound. "If we could just level the playing field—take away the regulations, the subsidies, and factor in the health care and environmental cleanup costs of cheap food—we could compete on price with anyone."

It's true that cheap industrial food is heavily subsidized in many ways such that its price in the supermarket does not reflect its real cost. But until the rules that govern our food system change, organic or sustainable food is going to cost more at the register, more than some people can afford. Yet for the great majority of us the story is not quite so simple. As a society we Americans spend only a fraction of our disposable income feeding ourselves—about a tenth, down from a fifth in the 1950s. Americans today spend less on food, as a percentage of disposable income, than any other industrialized nation, and probably less than any people in the history of the world. This suggests that there are many of us who could afford to spend more on food if *we chose to*. After all, it isn't only the elite who in recent years have found an extra fifty or one hundred dollars each month to spend on cell phones (now owned by more than half the U.S. population, children included) or television, which close to 90 percent of all U.S. households now pay for. Another formerly free good that more than half of us happily pay for today is water. So is the unwillingness to pay more for food really a matter of affordability or priority?

As things stand, artisanal producers like Joel compete not on price but quality, which, oddly enough, is still a somewhat novel idea when it comes to food. "When someone drives up to the farm in a BMW and asks me why our eggs cost more, . . . well, first I try not to get mad. Frankly, any city person who doesn't think I deserve a white-collar salary as a farmer doesn't deserve my special food. Let them eat E. coli. But I don't say that. Instead, I take him outside and point at his car. 'Sir, you clearly understand quality and are willing to pay for it. Well, food is no different: You get what you pay for.'

"Why is it that we exempt food, of all things, from that rule? Industrial agriculture, because it depends on standardization, has bombarded us with the message that all pork is pork, all chicken is chicken, eggs eggs, even though we all know that can't really be true. But it's downright un-American to suggest that one egg might be nutritionally superior to another." Joel recited the slogan of his local supermarket chain: "'We pile it high and sell it cheap.' What other business would ever sell its products that way?"

When you think about it, it is odd that something as important to our health and general well-being as food is so often sold strictly on the basis of price. The value of relationship marketing is that it allows many kinds of information besides price to travel up and down the food chain: stories as well as numbers, qualities as well as quantities, values rather than "value." And as soon as that happens people begin to make different kinds of buying decisions, motivated by criteria other than price. But instead of stories about how it was produced accompanying our food, we get bar codes—as inscrutable as the industrial food chain itself, and a fair symbol of its almost total opacity.

Not that a bar code *needs* to be so obscure or reductive. Supermarkets in Denmark have experimented with adding a second bar code to packages of meat that when scanned at a kiosk in the store brings up on a monitor images of the farm where the meat was raised, as well as detailed information on the particular animal's genetics, feed, medications, slaughter date, etc. Most of the meat in our supermarkets simply couldn't withstand that degree of transparency; if the bar code on the typical package of pork chops summoned images of the CAFO it came from, and information on the pig's diet and drug regimen, who could bring themselves to buy it? Our food system depends on consumers' not knowing much about it beyond the price disclosed by the checkout scanner. Cheapness and ignorance are mutually reinforcing. And it's a short way from not knowing who's at the other end of your food chain to not caring—to the carelessness of both producers and consumers. Of course, the global economy couldn't very well function without this wall of ignorance and the indifference it breeds. This is why the rules of world trade explicitly prohibit products from telling even the simplest stories—"dolphin safe," "humanely slaughtered," etc.—about how they were produced.

For his part, Joel would just as soon build local economies in which bar codes are unnecessary rather than attempt to enhance them—to use technology or labeling schemes to make the industrial food chain we have more transparent. I realized with a bit of a jolt that his pastoral, or agrarian, outlook doesn't adequately deal with the fact that so many of us now live in big cities far removed from the places where our food is grown and from opportunities for relationship marketing. When I asked how a place like New York City fit into his vision of a local food economy he startled me with his answer: "Why do we have to have a New York City? What good is it?"

If there was a dark side to Joel's vision of the postindustrial food chain, I realized, it was the deep antipathy to cities that has so often shadowed rural populism in this country. Though when I pressed him, pointing out that New York City, den of pestilence and iniquity though it might be, was probably here to stay and would need to eat, he allowed that farmer's markets and CSAs—"community supported agriculture," schemes in which customers "subscribe" to a farm, paying a few hundred dollars at the start of the growing season in exchange for a weekly box of produce through the summer—might be a good way for urbanites to connect with distant farmers. For my own part, this taut little exchange made me appreciate what a deep gulf of culture and experience separates me from Joel—and yet at the same time, what a sturdy bridge caring about food can sometimes provide.

(Sometimes, but not always, for the antipathy of city and country still runs deep—and in both directions. I once encouraged a food writer from a big city newspaper to pay a visit to Polyface. The day she got back she telephoned me, all in a lather about the alien beings she'd had to spend the day with in Swoope: *"You never warned me he had a Jesus fish on his front door!"*)

WHEN JOEL AND I arrived at Bev's office that afternoon, we were greeted by an intense, wiry, blue-eyed fellow in his forties wearing shorts and a Polyface baseball cap, and talking a mile a minute. Joel had explained on the way down that Bev was at the moment operating under excruciating financial pressure: He had mortgaged his family's farm to build a small meat-processing plant. Bev's experience at the farmer's markets had convinced him of the growing demand for pastured meat, but supply was limited by the shortage of small processing plants willing to work with the state's grass farmers. So he'd decided to build one himself.

Bev was nearing the end of his financial rope while the USDA dilly-dallied on the approvals he needed to open. Yet when he'd finally secured the necessary permits, hired a crew, and begun killing animals, the USDA abruptly pulled its inspector, effectively shutting him down. They explained that Bev wasn't processing enough animals fast enough to justify the inspector's time—in other words, he wasn't sufficiently industrial, which of course was precisely the point of the whole venture. I realized Joel had wanted me to see Bev's predicament as proof of his contention that the government is putting obstacles in the path of an alternative food system.

Remarkably under the circumstances, Bev—whose business card gives his full name as Beverly P. Eggleston IV—had not lost his sense of humor or weakness for bad puns and high-velocity patter. When I told him what I'd been up to on the farm all week he cautioned me that "trying to follow Joel around will give you carpool tunnels and old-timer's disease." Joel thinks Bev is the funniest man alive. He also fervently wants him to succeed and has been advancing him thousands of dollars' worth of Polyface product to help float him while he does battle with the bureaucrats.

After Bev took us on a tour through the shiny new processing facility, a million dollars' worth of stainless steel and white tile built to exacting USDA specifications and sitting idle, we repaired to the trailer home parked out back, where Bev appeared to be living on potato chips and caffeinated soda. Every weekend he drives the three hundred miles up to Washington with a truckload of product from Joel and other grass farmers from all over Virginia. I asked him about selling pastured meats at farmer's markets, about exactly what it took to get people to pay the extra money.

"What I sell them on is where they're coming from," Bev explained. "There's a whole wheel of reasons to work with, and you've got three seconds to figure out what their issue is. Animal cruelty? Pesticides? Nutrition? Taste?" Joel had told me Bev is a born salesman ("He could sell a hat rack to a moose"), and it wasn't hard to imagine him working the Saturday crowd, hitting the adjacent chords of fear and pleasure and health, all the while barbecuing free samples and unspooling his high-speed shtick. "This is food for folks whose faces itch when the wool's being pulled over their eyes," Bev said, giving me a taste of his spiel. "Instead of mad cow disease, we've got glad cows at ease."

Not many farmers can do this; indeed, many farmers become farmers precisely so they don't have to do any such thing. They'd much prefer to work with animals or plants rather than human strangers, and for these farmers direct relationship marketing isn't an option, which is why they're happy to have someone like Bev work the farmer's markets for them, even if that means giving him a cut on top of the 6 percent the farmer's market already takes from every dollar of sales. It's still a vastly better deal than wholesaling commodities.

Sitting around the trailer's tiny kitchen table drinking sodas, Bev and Joel talked about the economics of selling food locally. Joel said the farmer's market was his least profitable outlet, which is why he had stopped doing them himself a few years ago. All the same, farmer's markets have blossomed in recent years, their number increasing from

1,755 a decade ago to 3,137 at last count. Joel was even higher on metropolitan buying clubs, a scheme with which I was not familiar. A group of families gets together to place a big order once or twice a month; a lead person organizes everything, and offers her home as a pickup site, usually in exchange for free product. The size of the order makes it worth the farmer's while to deliver, in Joel's case sometimes as far as Virginia Beach or Bethesda—half a day's drive. Metropolitan buying clubs represent the fastest-growing segment of Joel's market.

Who were these consumers? In Joel's case mostly young mothers concerned about the health of their children, many of them drawn from the homeschooling community ("People who have already opted out once") or from an organization called the Weston Price Foundation. Dr. Weston Price was a dentist who in the 1930s began to wonder why isolated "primitive" tribes had so much better teeth and general good health than people living in industrialized countries. He traveled all over the world researching the diets of the healthiest, longest-lived populations, and found certain common denominators in their diets: They ate lots of meat and fats from wild or pastured animals; unpasteurized dairy products; unprocessed whole grains; and foods preserved by fermentation. Today the foundation, which is run by a nutritionist and cookbook author named Sally Fallon, promotes these traditional diets in books and conferences, as well as on its Web site, where Joel is one of the producers often cited.

"The beauty of the Internet is that it allows like-minded people to find their tribes, and then for the tribes to find their way to us"—all without the expense of marketing or a storefront. Eatwild.com, a site that promotes the benefits of pastured meat and dairy, is another route by which consumers find their way to Polyface. "It's never been easier for people to opt out."

"Opting out" is a key term for Joel, who believes that it would be a fatal mistake to "try to sell a connected, holistic, ensouled product through a Western, reductionist, Wall Street sales scheme"—by which (I think) he means selling to Whole Foods. As far as both Joel and Bev are concerned there isn't a world of difference between Whole Foods and Wal-Mart. Both are part of an increasingly globalized economy that turns anything it touches into a commodity, reaching its tentacles wherever in the world a food can be produced most cheaply, and then transporting it wherever it can be sold most dearly.

Late in our conversation, Joel asked Bev and me if we'd seen a recent column by Allan Nation in *Stockman Grass Farmer* about "artisanal economics." Drawing on the theories of Harvard Business School professor Michael Porter, Nation had distinguished between industrial and artisanal enterprises to demonstrate why attempts to blend the two modes seldom succeed. Industrial farmers are in the business of selling commodities, he explained, a business where the only viable competitive strategy is to be the least-cost producer. The classic way any industrial producer lowers the costs of his product is by substituting capital—new technologies and fossil-fuel energy—for skilled labor and then stepping up production, exploiting the economies of scale to compensate for shrinking profit margins. In a commodity business a producer must sell ever more cheaply and grow ever bigger or be crushed by a competitor who does.

Nation contrasted this industrial model with its polar opposite, what he calls "artisanal production," where the competitive strategy is based on selling something special rather than being the least-cost producer of a commodity. Stressing that "productivity and profits are two entirely different concepts," Nation suggests that even a small producer can be profitable so long as he's selling an exceptional product and keeping his expenses down. Yet this artisanal model works only so long as it doesn't attempt to imitate the industrial model in any respect. It must not try to replace skilled labor with capital; it must not grow for the sake of growth; it should not strive for uniformity in its products but rather make a virtue of variation and seasonality; it shouldn't invest capital to reach national markets but rather should focus on local markets, relying on reputation and word of mouth rather than on advertising; and lastly, it should rely as much as possible on free solar energy rather than costly fossil fuels.

"The biggest problem with alternative agriculture today," Nation writes, "is that it seeks to incorporate bits and pieces of the industrial

model and bits and pieces of the artisanal model. This will not work. . . .
In the middle of the road, you get the worst of both worlds."

Nation's column had helped Joel understand why his broiler business was more profitable than his beef or pork business. Since he could
process the chickens himself, the product was artisanal from start to
finish; his beef and pork, on the other hand, had to pass through an industrial processing plant, adding to his costs and shrinking his margins.

No one needed to spell it out, but the Porter/Nation theory also
helped explain Bev's current predicament. He had built an artisanal
meat plant, designed to custom-process pastured livestock humanely
and scrupulously, no more than a few dozen animals a day. But his artisanal enterprise was being forced to conform to a USDA regulatory system that is based on an industrial model—indeed, that was created in
response to the industrial abuses Upton Sinclair chronicled in *The Jungle*.
The federal regulatory regime is expressly designed for a large slaughterhouse operated by unskilled and indifferent workers killing and cutting as many as four hundred feedlot animals an hour. The volume of
such an operation can easily cover the costs of things like a dedicated
restroom for the inspector, or elaborate machinery to steam clean (or
irradiate) carcasses presumed to carry *E. coli*. The specifications and
costly technologies implicitly assume that the animals being processed
have been living in filth and eating corn rather than grass. The industrial
packing plant where number 534 met his end can take a steer from
knocker to boxed beef for about fifty dollars a head; it would cost
nearly ten times as much to process him in a custom facility like this.
The industrial and artisanal economies clash right here in Bev's packing
plant, and sadly, it's not hard to guess which one will ultimately prevail.

2. THURSDAY MORNING

I woke to the sound of Joel's brother Art's panel truck noisily backing
up to the salesroom door. The clock said 5:45 A.M. Thursday is delivery
day, and Art likes to start staging orders and organizing his truck before

any of the other farmers he makes deliveries for shows up. I threw on some clothes and dashed out to meet him. Art is five years older than Joel and, on first impression, a very different sort of character: not nearly so sunny or expansive, more grounded in the world as it is and perhaps as a result, prone to flashes of crankiness I've never observed in Joel. But then, Art works in a less pastoral world, one in which he has to contend with city traffic and overzealous meter maids as well as the occasional temperamental chef. Compared to his brother's revolutionary zeal, Art seems to have passed the point of believing that this world, or for that matter the human soul, is ever going to be substantially different than we find it.

Every Thursday Art mounts a scrupulously planned military operation to supply Charlottesville's white-tablecloth restaurants with pastured meat and eggs from Polyface, as well as produce, dairy products, and mushrooms from a half dozen other small producers in the Shenandoah Valley. He phones his farmers on Monday night to find out what they've got, faxes his chefs an order list Tuesday morning, sells and writes orders all day Tuesday, Tuesday night faxes them to the farmers so they can harvest Wednesday, and then reconnoiter with Art in the Polyface parking lot shortly after dawn on Thursday.

I spent the better part of Thursday riding shotgun in Art's panel truck, an old orange Dodge Caravan with a cranky compressor on the roof and a sign on the side that says On Delivery From Polyface Inc. Follow me to the Best Restaurants in Town. Which seemed to be more or less the case: Most of Charlottesville's best chefs buy from Polyface, primarily chickens and eggs, but also lots of pork and as many rabbits as Daniel can raise. We made most of our deliveries after lunch, when the kitchens were prepping dinner and relatively quiet. After Art nailed down a quasi-legal parking spot, I'd help him haul in plastic totes the size of laundry baskets laden with meat and produce. The chefs had uniformly high praise for the quality of Polyface produce, and clearly felt good about supporting a local farm, which many of them had visited on one of the Chef Appreciation Days that Polyface holds each summer. I could have filled a notebook with their encomiums. A few:

"Okay, a happier chicken, great, but frankly for me it's all about the taste, which is just so different—this is a chickenier chicken."

"Art's chickens just taste cleaner, like the chicken I remember when I was a kid. I try to buy from people who are in my community and stand by their food. Don Tyson, on the other hand, stands behind a bunch of lawyers."

"Oh, those beautiful eggs! The difference is night and day—the color and richness and fat content. There's just no comparison. I always have to adjust my recipes for these eggs—you never need as many as they call for."

Between stops, Art mentioned that Joel's eggs usually gave him his foot in the door when trying to land a new account. We stopped in at one such prospect, a newly opened restaurant called the Filling Station. Art introduced himself and presented the chef with a brochure and a dozen eggs. The chef cracked one into a saucepan; instead of spreading out flabbily, the egg stood up nice and tall in the pan. Joel refers to this as "muscle tone." When he first began selling eggs to chefs, he'd crack one right into the palm of his hand, and then flip the yolk back and forth from one hand to another to demonstrate its integrity. The Filling Station chef called his staff over to admire the vibrant orange color of the yolk. Art explained that it was the grass diet that gave the eggs their color, indicating lots of beta-carotene. I don't think I'd ever seen an egg yolk rivet so many people for so long. Art beamed; he was in.

At one restaurant, the chef inquired if Art could find him some game birds; maybe in the fall, Art offered. Later, back in the truck, Art launched into a little diatribe about seasonality—one of the stiffest challenges facing the development of a local food economy.

"We have to battle the idea that you can have anything you want any time you want it. Like 'spring lamb.' What the hell does that mean? That's not its natural cycle. You want lambs to hit the ground when the grass is lush, in April. They won't be ready for eight to ten months after that—not till early winter. But the market's become totally out of sync with nature. We should eat red meat when it's cold, but people want chicken in the winter, when we don't have it."

A global food market, which brings us New Zealand lamb in the spring, Chilean asparagus in December, and fresh tomatoes the year round, has smudged the bright colors of the seasonal food calendar we all once knew by heart. But for local food chains to succeed, people will have to relearn what it means to eat according to the seasons. This is especially true in the case of pastured animals, which can be harvested only after they've had several months on rapidly growing grass. Feeding animals corn in CAFOs has accustomed us to a year-round supply of fresh meats, many of which we forget were once eaten as seasonally as tomatoes or sweet corn: People would eat most of their beef and pork in late fall or winter, when the animals were fat, and eat chicken in the summer.

Joel told me that when he first began selling eggs to chefs, he found himself apologizing for their pallid hue in winter; the yolks would lose their rich orange color when the chickens came in off the pasture in November. Then he met a chef who told him not to worry about it. The chef explained that in cooking school in Switzerland he'd been taught recipes that specifically called for April eggs, August eggs, and December eggs. Some seasons produce better yolks, others better whites, and chefs would adjust their menus accordingly.

Both Joel and Art evinced the deepest respect for their chefs, who not only seldom argued price and wrote checks right on the spot, but clearly appreciated their work and, very often, acknowledged it right on their menus: "Polyface Farm Chicken" is something I saw on menus and specials boards all over Charlottesville.

This informal alliance of small farmers and local chefs is something you find in many cities these days. Indeed, ever since Alice Waters opened Chez Panisse in Berkeley in 1973, chefs have been instrumental in helping rebuild local food economies all over America. Waters made a point of sourcing much of her food from local organic growers, cooked only what was in season, and shone the bright light of glamour on the farmers, turning many of them into menu celebrities. Chefs like Waters have also done much to educate the public about the virtues of local agriculture, the pleasures of eating by the season, and the superior qualities of exceptionally fresh food grown with care and without chemicals. The Roman writer Livy once warned that when a society's chefs come to be regarded as consequential figures, it is a sure sign that society is well down the road to decadence. Livy's rule might have held up until the 1960s in America, but clearly no longer. Who would ever have guessed before then that America's chefs would be leading a movement to save small farmers and reform America's food system?

To talk to the chefs, customers, and farmers working together in this one corner of the country to rebuild a local food chain is to appreciate that it is a movement, and not merely a market. Or rather it is a novel hybrid, a market as movement, for at its heart is a new conception of what it means to be a consumer—an attempt to redeem that ugly word, with its dismal colorings of selfishness and subtraction. Many of the Polyface customers I met (though by no means all of them) had come to see their decision to buy a chicken from a local farmer rather than from Wal-Mart as a kind of civic act, even a form of protest. A protest of *what* exactly is harder to pin down, and each person might put it a little differently, but the customers I met at Polyface had gone to some trouble and expense to opt out—of the supermarket, of the fast-food nation, and, standing behind that, of a globalized industrial agriculture. Their talk of distrusting Wal-Mart, resenting the abuse of animals in farm factories, insisting on knowing who was growing their food, and wanting to keep their food dollars in town—all this suggested that for many of these people spending a little more for a dozen eggs was a decision inflected by a politics, however tentative or inchoate.

Shortly before I traveled to Virginia I'd read an essay by Wendell Berry called "The Whole Horse" in which he argued that reversing the damage done to local economies and the land by the juggernaut of world trade would take nothing less than "a revolt of local small producers and local consumers against the global industrialism of the corporation." He detected the stirrings of such a rebellion in the rise of local food systems, and the growing market "for good, fresh, trustworthy food, food from producers known and trusted by consumers." Berry would have us believe that what I was seeing in the Polyface salesroom represented a local uprising in a gathering worldwide rebellion against what he calls "the total economy."

Why should food, of all things, be the linchpin of that rebellion? Perhaps because food is a powerful metaphor for a great many of the values to which people feel globalization poses a threat, including the distinctiveness of local cultures and identities, the survival of local landscapes, and biodiversity. When José Bové, the French antiglobalization activist (and Roquefort farmer), wanted to make his stand against globalization, he drove his tractor through the plate glass not of a bank or insurance company, but of a McDonald's. Indeed, the most powerful protests against globalization to date have all revolved around food: I'm thinking of the movement against genetically modified crops, the campaign against patented seeds in India (which a few years ago brought four hundred thousand Indians into the streets to protest WTO intellectual property rules), and Slow Food, the Italian-born international movement that seeks to defend traditional food cultures against the global tide of homogenization.

Even for people who find the logic of globalization otherwise compelling, the globalization of food often stops them short. That logic treats food as a commodity like any other, and that simply doesn't square with people's beliefs or experience. Once the last barrier to free trade comes down, and the last program of government support for farmers ends, our food will come from wherever in the world it can be produced most cheaply. The iron law of competitive advantage dictates that if another country can grow something more efficiently—whether because its land or labor is cheaper or its environmental laws more lax—we will no longer grow it here. What's more, under the global economic dispensation, this is an outcome to be wished for, since it will free our land for more productive uses—more houses, say. Since land in the United States is relatively expensive, and our tolerance for agricultural pollution and animal cruelty is wearing thin, in the future all our food may come from elsewhere. This argument has been made by, among others, economist Steven Blank, in a book rather bloodlessly titled *The End of Agriculture in the American Portfolio*.

And why *should* a nation produce its own food when others can produce it more cheaply? A dozen reasons leap to mind, but most of them the Steven Blanks of the world—and they are legion—are quick to dismiss as sentimental. I'm thinking of the sense of security that comes from knowing that your community, or country, can feed itself; the beauty of an agricultural landscape; the outlook and kinds of local knowledge that farmers bring to a community; the satisfactions of buying food from a farmer you know rather than the supermarket; the locally inflected flavor of a raw-milk cheese or honey. All those things—all those pastoral values—globalization proposes to sacrifice in the name of efficiency and economic growth.

Though you do begin to wonder who is truly the realist in this debate, and who the romantic. We live, as Berry has written (in an essay called "The Total Economy"), in an era of "sentimental economics," since the promise of global capitalism, much like the promise of communism before it, ultimately demands an act of faith: that if we permit the destruction of certain things we value here and now we will achieve a greater happiness and prosperity at some unspecified future time. As Lenin put it, in a sentiment the WTO endorses in its rulings every day, you have to break a few eggs to make an omelet.

Perhaps it is no accident that sentimental communism foundered precisely on the issue of food. The Soviets sacrificed millions of small farms and farmers to the dream of a collectivized industrial agriculture that never managed to do what a food system has to do: feed the nation. By the time of its collapse, more than half of the food consumed in the Soviet Union was being produced by small farmers and home gardeners operating without official sanction, on private plots tucked away in the overlooked corners and cracks of the crumbling Soviet monolith. George Naylor, speaking from deep inside the American monolith, might be onto something when, during our conversations about industrial agriculture, he likened the rise of alternative food chains in America to "the last days of Soviet agriculture. The centralized food system wasn't serving the people's needs, so they went around it. The rise of farmer's markets and CSAs is sending the same signal today." Of course the problems of our food system are very different—if anything, it produces too much food, not too little, or too much of the wrong food. But there's no question that it is failing many consumers and producers, which is why they are finding creative ways around it.

So much about life in a global economy feels as though it has passed beyond the individual's control—what happens to our jobs, to the prices at the gas station, to the vote in the legislature. But somehow food still feels a little different. We can still decide, every day, what we're going to put into our bodies, what sort of food chain we want to participate in. We can, in other words, reject the industrial omelet on offer and decide to eat another. This might not sound like a big deal, but it could be the beginnings of one. Already the desire on the part of consumers to put something different into their bodies has created an $11 billion market in organic food. That marketplace was built by consumers and farmers working informally together outside the system, with exactly no help from the government.

Today the total economy, astounding in its ability to absorb every challenge, is well on its way to transforming organic from a reform movement into an industry—another flavor in the global supermarket. It took capitalism less than a quarter century to turn even something as ephemeral as bagged salads of cut and washed organic mesclun, of all things, into a cheap international commodity retailed in a new organic supermarket. Whether this is a good or bad thing people will disagree.

Joel Salatin and his customers want to be somewhere that that juggernaut can't go, and it may be that by elevating local above organic, they have found exactly that place. By definition local is a hard thing to sell in a global marketplace. Local food, as opposed to organic, implies a new economy as well as a new agriculture—new social and economic relationships as well as new ecological ones. It's a lot more complicated.

Of course, just because food is local doesn't necessarily mean it will be organic or even sustainable. There's nothing to stop a local farmer from using chemicals or abusing animals—except the gaze or good word of his customers. Instead of looking at labels, the local food customer will look at the farm for himself, or look the farmer in the eye and ask him about how he grows his crops or treats his animals. That said, there are good reasons to think a genuinely local agriculture will tend to be a more sustainable agriculture. For one thing, it is much less likely to rely on monoculture, the original sin from which almost every other problem of our food system flows. A farmer dependent on a local market will, perforce, need to grow a wide variety of things rather than specialize in the one or two plants or animals that the national market (organic or otherwise) would ask from him.

The supermarket wants all its lettuce from the Salinas Valley, all its apples from Washington State, and all its corn from Iowa. (At least until the day it decides it wants all its corn from Argentina, all its apples from China, and all its lettuce from Mexico.) People in Iowa can eat only so much corn and soybeans themselves. So when Iowans decide to eat locally, rather than from the supermarket, their farmers will quickly learn to grow a few other things besides. And when they do, they'll probably find that they can give up most of their fertilizers and pesticides, since a diversified farm will produce much of its own fertility and its own pest control.

Shopping in the organic supermarket underwrites important values on the farm; shopping locally underwrites a whole set of other values as well. That's because farms produce a lot more than food; they also produce a kind of landscape and a kind of community. Whether Polyface's customers spend their food dollars here in Swoope or in the Whole Foods in Charlottesville will have a large bearing on whether this lovely valley—this undulating checkerboard of fields and forests—will endure, or whether the total economy will find a "higher use" for it. "Eat your view!" is a bumper sticker often seen in Europe these days; as it implies, the decision to eat locally is an act of conservation, too, one that is probably more effective (and sustainable) than writing checks to environmental organizations.

"Eat your view!" takes work, however. To participate in a local food economy requires considerably more effort than shopping at the Whole Foods. You won't find anything microwaveable at the farmer's market or in your CSA box, and you won't find a tomato in December. The local food shopper will need to put some work into sourcing his food—into learning who grows the best lamb in his area, or the best sweet corn. And then he will have to become reacquainted with his kitchen. Much of the appeal of the industrial food chain is its convenience; it offers busy people a way to delegate their cooking (and food preservation) to others. At the other end of the industrial food chain that begins in a cornfield in Iowa sits an industrial eater at a table. (Or, increasingly, in a car.) The achievement of the industrial food system over the past half century has been to transform most of us into precisely that creature.

All of which is to say that a successful local food economy implies not only a new kind of food producer, but a new kind of eater as well, one who regards finding, preparing, and preserving food as one of the pleasures of life rather than a chore. One whose sense of taste has ruined him for a Big Mac, and whose sense of place has ruined him for shopping for groceries at Wal-Mart. This is the consumer who understands— or remembers—that, in Wendell Berry's memorable phrase, "eating is an agricultural act." He might have added that it's a political act as well.

This is precisely the mission that Slow Food has set for itself: to remind a generation of industrial eaters of their connections to farmers and farms, and to the plants and animals we depend on. The movement, which began in 1989 as a protest against the opening of a McDonald's in Rome, recognizes that the best way to fight industrial eating is by simply recalling people to the infinitely superior pleasures of traditional foods enjoyed communally. The consumer becomes, in founder Carlo Petrini's phrase, a "coproducer"—his eating contributes to the survival of landscapes and species and traditional foods that would otherwise succumb to the fast-food ideal of "one world, one taste." Even connoisseurship can have a politics, Slow Food wagers, since an eater in closer touch with his senses will find less pleasure in a box of Chicken McNuggets than in a pastured chicken or a rare breed of pig. It's all very

Italian (and decidedly un-American): to insist that doing the right thing is the most pleasurable thing, and that the act of consumption might be an act of addition rather than subtraction.

ON MY LAST DAY on the farm, a soft June Friday afternoon, Joel and I sat talking at a picnic table behind the house while a steady stream of customers dropped by to pick up their chickens. I asked him if he believed the industrial food chain would ever be overturned by an informal, improvised movement made up of farmer's markets, box schemes, metropolitan buying clubs, slow-foodies, and artisanal meat-processing plants like Bev Eggleston's. Even if you count the organic supermarket, the entire market for all alternative foods remains but a flea on the colossus of the industrial food economy, with its numberless fast-food outlets and supermarkets backed by infinite horizons of corn and soybeans.

"We don't have to beat them," Joel patiently explained. "I'm not even sure we should try. We don't need a law against McDonald's or a law against slaughterhouse abuse—we ask for too much salvation by legislation. All we need to do is empower individuals with the right philosophy and the right information to opt out en masse.

"And make no mistake: It's happening. The mainstream is splitting into smaller and smaller groups of like-minded people. It's a little like Luther nailing his ninety-five theses up at Wittenberg. Back then it was the printing press that allowed the Protestants to break off and form their own communities; now it's the Internet, splintering us into tribes that want to go their own way."

Of course! Joel saw himself as more of a Luther than a Lenin; the goal wasn't to blow up the Church, but simply to step around it. Protestantism also comes in many denominations, as I suspect will the future of food. Deciding whether that future should more closely resemble Joel's radically local vision or Whole Foods' industrial organic matters less than assuring that thriving alternatives exist; feeding the cities may require a different sort of food chain than feeding the countryside. We may need a great many different alternative food chains, organic and local, biodynamic and slow, and others yet undreamed of. As in the fields, nature provides the best model for the marketplace, and nature never puts all her eggs in one basket. The great virtue of a diversified food economy, like a diverse pasture or farm, is its ability to withstand any shock. The important thing is that there be multiple food chains, so that when any one of them fails—when the oil runs out, when mad cow or other food-borne diseases become epidemic, when the pesticides no longer work, when drought strikes and plagues come and soils blow away—we'll still have a way to feed ourselves. It is because some of those failures are already in view that the salesroom at Polyface Farm is buzzing with activity this afternoon, and why farmer's markets in towns and cities all across America are buzzing this afternoon, too.

"An alternative food system is rising up on the margins," Joel continued. "One day Frank Perdue and Don Tyson are going to wake up and find that their world has changed. It won't happen overnight, but it will happen, just as it did for those Catholic priests who came to church one Sunday morning only to find that, my goodness, there aren't as many people in the pews today. Where in the world has everybody gone?"

FOURTEEN

THE MEAL

Grass Fed

Before I left the farm Friday, I gathered together the makings for that evening's dinner, which I'd arranged to cook for some old friends who lived in Charlottesville. I had originally thought about filling a cooler with Polyface meat and bringing it home with me to California to cook there, but decided it would be more in keeping with the whole local food chain concept to eat this particular meal within a leisurely drive of the farm where it had been grown. After all, it was the sin of flying meat across the country that had brought me to Swoope in the first place, and I hated for Joel to think that an entire week of his instruction had left me unimproved.

From the walk-in, I picked out two of the chickens we had slaughtered on Wednesday and a dozen of the eggs I'd helped gather Thursday evening. I also stopped by the hoop house and harvested a dozen ears of sweet corn. (In consideration of my week's labors, Joel refused to accept payment for the food, but had I paid for it, the chicken would have cost $2.05 a pound, and the eggs $2.20 a dozen—prices that compare very favorably with Whole Foods's. This is not boutique food.)

On the way into Charlottesville, I stopped to pick up a few other in-gredients, trying as best as I could to look for local produce and pre-serve the bar code virginity of this meal. For my salad, I found some nice-looking locally grown rocket. At the wine shop I found a short, chauvinistic shelf of Virginia wines, but here I hesitated. How far could I take this local conceit before it ruined my meal? I hadn't had a sip of wine all week and was really looking forward to a decent one. I'd read somewhere that wine-making in Virginia was "coming into its own," but isn't that what they always say? Then I spotted a Viognier for twenty-five bucks—the priciest Virginia wine I'd ever seen. I took this as a sign of genuine confidence on somebody's part, and added the bot-tle to my cart.

I also needed some chocolate for the dessert I had in mind. Fortu-nately the state of Virginia produces no chocolate to speak of, so I was free to go for the good Belgian stuff, panglessly. In fact, even the most fervent eat-local types say it's okay for a "foodshed" (a term for a re-gional food chain, meant to liken it to a watershed) to trade for goods it can't produce locally—coffee, tea, sugar, chocolate—a practice that predates the globalization of our food chain by a few thousand years. (Whew . . .)

During the week I'd given some thought to what I should make; the farm's varied offerings certainly gave me plenty of choices. Working backward, I knew I wanted to make a dessert that would prominently feature Polyface eggs, having heard so much from the chefs about their magical properties. A chocolate soufflé, since it calls for a certain degree of magic, seemed the obvious choice. For a side dish, sweet corn was a no-brainer; there'd be kids at the table and no one had tasted corn yet this summer. But what meat to serve? Because it was only June, Polyface had no fresh beef or pork or turkey; Joel wouldn't begin slaughtering beeves and turkeys till later in the summer, hogs not till the fall. There was frozen beef and pork in the walk-in, last season's, but I preferred to make something fresh. Rabbit seemed risky; I had no idea whether Mark and Liz liked it, and the chances that their boys would eat bunny were slim. So that had left chicken, the animal with which I'd been most intimate this week. Which, truth to tell, left me feeling vaguely queasy. Was I going to be able to enjoy eating chicken so soon after my stint in the processing shed and gut-composting pile?

That queasiness perhaps explains the multistep preparation I finally settled on. When I got to Mark and Liz's house, there were still several hours before dinner, which meant there was enough time for me to brine the chicken. So I cut each of the two birds into eight pieces and immersed them in a bath consisting of water, kosher salt, sugar, a bay leaf, a splash of soy sauce, a garlic clove, and a small handful of peppercorns and coriander seeds. My plan was to slow roast the chicken pieces on a wood fire, and brining—which causes meat to absorb moisture and breaks down the proteins that can toughen it on the grill—would keep the chicken from drying out.

But the brining (like the carving of the birds into pieces) promised to do something else, too, something for me as much as the meat: It would put a little distance between the meal and Wednesday's kill, certain aromas of which were still lodged in my nostrils. One of the reasons we cook meat (besides making it tastier and easier to digest) is to civilize, or sublimate, what is at bottom a fairly brutal transaction between animals. The anthropologist Claude Lévi-Strauss described the work of civilization as the process of transforming the raw into the cooked—nature into culture. For these particular chickens, which I had personally helped to kill and eviscerate, the brining would make a start on that transformation even before the cooking fire was lit. Both literally and metaphorically, a saltwater bath cleanses meat, which perhaps explains why the kosher laws—one culture's way of coming to terms with the killing and eating of animals—insist on the salting of meat.

After a few hours, I removed and rinsed the chicken pieces, and then spread them out to dry for an hour or two, so that the skin, now slightly waterlogged, would brown nicely. Since Mark and Liz had a gas barbecue, I'd have to simulate my wood fire. So I snipped a couple of twigs off their apple tree, stripped the leaves, and placed the twigs on top of the grill, where the green wood would smolder rather than burn. I turned the gas down low and, after rubbing a little olive oil on the chicken pieces, arranged them on the grill among the apple branches, leaving some room to add the corn later.

While the chicken roasted slowly outside, I got to work in the kitchen preparing the soufflé with Willie, Mark and Liz's twelve-year-old. While Willie melted the chocolate in a saucepan, I separated the eggs. The yolks were a gorgeous carroty shade of orange and they did seem to possess an unusual integrity; separating them from the whites was a cinch. After adding a pinch of salt, I began beating the egg whites; within minutes they turned from translucent to bright white and formed soft, rounded peaks, which is when Julia Child says to begin adding sugar, and to turn the beater on high. Now the egg whites rapidly doubled in volume, then doubled again, as billions of microscopic air pockets formed amid the stiffening egg proteins. When the heat of the oven caused these air pockets to expand, the soufflé would rise, assuming everything went according to plan. Once the egg whites formed a stiff, spiky snowscape, I stopped. Willie had already blended the yolks into his melted chocolate, so we now gently folded that thick syrup into my egg whites, then poured the airy, toast-colored mixture into a soufflé dish and put it aside. I could see why pastry chefs in Charlottesville swore by Polyface eggs: What Joel had called their "muscle tone" made baking with them a breeze.

Willie and I brought the corn out on the deck to shuck. The ears were so fresh that the husks squealed as you peeled them back. I mentioned to Willie that our entire meal would be a celebration of the chicken—not only the eponymous entrée, which we could smell sweetly roasting on the grill, but the soufflé with its half-dozen eggs, and even this corn, which I explained had grown in a deep bed of composted chicken manure. Probably not the sort of detail you'd want to mention on a menu, but Willie agreed there was something pretty neat about the alchemy involved, how a plant could transform chicken crap into something as sweet and tasty and golden as an ear of corn.

Golden Bantam, the corn in question, is an heirloom variety introduced in 1902, long before the hybridizers figured out how to amp up the sweetness in sweet corn. This momentous change in the genetics of our corn is an artifact of an industrial food chain, which demands that vegetables be able to endure a cross-country road trip after picking so that they might be available everywhere the year round. This was a particular problem for corn, the sugars of which begin turning to starch the moment it is picked. So in the early sixties the breeders figured out a way to breed in extra copies of the genes responsible for producing sugars. But something was lost in the translation from local to cosmopolitan corn: The kernels lost much of their creaminess, and the specific taste of corn was overwhelmed by a generic, one-dimensional sweetness. The needs of a long industrial food chain might justify such a trade-off, but when you can eat corn picked a few hours before dinner, there's no reason for it. Unless of course an industrial diet of easy sugars has dulled your taste for the earthy sweetness of corn, now that it has to compete with things like soda.

I HAD MADE pretty much the same meal on several occasions at home, using the same basic foodstuffs, yet in certain invisible ways this wasn't the same food at all. Apart from the high color of the egg yolks, these eggs looked pretty much like any other eggs, the chicken like chicken, but the fact that the animals in question had spent their lives outdoors on pastures rather than in a shed eating grain distinguished their flesh and eggs in important, measurable ways. A growing body of scientific research indicates that pasture substantially changes the nutritional profile of chicken and eggs, as well as of beef and milk. The question we asked about organic food—is it any better than the conventional kind?—turns out to be much easier to answer in the case of grass-farmed food.

Perhaps not surprisingly, the large quantities of beta-carotene, vitamin E, and folic acid present in green grass find their way into the flesh of the animals that eat that grass. (It's the carotenoids that give these egg yolks their carroty color.) That flesh will also have considerably less fat in it than the flesh of animals fed exclusively on grain—also no surprise, in light of what we know about diets high in carbohydrates. (And about exercise, something pastured animals actually get.) But all fats are not created equal—polyunsaturated fats are better for us than saturated ones, and certain unsaturated fats are better than others. As it turns out, the fats created in the flesh of grass eaters are the best kind for us to eat.

This is no accident. Taking the long view of human nutrition, we evolved to eat the sort of foods available to hunter-gatherers, most of whose genes we've inherited and whose bodies we still (more or less) inhabit. Humans have had less than ten thousand years—an evolutionary blink—to accustom our bodies to agricultural food, and as far as our bodies are concerned, industrial agricultural food—a diet based largely on a small handful of staple grains, like corn—is still a biological novelty. Animals raised outdoors on grass have a diet much more like that of the wild animals humans have been eating at least since the Paleolithic era than that of the grain-fed animals we only recently began to eat.

So it makes evolutionary sense that pastured meats, the nutritional profile of which closely resembles that of wild game, would be better for us. Grass-fed meat, milk, and eggs contain less total fat and less saturated fats than the same foods from grain-fed animals. Pastured animals also contain conjugated linoleic acid (CLA), a fatty acid that some recent studies indicate may help reduce weight and prevent cancer, and which is absent from feedlot animals. But perhaps most important, meat, eggs, and milk from pastured animals also contain higher levels of omega-3s, essential fatty acids created in the cells of green plants and algae that play an indispensable role in human health, and especially in the growth and health of neurons—brain cells. (It's important to note that fish contain higher levels of the most valuable omega-3s than land animals, yet grass-fed animals do offer significant amounts of such important omega-3s as alpha linolenic acid—ALA.) Much research into the role of omega-3s in the human diet remains to be done, but the preliminary findings are suggestive: Researchers report that pregnant women who receive supplements of omega-3s give birth to babies with higher IQs; children with diets low in omega-3s exhibit more behavioral and learning problems at school; and puppies eating diets high in omega-3s prove easier to train. (All these claims come from papers presented at a 2004 meeting of the International Society for the Study of Fatty Acids and Lipids.)

One of the most important yet unnoticed changes to the human diet in modern times has been in the ratio between omega-3 and omega-6, the other essential fatty acid in our food. Omega-6 is produced in the seeds of plants; omega-3 in the leaves. As the name indicates, both kinds of fat are essential, but problems arise when they fall out of balance. (In fact, there's research to suggest that the ratio of these fats in our diet may be more important than the amounts.) Too high a ratio of omega-6 to omega-3 can contribute to heart disease, probably because omega-6 helps blood clot, while omega-3 helps it flow. (Omega-6 is an inflammatory; omega-3 an anti-inflammatory.) As our diet—and the diet of the animals we eat—shifted from one based on green plants to one based on grain (from grass to corn), the ratio of omega-6 to omega-3 has gone from roughly one to one (in the diet of hunter-gatherers) to more than ten to one. (The process of hydrogenating oil also eliminates omega-3s.) We may one day come to regard this shift as one of the most deleterious dietary changes wrought by the industrialization of our food chain. It was a change we never noticed, since the importance of omega-3s was not recognized until the 1970s. As in the case of our imperfect knowledge of soil, the limits of our knowledge of nutrition have obscured what the industrialization of the food chain is doing to our health. But changes in the composition of fats in our diet may account for many of the diseases of civilization—cardiac, diabetes, obesity, etc.—that have long been linked to modern eating habits, as well as for learning and behavioral problems in children and depression in adults.

Research in this area promises to turn a lot of conventional nutritional thinking on its head. It suggests, for example, that the problem with eating red meat—long associated with cardiovascular disease—may owe less to the animal in question than to that animal's diet. (This might explain why there are hunter-gatherer populations today who eat far more red meat than we do without suffering the cardiovascular consequences.) These days farmed salmon are being fed like feedlot cattle, on grain, with the predictable result that their omega-3 levels fall well below those of wild fish. (Wild fish have especially high levels of omega-3 because the fat concentrates as it moves up the food chain from the algae and phytoplankton that create it.) Conventional nutritional wisdom holds that salmon is automatically better for us than beef, but that judgment assumes the beef has been grain fed and the salmon krill fed; if the steer is fattened on grass and the salmon on grain, we might actually be better off eating the beef. (Grass-finished beef has a two-to-one ratio of omega-6 to -3 compared to more than ten to one in corn-fed beef.) The species of animal you eat may matter less than what the animal you're eating has itself eaten.

The fact that the nutritional quality of a given food (and of that food's food) can vary not just in degree but in kind throws a big wrench into an industrial food chain, the very premise of which is that beef is beef and salmon salmon. It also throws a new light on the whole question of cost, for if quality matters so much more than quantity, then the price of a food may bear little relation to the value of the nutrients in it. If units of omega-3s and beta-carotene and vitamin E are what an egg shopper is really after, then Joel's $2.20 a dozen pastured eggs actually represent a much better deal than the $0.79 a dozen industrial eggs at the supermarket. As long as one egg looks pretty much like another, all the chickens like chicken, and beef beef, the substitution of quantity for quality will go on unnoticed by most consumers, but it is becoming increasingly apparent to anyone with an electron microscope or a mass spectrometer that, truly, this is not the same food.

OKAY, but what about to someone equipped with a more or less standard-issue set of human taste buds? How different does a pastured chicken actually taste? It certainly smelled wonderful when I raised the lid on the barbecue to put the corn on. The chicken was browning nicely, the skin beginning to crisp and take on the toasty tones of oiled wood. The corn, on which I'd rubbed some olive oil and sprinkled salt and pepper, would take only a few minutes—all it needed was to heat up and for a scattering of kernels to brown. The browning of the chicken skin and the corn looked similar but in fact it owed to completely different chemical reactions, reactions that were contributing to their flavors and smells. The corn was caramelizing, as its sugars broke apart under the heat and formed into hundreds of more complicated aromatic compounds, giving a smoky dimension to the corny sweetness. Meanwhile, the chicken skin was undergoing what chemists called the Maillard reaction, in which carbohydrates in the chicken react in dry heat with certain amino acids to create an even larger and more complicated set of compounds that, because they include atoms of sulfur and nitrogen, give a richer, meatier aroma and taste to the meat than it would otherwise possess. This, at least, is how a chemist would explain what I was seeing and smelling on the grill, as I turned the corn and chicken pieces and felt myself growing hungrier.

While the corn finished roasting, I removed the chicken from the grill and set it aside to rest. A few minutes later I called everyone to the table. Ordinarily I might have felt a little funny serving as both dinner host and guest, but Mark and Liz are such close friends it seemed perfectly natural to be cooking for them in their home. That's not to say I didn't feel the cook's customary preprandial apprehension, compounded in this instance by the fact that Liz herself is such a good cook, and holds very definite opinions about food. I certainly hadn't forgotten the time she'd wrinkled her nose and pushed away a Polyface steak I'd served her. Grass-fed beef is flavored by the pastures it grows on, usually but not always for the best. It had tasted fine to me.

I passed the platters of chicken and corn and proposed a toast. I offered thanks first to my hosts-cum-guests, then to Joel Salatin and his family for growing the food before us (and for giving it to us), and then finally to the chickens, who in one way or another had provided just about everything we were about to eat. My secular version of grace, I suppose, acknowledging the various material and karmic debts incurred by this meal, debts which I felt more keenly than usual.

"At the beginning of the meal," Brillat-Savarin writes in his chapter "On the Pleasures of the Table" in *The Physiology of Taste*, "each guest eats steadily, without speaking or paying attention to anything which may be said." And so we did, aside from a few sublingual murmurs of satisfaction. I don't mind saying the chicken was out of this world. The skin had turned the color of mahogany and the texture of parchment, almost like a Peking duck, and the meat itself was moist, dense, and almost shockingly flavorful. I could taste the brine and apple wood, of course, but also the chicken itself, which more than held its own against those strong flavors. This may not sound like much of a compliment, but to me the chicken smelled and tasted exactly like chicken. Liz voiced her approval in similar terms, pronouncing it a more chickeny chicken. Which is to say, I suppose, that it chimed with that capitalized idea of Chicken we hold in our heads but seldom taste anymore. So what accounted for it? The grass? The grubs? The exercise? I know what Joel would have said: When chickens get to live like chickens, they'll taste like chickens, too.

The flavors of everything else on the table had a similarly declarative quality: the roasted corn and lemony rocket salad, and even the peachy Viognier, all of them tasting almost flamboyantly themselves, their flavors forming a bright sequence of primary colors. There was nothing terribly subtle about this meal, but everything about it tasted completely in character.

Everyone was curious to hear about the farm, especially after tasting the food that had come off it. Matthew, who's fifteen and currently a vegetarian (he confined himself to the corn), had many more questions about killing chickens than I thought it wise to answer at the dinner table. But I did talk about my week on the farm, about the Salatins and their animals. I explained the whole synergistic ballet of chickens and cows and pigs and grass, without getting into specifics about the manure and grubs and composted guts that made the whole dance work. Thankfully all of that, the killing cones, too, had retreated to the mental background for me, chased by the smoky-sweet aromas of the meal, which I found myself able to thoroughly enjoy.

The unexpectedly fine wine helped too, as did the fact that the dinner table conversation drifted off as it will do, from my Paris Hilton adventures as a farmhand to Willie's songwriting (he is, mark my words, the next Bob Dylan), Matthew's summer football camp, Mark and Liz's books-in-progress, school, politics, war, and on and on, the topics spiraling away from the table like desultory rings of smoke. Being a Friday late in June, this was one of the longest evenings of the year, so no one felt in a rush to finish. Besides, I'd just put the soufflé in to bake when we sat down, so dessert was still a ways off.

In his chapter Brillat-Savarin draws a sharp distinction between the pleasures of eating—"the actual and direct sensation of a need being satisfied," a sensation we share with the animals—and the uniquely human "pleasures of the table." These consist of "considered sensations born of the various circumstances of fact, things, and persons accompanying the meal"—and comprise for him one of the brightest fruits of civilization. Every meal we share at a table recapitulates this evolution from nature to culture, as we pass from satisfying our animal appetites in semisilence to the lofting of conversational balloons. The pleasures of the table begin with eating (and specifically with eating meat, in Brillat-Savarin's view, since it was the need to cook and apportion meat that first brought us together to eat), but they can end up anywhere human talk cares to go. In the same way that the raw becomes cooked, eating becomes dining.

All such transformations were very much on my mind that evening, coming at the end of a week of farmwork that had put me in much closer touch with the biology of eating than the art. The line from composting chicken guts to gastronomy is almost unimaginably long, but there is a line. While we talked and waited for the soufflé to complete its magic rise, the smell of baking chocolate seeped out of the kitchen and filled the house. When at last I told Willie the time had come to open the oven, cross your fingers, I saw his smile blossom first, then the great crown of soufflé puffing out from the cinched white waist of its dish. Triumph!

Here was the most improbable transformation of all. There's something wondrous about any soufflé, how a half dozen eggs flavored by nothing more than sugar and chocolate can turn into something so ethereally Other. Soufflé, "to blow," comes from the Latin word for breath, of course, in recognition of the air that a soufflé mostly is. But soufflé has a spiritual sense, too, as in the breath of life (in English the word "spirit" comes from breath), which seems fitting, for isn't the soufflé as close as cookery ever comes to elevating matter into spirit?

This particular soufflé was good, not great; its texture was slightly grainier than it should be, which makes me think I may have beaten the eggs a little too long. But it tasted wonderful, everyone agreed, and as I rolled the rich yet weightless confection on my tongue, I closed my eyes and suddenly there they were: Joel's hens, marching down the gangplank from out of their Eggmobile, fanning out across the early morning pasture, there in the grass where this sublime bite began.

III

PERSONAL

THE
FOREST

FIFTEEN

THE FORAGER

1. SERIOUS PLAY

There was one more meal I wanted to make, and that was the meal at the end of the shortest food chain of all. What I had in mind was a dinner prepared entirely from ingredients I had hunted, gathered, and grown myself. Now, there are some people (though not all that many of them anymore) for whom such a radically self-made meal exists firmly in the realm of possibility. I am not one of them. The growing part was the only part I knew I could handle. I've been a gardener most of my life, and have made countless meals from my garden. These included no animal protein, however, and I had decided that this meal should feature representatives of all three edible kingdoms: animal, vegetable, and fungi. I was about as ill prepared to hunt the former and gather the latter as an eater could possibly be.

I had never hunted in my life. Indeed, I had never fired a gun loaded with anything more lethal than caps. Being a somewhat accident-prone individual (childhood mishaps included getting bitten in the cheek by a seagull and breaking my nose falling out of bed), I have always thought it wise to maintain a healthy distance between me and firearms. Besides, you have to have had a certain kind of dad in order to join the culture of hunting in America, and mine, one of the great indoorsmen, was emphatically not that dad. My father looked upon hunting as a human activity that had stopped making sense with the invention of the steakhouse. As a recreational pursuit that involved the certainty of going outside and the possibility of the sight of blood, hunting in his opinion was something best left to the gentiles. So in hunting my own dinner I would be starting very much from scratch.

Thanks to my mother's more extensive engagement with the natural world, I did have some childhood experience as a gatherer. During the summer she would take us to the beach at low tide to dig for steamers, excavating by hand the airholes the clams had made in the sand flats until they squirted us in self-defense. In the waning days of summer we would pick beach plums that she would transform into a deliciously tart jelly the brilliant color of rubies. All winter long her beach plum jelly summoned memories of summer vacation: August on toast. My sisters and I also filled bowls with wild blackberries and huckleberries for dessert. Once as a teenager I gathered enough wild grapes to attempt to make wine. My understanding of fermentation was shaky, however, and after a week or so the sealed container I had put the crushed fruit in exploded, splattering the ceiling and all four walls of the living room where I had left it with grape-skin confetti. Another time I tried to make root beer from the roots of a sassafras sapling. The resulting concoction smelled right, but that was about it.

These elementary foraging expeditions were always accompanied by scary surgeon general–like warnings from my mother about the deadly poisons lurking in berries and mushrooms growing in the wild; she made it sound like it wouldn't take much for a kid to get himself killed snacking in the woods. So I never picked any but the most iconic fruits, and while I enjoyed eating store-bought mushrooms, I never so much as touched one in the woods. My mother had inculcated a fear of fungi in me that put picking a wild mushroom in the same class of certain-death behaviors as touching downed power lines or climbing into the cars of strangers proffering candy.

So my fungiphobia was another thing I'd have to overcome if I hoped to ever serve a personally hunted and gathered meal, because wild mushrooms had to be on the menu. Mushroom hunting seems to me the very soul of foraging, throwing both the risks and rewards of eating from the wild into the sharpest possible relief. If I hoped to host representatives of all three kingdoms on my plate, learning to distinguish the delicious from the deadly among the fungi was a necessity. (Actually I hoped to wangle a fourth kingdom in there—a mineral—if I could manage to locate a salt flat within driving distance of my house.)

Why go to all this trouble? It's not as though the forager food chain represents a viable way for us to eat at this point in history; it doesn't. For one thing, there is not enough game left to feed us all, and probably not enough wild plants and mushrooms either. The prevailing theory as to why, as a species, we left off hunting and gathering is that we had ruined that perfectly good lifestyle by overdoing it, killing off the megafauna on which we depended. Otherwise it's hard to explain why humans would ever have traded such a healthy and comparatively pleasant way of life for the backbreaking, monotonous work of agriculture. Agriculture brought humans a great many blessings, but it also brought infectious disease (from living in close quarters with one another and our animals) and malnutrition (from eating too much of the same thing when crops were good, and not enough of anything when they weren't). Anthropologists estimate that typical hunter-gatherers worked at feeding themselves no more than seventeen hours a week, and were far more robust and long-lived than agriculturists, who have only in the last century or two regained the physical stature and longevity of their Paleolithic ancestors.

So even if we wanted to go back to hunting and gathering wild species, it's not an option: There are far too many of us and not nearly enough of them. Fishing is the last economically important hunter-gatherer food chain, though even this foraging economy is rapidly giving way to aquaculture, for the same reasons hunting wild game succumbed to raising livestock. It is depressing though not at all difficult to imagine our grandchildren living in a world in which fishing for a living is history.

For most of us today hunting and gathering and growing our own food is by and large a form of play. That's not to say there aren't still subcultures of people, especially in rural places, who hunt some portion of the protein in their diet, feed themselves out of their gardens, and even earn an income foraging for wild delicacies such as morels or ramps or abalone. But the exorbitant price these wild tastes bring in the marketplace is only proof that very few of us can be serious foragers anymore.

So though a hunter-gatherer food chain still exists here and there to one degree or another, it seems to me its chief value for us at this point is not so much economic or practical as it is didactic. Like other important forms of play, it promises to teach us something about who we are beneath the crust of our civilized, practical, grown-up lives. Foraging for wild plants and animals is, after all, the way the human species has fed itself for 99 percent of its time on earth; this is precisely the food chain natural selection designed us for. Ten thousand years as agriculturists has selected for a small handful of new traits suited to our new existence (a tolerance for lactose in adults is one example), but for the most part we still, somewhat awkwardly, occupy the bodies of foragers and look out at the world through the hunter's eye.

"We don't have to go back to the Pleistocene," wrote Paul Shepard, an environmental philosopher who exalted wildness and deplored modernity, "because our bodies never left." Somehow I doubted I would feel quite that at home stalking game in the woods, but it was reassuring to think that in doing so I would be contesting only my upbringing, not my genes.

My wager in undertaking this experiment is that hunting and gathering (and growing) a meal would perforce teach me things about the ecology and ethics of eating that I could not get in a supermarket or fast-food chain or even on a farm. Some very basic things: about the ties between us and the species (and natural systems) we depend upon;

about how we decide what in nature is good to eat and what is not; and about how the human body fits into the food chain, not only as an eater but as a hunter and, yes, a killer of other creatures. For one of the things I was hoping to accomplish by rejoining, however briefly, this shortest and oldest of food chains was to take some more direct, conscious responsibility for the killing of the animals I eat. Otherwise, I felt, I really shouldn't be eating them. While I'd already slaughtered a handful of chickens in Virginia, the experience had disconcerted me and left the hardest questions untouched. Killing doomed domesticated animals on an assembly line, where you have to keep pace with the expectations of others, is an excellent way to remain only semiconscious about what it is you're really doing. By contrast the hunter, at least as I imagined him, is alone in the woods with his conscience.

And this, I suppose, points to what I was really after in taking up hunting and gathering: to see what it'd be like to prepare and eat a meal in full consciousness of what was involved. I realized that this had been the ultimate destination of the journey I'd been on since traveling to an Iowa cornfield: to look as far into the food chains that support us as I could look, and recover the fundamental biological realities that the complexities of modern industrialized eating keep from our view.

"[T]here is value in any experience that reminds us of our dependency on the soil-plant-animal-man food chain, and of the fundamental organization of the biota," Aldo Leopold wrote in *A Sand County Almanac*. He was talking specifically about hunting, but the same might be said of gardening or hunting for mushrooms. "Civilization has so cluttered this elemental man-earth relationship with gadgets and middlemen that awareness of it is growing dim. We fancy that industry supports us, forgetting what supports industry."

Leopold's injunction was back there somewhere behind my desire to do some hunting and gathering, as no doubt was a line of Henry David Thoreau's that had irritated me when I first came across it years ago. "We cannot but pity the boy who has never fired a gun," he wrote in *Walden*. "He is no more humane, while his education has been sadly neglected." That pitiable, uneducated boy was me. But this boy was determined now to take up Thoreau's and Leopold's challenges: to personally hunt down and drive into a corner that whole charged web of relations with other species we so blandly call "eating," reduce it to its lowest denominator, look at it squarely, and see whatever there was to see.

2. MY FORAGER VIRGIL

The desire was one thing, fulfilling it quite another. A whole host of hard questions now strode into view. How was I going to learn to fire a gun, let alone hunt? Did I need a license? What if I actually managed to kill something—then what? How do you "dress" an animal you've killed? (And what kind of euphemism is that, anyway?) Was it realistic to think I could learn to identify mushrooms with enough confidence to actually eat them?

What I badly needed, I realized, was my own personal foraging Virgil, a fellow not only skilled in the arts of hunting and gathering (and butchering), but also well versed in the flora, fauna, and fungi of Northern California, about which I knew approximately nothing. You see, there was this whole other complication I've neglected to mention: On the eve of this experiment I had just moved to Northern California, a place that is an ecological world away from the New England woods and fields that I knew my way around—a little. I was going to have to learn to hunt and gather and garden on what amounted to a different planet, for it was inhabited by dozens of exotic species about which I possessed not the first useful fact. What did people hunt here, anyway, and when did they hunt it? Which Plant Hardiness Zone is Berkeley in? What time of year do the mushrooms mushroom around here, and where?

As serendipity would have it, a foraging Virgil appeared in my life at exactly the right moment, though it took me a while to recognize him. Angelo Garro is a stout, burly Italian with a five-day beard, sleepy brown eyes, and a passion verging on obsession about the getting and preparing of food. Shortly after we moved to California, I started running into Angelo at dinners to which we'd been invited, though I noticed that only rarely did he play the typical, more or less passive, role of guest. No, Angelo was always intimately involved in the story of the meal. He'd gotten the halibut from a friend at the dock in Bolinas that morning, picked the fennel along the highway on the drive over, made the wine on the table, and brined the olives and personally cured the prosciutto being served. Inevitably he wound up in the kitchen cooking the dinner or passing platters of his famous fennel cakes to whet our appetites while he explained the proper way to make farro pasta or boar salami or balsamic vinegar, this last assuming you had ten or twelve years and the right kind of barrels. The guy was a one-man traveling food network, a poster boy for the Slow Food movement.

Eventually I pieced together Angelo's story. He is a fifty-eight-year-old Sicilian, from the town of Provencia, who left home at eighteen, following a girl to Canada; twenty years later he followed a different girl to San Francisco, where he has lived ever since. He makes his living designing and forging architectural wrought iron; he lives in a forge that has been a blacksmith shop since the time of the gold rush. But Angelo will tell you his consuming passion is food and, specifically, recapturing the flavors and the foodways of his childhood, which he sometimes gives the impression was prematurely interrupted. A particularly successful dish, he will say, is one that "tastes like my mother."

"When I moved away I would call for recipes and for memory of smell and taste, and now I'm trying to replicate what I left behind."

Several months after meeting Angelo he appeared again, this time, strangely enough, on my car radio. He was being interviewed on public radio for a segment on foraging produced by the Kitchen Sisters. Their microphones followed Angelo on a porcini hunt and then into a duck blind at dawn. While he waited for the sun and the ducks to rise, Angelo spoke in an accented whisper about his past and his passions. "In Sicily I could tell by the smell what time of the year it was," he said. "Orange season, oranges, persimmons, olives, and olive oil."

Angelo spends many of his days in California re-creating the calendar of life in Sicily, a calendar that is strictly organized around seasonal foods. "You know, food in Sicily doesn't come from the Safeway," he will say. "It comes from the garden, it comes from nature." So there are eels to catch for the traditional seven-fish dinner on Christmas Eve ("You almost can't have Christmas without eel"); chanterelles to hunt in January; wild fennel to gather in April; olives to pick and cure in August; grapes to harvest and crush in September; game to hunt and cure in October; and porcini to hunt after the first rains in November. Each of these rites is performed in the company of friends—and is accompanied by a good meal, homemade wine, and conversation.

"I have the passions of foraging, passion of hunting, opera, my work," he told the Kitchen Sisters. "I have the passion of cooking, pickling, curing salamis, sausage, making wine in the fall. This is my life. I do this with my friends. It's to my heart."

Even before the radio segment ended I knew I had found my Virgil. The next time I bumped into Angelo I asked him if I could tag along on his next foraging expedition. "Sure, okay, we go hunt chanterelle in Sonoma. I call you when it's time." Emboldened, I asked about going hunting too. "Okay, we could hunt one day, maybe some duck, maybe the pig, but first you need license and learn to shoot."

The pig? Clearly there was even more to learn than I had thought.

3. HUNTER ED.

It took me a couple of months to sort out the procedures for securing a hunter's license, which involved enrolling in a hunter education course and taking a test. It seems they'll sell a high-powered rifle to just about anybody in California, but it's against the law to aim the thing at an animal without first enduring a fourteen-hour class and a one-hundred-question multiple-choice exam that demands some study. The next scheduled session was on a Saturday two months off.

Yet now that I knew I would be going hunting eventually, for game as well as for mushrooms, something peculiar happened. I became an incipient forager, a forager-in-waiting. The mere expectation of hunting and gathering abruptly changed what it meant—and what it felt like— to take a walk in the woods. All at once I started looking at, and think- ing about, everything in the landscape in terms of its potential as a source of food. "Nature," as the Woody Allen character says in *Love and Death*, "is like an enormous restaurant."

It was almost as if I had donned a new pair of glasses that divided the natural world into the possibly good to eat and the probably not. Though in most cases of course I had no idea which was which; being so new to this, and to this place, my forager vision was far from perfect. Still, I began to notice things. I noticed the soft yellow globes of chamomile edging the path I hiked most afternoons, and spotted clumps of miner's lettuce off in the shade (*Claytonia*, a succulent coin- shaped green I had once grown in my Connecticut garden) and wild mustard out in the sun. (Angelo called it *rapini*, and said the young leaves were delicious sautéed in olive oil and garlic.) There were black- berries in flower and the occasional edible bird: a few quail, a pair of doves. Okay, this might not have been the most exalted way of experi- encing nature, but it did sharpen my eye and engage my attention in a way it hadn't been engaged in years. I began consulting field guides to help me identify the many unfamiliar species I'd been content to treat as leafy, fungal, and feathery background noise.

Hiking in the Berkeley Hills one afternoon in January I noticed a narrow shady path dropping off the main trail into the woods, and I followed it down into a grove of big oaks and bay laurel trees. I'd read that chanterelles came up this time of year around old live oak trees, so I kept an eye out. The only place I'd seen a chanterelle before was over pasta or in the market, but I knew I was looking for a yellowish-orange and thickly built trumpet. I scanned the leaf litter around a couple of oaks but saw nothing. Just when I'd given up and turned to head back, however, I noticed a bright, yolky glimmer of something pushing up the carpet of leaves not two feet from where I'd just stepped. I brushed away the leaves and there it was, this big, fleshy, vase-shaped mushroom that I was dead certain had to be a chanterelle.

Or was it?

How certain was that?

I took the mushroom home, brushed off the soil, and put it on a plate, then pulled out my field guides to see if I could confirm the identification. Everything matched up: the color, the faint apricot smell, the asymmetrical trumpet shape on top, the underside etched in a shallow pattern of "false" gills. I felt fairly confident. But confident enough to eat it? Not quite. The field guide mentioned something called a "false chanterelle" that had slightly "thinner" gills. Uh oh. Thinner, thicker: These were relative terms; how could I tell if the gills I was looking at were thin or thick ones? Compared to what? My mother's mycophobic warnings rang in my ears. I couldn't trust my eyes. I couldn't quite trust the field guide. So whom could I trust? Angelo! But that meant driving my lone mushroom across the bridge to San Francisco, which seemed excessive. My desire to sauté and eat my first-found chanterelle squabbled with my doubts about it, slender as they were. But by now I had passed the point of being able to enjoy this putative chanterelle without anxiety, so I threw it out.

I didn't realize it at the time, but I had impaled myself that afternoon on the horns of the omnivore's dilemma.

SIXTEEN

THE OMNIVORE'S DILEMMA

1. GOOD TO EAT, GOOD TO THINK

My encounter with the chanterelle—or was it a false chanterelle?—put me in touch with one of the most elemental facts about human eating: It can be dangerous, and even when it isn't dangerous, it is fraught. The blessing of the omnivore is that he can eat a great many different things in nature. The curse of the omnivore is that when it comes to figuring out which of those things are safe to eat, he's pretty much on his own.

As noted at the beginning of this book, the omnivore's dilemma, or paradox, was first described in the 1976 paper, "The Selection of Foods by Rats, Humans, and Other Animals," by University of Pennsylvania psychologist Paul Rozin. Rozin studied food selection behavior in rats, which are omnivores, in the hopes of understanding something about food selection in people. Like us, rats daily confront the bounty of nature and its manifold perils—perils designed to protect plants, animals, and microbes from being eaten. To defend themselves from predation, plants and fungi produce a great many poisons, everything from cyanide and oxalic acid to a wide variety of toxic alkaloids and gluco-sides; similarly, bacteria colonizing dead plants and animals produce toxins to keep other potential eaters at bay. (Also similarly, we humans manufacture toxins to keep rats from eating our food.)

Among the more specialized eaters, natural selection takes care of the whole problem of food selection, hardwiring the monarch butter-fly, say, to regard the milkweed as food and everything else in nature as not food. No thought or emotion need go into deciding whether to eat any given thing. This approach works for the monarch because its di-gestion can wring everything it needs for its survival from milkweed leaves (including a toxin that makes the butterfly itself unappetizing to birds). But rats and humans require a wider range of nutrients and so must eat a wider range of foods, some of them questionable. Whenever they encounter a potential new food they find themselves torn between two conflicting emotions unknown to the specialist eater, each with its own biological rationale: neophobia, a sensible fear of ingesting any-thing new, and neophilia, a risky but necessary openness to new tastes.

Rozin found that the rat minimizes the risk of the new by treating its digestive tract as a kind of laboratory. It nibbles a very little bit of the new food (assuming it is food) and then waits to see what happens. The animal evidently has a good enough grasp of causality ("delayed learn-ing," as the social scientists call it) to link a stomachache in the present to something it ingested a half hour before, and a good enough mem-ory to store that finding as a lifelong aversion to that particular sub-stance. (This is what makes poisoning rats so difficult.) I might have used the same strategy to test my chanterelle, eating a tiny bite of it and waiting to see what happened.

Rozin's early work on food selection behavior postulated that the "omnivoral problem" would explain a great deal, not only about how and what we eat, but who we are as a species, and subsequent research by him and others, in anthropology as well as psychology, has done much to confirm his hunch. The concept of the omnivore's dilemma helps unlock not only simple food-selection behaviors in animals, but much more complex "biocultural" adaptations in primates (humans included) as well as a wide range of otherwise baffling cultural practices in humans, the species for whom, as Claude Lévi-Strauss famously said, food must be "not only good to eat, but also good to think."

The omnivore's dilemma is replayed every time we decide whether or not to ingest a wild mushroom, but it also figures in our less primordial encounters with the putatively edible: when we're deliberating the nutritional claims on the boxes in the cereal aisle; when we're settling on a weight-loss regimen (low fat or low carb?); or deciding whether to sample McDonald's' newly reformulated chicken nugget; or weighing the costs and benefits of buying the organic strawberries over the conventional ones; or choosing to observe (or flout) kosher or halal rules; or determining whether or not it is ethically defensible to eat meat—that is, whether meat, or any other of these things, is not only good to eat, but good to think as well.

2. HOMO OMNIVOROUS

The fact that we humans are indeed omnivorous is deeply inscribed in our bodies, which natural selection has equipped to handle a remarkably wide-ranging diet. Our teeth are omnicompetent—designed for tearing animal flesh as well as grinding plants. So are our jaws, which we can move in the manner of a carnivore, a rodent, or an herbivore, depending on the dish. Our stomachs produce an enzyme specifically designed to break down elastin, a type of protein found in meat and nowhere else. Our metabolism requires specific chemical compounds that, in nature, can be gotten only from plants (like vitamin C) and others that can be gotten only from animals (like vitamin B-12). More than just the spice of human life, variety for us appears to be a biological necessity.

By comparison, nature's specialists can get everything they need from a small number of foods and, very often, a highly specialized digestive system, freeing them from the need to devote a lot of brainpower to the challenges of omnivorousness. The ruminant, for example, specializes in eating grass, even though the grasses by themselves don't supply all the nutrients the animal needs. What they do supply is food for the microbes living in the animal's rumen, which in turn supply the other nutrients the animal needs to survive. The ruminant's genius for keeping itself well fed resides in its gut rather than its brain.

There does seem to be an evolutionary trade-off between big brains and big guts—two very different evolutionary strategies for dealing with the question of food selection. The case of the koala, one of nature's pickiest eaters, exemplifies the small-brain strategy. You don't need a lot of brain circuitry to figure out what's for dinner when all you ever eat is eucalyptus leaves. As it happens, the koala's brain is so small it doesn't even begin to fill up its skull. Zoologists theorize that the koala once ate a more varied and mentally taxing diet than it does now, and that as it evolved toward its present, highly circumscribed concept of lunch, its underemployed brain actually shrank. (Food faddists take note.) More important to the koala than brains is a gut big enough to break down all those fibrous leaves. By the same token, the digestive tract of primates like us has grown progressively shorter as we've evolved to eat a more varied, higher quality diet.

Eating might be simpler as a thimble-brained monophage, but it's also a lot more precarious, which partly explains why there are so many more rats and humans in the world than koalas. Should a disease or drought strike the eucalyptus trees in your neck of the woods, that's it for you. But the rat and the human can live just about anywhere on earth, and when their familiar foods are in short supply, there's always another they can try. Indeed, there is probably not a nutrient source on earth that is not eaten by some human somewhere—bugs, worms, dirt, fungi, lichens, seaweed, rotten fish; the roots, shoots, stems, bark, buds, flowers, seeds, and fruits of plants; every imaginable part of every imaginable animal, not to mention haggis, granola, and Chicken McNuggets. (The deeper mystery, only partly explained by neophobia, is why any given human group will eat so few of the numberless nutrients available to it.)

The price of this dietary flexibility is much more complex and metabolically expensive brain circuitry. For the omnivore a tremendous amount of mental wiring must be devoted to sensory and cognitive tools for figuring out which of all these questionable nutrients it is safe to eat. There's just too much information involved in food selection to encode every potential food and poison in the genes. So instead of genes to write our menus omnivores evolved a complicated set of sensory and mental tools to help us sort everything out. Some of these tools are fairly straightforward and we share them with many other mammals; others represent impressive feats of adaptation by primates; still others straddle the blurry line between natural selection and cultural invention.

The first tool is of course our sense of taste, which performs some of the basic work screening foods for value and safety. Or as Brillat-Savarin put it in *The Physiology of Taste*, taste "helps us to choose, from the various substances offered us by nature, those which are proper to be consumed." Taste in humans gets complicated, but it starts with two powerful instinctual biases, one positive, the other negative. The first bias predisposes us toward sweetness, a taste that signals a particularly rich source of carbohydrate energy in nature. Indeed, even when we're otherwise sated, our appetite for sweet things persists, which is probably why dessert shows up in the meal when it does. A sweet tooth represents an excellent adaptation for an omnivore whose big brain demands a tremendous amount of glucose (the only type of energy the brain can use), or at least it once did, when sources of sugar were few and far between. (The adult human brain accounts for 2 percent of our body weight but consumes 18 percent of our energy, all of which must come from a carbohydrate. Food faddists take note two.)

Our sense of taste's second big bias predisposes us against bitter flavors, which is how many of the defensive toxins produced by plants happen to taste. Pregnant women are particularly sensitive to bitter tastes, probably an adaptation to protect the developing fetus against even the mild plant toxins found in foods like broccoli. A bitter flavor on the tongue is a warning to exercise caution lest a poison pass what Brillat-Savarin called the sense of taste's "faithful sentries."

Disgust turns out to be another valuable tool for negotiating the omnivore's dilemma. Though the emotion has long since attached itself to a great many objects having nothing to do with food, food is where and why it began, as the etymology of the word indicates. (It comes from the Middle French verb *desgouster*, to taste.) Rozin, who has written or coauthored several fascinating articles about disgust, defines it as the fear of incorporating offending substances into one's body. Much of what people deem disgusting is culturally determined, but there are certain things that apparently disgust us all, and all these substances, Rozin notes, come from animals: bodily fluids and secretions, corpses, decaying flesh, feces. (Curiously, the one bodily fluid of other people that doesn't disgust us is the one produced by the human alone: tears. Consider the sole type of used tissue you'd be willing to share.) Disgust is an extremely useful adaptation, since it prevents omnivores from ingesting hazardous bits of animal matter: rotten meat that might carry bacterial toxins or infected bodily fluids. In the words of Harvard psychologist Steven Pinker, "Disgust is intuitive microbiology."

Yet helpful as it is, our sense of taste is not a completely adequate guide to what we can and cannot eat. In the case of plants, for instance, it turns out that some of the bitterest ones contain valuable nutrients, even useful medicines. Long before the domestication of plants (a process in which we generally selected for nonbitterness), early humans developed various other tools to unlock the usefulness of these foods, either by overcoming their defenses or overcoming our own aversion to how they taste.

That's precisely what people must have done in the case of the sap in the opium poppy or the bark of the willow, both of which taste extremely bitter—and both of which contain powerful medicines. Once humans discovered the curative properties of salicylic acid in willows (the active ingredient in aspirin) and the relief from pain offered by the poppy's opiates, our instinctive aversion to these plants' bitterness gave way to an even more convincing cultural belief that the plants were worth ingesting even so; basically, our powers of recognition, memory, and communication overcame the plants' defenses.

Humans also learned to overcome plant defenses by cooking or otherwise processing foods to remove their bitter toxins. Native Americans, for example, figured out that if they ground, soaked, and roasted acorns they could unlock the rich source of nutrients in the bitter nuts. Humans also discovered that the roots of the cassava, which effectively defends itself against most eaters by producing cyanide, could be made edible by cooking. By learning to cook cassava humans unlocked a fabulously rich source of carbohydrate energy, one that, just as important, they had all to themselves, since locusts, pigs, porcupines, and all the other potential cassava eaters haven't yet figured out how to overcome the plant's defense.

Cooking, one of the omnivore's cleverest tools, opened up whole new vistas of edibility. Indeed, in doing so it probably made us who we are. By making these foods more digestible, cooking plants and animal flesh vastly increased the amount of energy available to early humans, and some anthropologists believe this boon accounts for the dramatic increase in the size of the hominid brain about 1.9 million years ago. (Around the same time our ancestors' teeth, jaws, and gut slimmed down to their present proportions, since they were no longer needed to process large quantities of raw food.) By improving digestibility cooking also cut down on the time we had to spend foraging for plants and simply chewing raw meat, freeing that time and energy for other pursuits.

Last but not least, cooking abruptly changed the terms of the evolutionary arms race between omnivores and the species they would eat by allowing us to overcome their defenses. Apart from fruits, which have a declared interest in becoming another species' lunch (this being their strategy for spreading their seeds), and grasses, which welcome grazing as a strategy to keep their habitat free of shady competitors, most wild foods are parts of plants or animals that have no interest in being eaten; they evolved defenses to keep themselves whole. But evolution doesn't stand still, and eaters are constantly evolving counteradaptations to overcome the defenses of nutrient sources: a new digestive enzyme to detoxify a plant or fungal poison, say, or a new perceptual skill to overcome an edible creature's camouflage. In response, the plants, animals, and fungi evolved new defenses to make themselves either more difficult to catch or to digest. This arms race between the eaters and the potentially eaten unfolded at a stately pace until early humans came on the scene. For a countermeasure such as cooking bitter plants completely changed the rules of the game. All at once a species' painstakingly developed defense against being eaten had been breached and, assuming it could erect a new defense, that was going to take time—evolutionary time.

Cooking is often cited (along with tool making and a handful of other protohuman tricks) as evidence that the human omnivore entered a new kind of ecological niche in nature, one that some anthropologists have labeled "the cognitive niche." The term seems calculated to smudge the line between biology and culture, which is precisely the point. To these anthropologists the various tools humans have developed to overcome the defenses of other species—not only food-processing techniques but a whole gamut of hunting and gathering tools and talents—represent biocultural adaptations, so-called because they constitute evolutionary developments rather than cultural inventions that somehow stand apart from natural selection.

In this sense learning to cook cassava roots or disseminate the hard-won knowledge of safe mushrooms is not all that different from recruiting rumenal bacteria to nourish oneself. The cow depends on the ingenious adaptation of the rumen to turn an exclusive diet of grasses into a balanced meal; we depend instead on the prodigious powers of recognition, memory, and communication that allow us to cook cassava or identify an edible mushroom and share that precious information. The same process of natural selection came up with both strategies; one just happens to rely on cognition, the other goes with the gut.

3. THE ANXIETY OF EATING

Being an omnivore occupying a cognitive niche in nature is both a boon and a challenge, a source of tremendous power as well as anxiety. Omnivory is what allowed humans to adapt to a great many environments all over the planet, and to survive in them even after our favored foods were driven to extinction, whether by accident or because of our own too-great success in overcoming other species' defenses. After the mastodon there would be the bison and then the cow; after the sturgeon, the salmon, and then, perhaps, some novel mycoprotein like "quorn."

Being a generalist offers us deep satisfactions, too, enjoyments that flow equally from the omnivore's innate neophilia—the pleasure of variety—and neophobia—the comfort of the familiar. What began as a set of simple sensory responses to food (sweet, bitter, disgusting) we've elaborated into more complicated canons of taste that afford us aesthetic pleasures undreamed of by the koala or cow. Since "everything that is edible is at the mercy of his vast appetite," Brillat-Savarin writes, "the machinery of taste attains a rare perfection in man," making "man the only gourmand in the whole of nature." Taste in this more cultivated sense brings people together, not only in small groups at the table but as communities. For a community's food preferences—the strikingly short list of foods and preparations it regards as good to eat and think—represent one of the strongest social glues we have. Historically, national cuisines have been remarkably stable, and resistant to change, which is why the immigrant's refrigerator is the very last place to look for signs of assimilation.

Yet the surfeit of choice that confronts the omnivore brings stresses and anxieties also undreamed of by the cow or the koala, for whom the distinction between The Good Things to Eat and the Bad is second nature. And while our senses can help us draw the first rough distinctions between good and bad foods, we humans have to rely on culture to remember and keep it all straight. So we codify the rules of wise eating in an elaborate structure of taboos, rituals, manners, and culinary traditions, covering everything from the proper size of portions to the order in which foods should be consumed to the kinds of animals it is and is not okay to eat. Anthropologists argue over whether all these rules make biological sense—some, like the kosher rules, are probably designed more to enforce group identity than to protect health. But certainly a great many of our food rules do make biological sense, and they keep each of us from having to confront the omnivore's dilemma every time we visit the supermarket or sit down to eat.

That set of rules for preparing food we call a cuisine, for example, specifies combinations of foods and flavors that on examination do a great deal to mediate the omnivore's dilemma. The dangers of eating raw fish, for example, are minimized by consuming it with wasabi, a potent antimicrobial. Similarly, the strong spices characteristic of many cuisines in the tropics, where food is quick to spoil, have antibacterial properties. The meso-American practice of cooking corn with lime and serving it with beans, like the Asian practice of fermenting soy and serving it with rice, turn out to render these plant species much more nutritious than they otherwise would be. When not fermented, soy contains an antitrypsin factor that blocks the absorption of protein, rendering the bean indigestible; unless corn is cooked with an alkali like lime its niacin is unavailable, leading to the nutritional deficiency called pellagra. Corn and beans each lack an essential amino acid (lysine and methionine, respectively); eat them together and the proper balance is restored. Similarly, a dish that combines fermented soy with rice is nutritionally balanced. As Rozin writes, "[C]uisines embody some of a culture's accumulated wisdom about food." Often when one culture imports another's food species without importing the associated cuisine, and its embodied wisdom, they make themselves sick.

Rozin suggests that cuisines also help negotiate the tension between the omnivore's neophilia and neophobia. By preparing a novel kind of food using a familiar complex of flavors—by cooking it with traditional spices, say, or sauces—the new is rendered familiar, "reducing the tension of ingestion."

ANTHROPOLOGISTS MARVEL at just how much cultural energy goes into managing the food problem. But as students of human nature have long suspected, the food problem is closely tied to . . . well, to several other big existential problems. Leon Kass, the ethicist, wrote a fascinating book called *The Hungry Soul: Eating and the Perfection of Our Nature* in which he teases out the many philosophical implications of human eating. In a chapter on omnivorousness Kass quotes at length from Jean-Jacques Rousseau, who in his Second Discourse on man draws a connection between our freedom from instinct in eating and the larger problem of free will. Rousseau is after somewhat bigger game in this passage, but along the way he offers as good a statement of the omnivore's dilemma as you're likely to find:

> . . . nature does everything in the operations of a beast, whereas man contributes to his operations by being a free agent. The former chooses or rejects by instinct and the latter by an act of freedom, so that a beast cannot deviate from the rule that is prescribed to it even when it would be advantageous to do so, and a man deviates from it often to his detriment. Thus a pigeon would die of hunger near a basin filled with the best meats, and a cat upon heaps of fruits or grain, although each could very well nourish itself on the food it disdains if it made up its mind to try some. Thus dissolute men abandon themselves to the excesses which cause them fever and death, because the mind depraves the senses and because the will still speaks when nature is silent.

Guided by no natural instinct, the prodigious and open-ended human appetite is liable to get us into all sorts of trouble, well beyond the stomachache. For if nature is silent what's to stop the human omnivore from eating anything—including, most alarmingly, other human omnivores? A potential for savagery lurks in a creature capable of eating anything. If nature won't draw a line around human appetite, then human culture must step in, as indeed it has done, bringing the omnivore's eating habits under the government of all the various taboos (foremost the one against cannibalism), customs, rituals, table manners, and culinary conventions found in every culture. There is a short and direct path from the omnivore's dilemma to the astounding number of ethical rules with which people have sought to regulate eating for as long as they have been living in groups.

"Without virtue" to govern his appetites, Aristotle wrote, man of all the animals "is most unholy and savage, and worst in regard to sex and eating." Paul Rozin has suggested, only partly in jest, that Freud would have done well to build his psychology around our appetite for food rather than our appetite for sex. Both are fundamental biological drives necessary to our survival as a species, and both must be carefully channeled and socialized for the good of society. ("You can't just grab any tasty-looking morsel," he points out.) But food is more important than sex, Rozin contends. Sex we can live without (at least as individuals), and it occurs with far less frequency than eating. Since we also do rather more of our eating in public there has been "a more elaborate cultural transformation of our relationship to food than there is to sex."

4. AMERICA'S NATIONAL EATING DISORDER

Rozin doesn't say as much, but all the customs and rules culture has devised to mediate the clash of human appetite and society probably bring greater comfort to us as eaters than as sexual beings. Freud and others lay the blame for many of our sexual neuroses at the door of an overly repressive culture, but that doesn't appear to be the principal culprit in our neurotic eating. To the contrary, it seems as though our eating tends to grow *more* tortured as our culture's power to manage our relationship to food weakens.

This seems to me precisely the predicament we find ourselves in today as eaters, particularly in America. America has never had a stable national cuisine; each immigrant population has brought its own food-ways to the American table, but none has ever been powerful enough to hold the national diet very steady. We seem bent on reinventing the American way of eating every generation, in great paroxysms of neophilia and neophobia. That might explain why Americans have been such easy marks for food fads and diets of every description.

This is the country, after all, where at the turn of the last century Dr. John Harvey Kellogg persuaded great numbers of the country's most affluent and best educated to pay good money to sign themselves into his legendarily nutty sanitarium at Battle Creek, Michigan, where they submitted to a regime that included all-grape diets and almost hourly enemas. Around the same time millions of Americans succumbed to the vogue for "Fletcherizing"—chewing each bite of food as many as one hundred times—introduced by Horace Fletcher, also known as the Great Masticator.

This period marked the first golden age of American food faddism, though of course its exponents spoke not in terms of fashion but of "scientific eating," much as we do now. Back then the best nutritional science maintained that carnivory promoted the growth of toxic bacteria in the colon; to battle these evildoers Kellogg vilified meat and mounted a two-fronted assault on his patients' alimentary canals, introducing quantities of Bulgarian yogurt at both ends. It's easy to make fun of people who would succumb to such fads, but it's not at all clear that we're any less gullible. It remains to be seen whether the current Atkins school theory of ketosis—the process by which the body resorts to burning its own fat when starved of carbohydrates—will someday seem as quaintly quackish as Kellogg's theory of colonic autointoxication.

What is striking is just how little it takes to set off one of these applecart-toppling nutritional swings in America; a scientific study, a new government guideline, a lone crackpot with a medical degree can alter this nation's diet overnight. One article in the *New York Times Magazine* in 2002 almost single-handedly set off the recent spasm of carbophobia in America. But the basic pattern was fixed decades earlier, and suggests just how vulnerable the lack of stable culinary traditions leaves us to the omnivore's anxiety, and the companies and quacks who would prey on it. So every few decades some new scientific research comes along to challenge the prevailing nutritional orthodoxy; some nutrient that Americans have been happily chomping for decades is suddenly found to be lethal; another nutrient is elevated to the status of health food; the industry throws its weight behind it; and the American way of dietary life undergoes yet another revolution.

Harvey Levenstein, a Canadian historian who has written two fascinating social histories of American foodways, neatly sums up the beliefs that have guided the American way of eating since the heyday of John Harvey Kellogg: "that taste is not a true guide to what should be eaten; that one should not simply eat what one enjoys; that the important components of food cannot be seen or tasted, but are discernible only in scientific laboratories; and that experimental science has produced rules of nutrition that will prevent illness and encourage longevity." The power of any orthodoxy resides in its ability not to seem like one and, at least to a 1906 or 2006 genus American, these beliefs don't seem in the least bit strange or controversial.

It's easy, especially for Americans, to forget just how novel this nutritional orthodoxy is, or that there are still cultures that have been eating more or less the same way for generations, relying on such archaic criteria as taste and tradition to guide them in their food selection. We Americans are amazed to learn that some of the cultures that set their culinary course by the lights of habit and pleasure rather than nutritional science and marketing are actually healthier than we are—that is, suffer a lower incidence of diet-related health troubles.

The French paradox is the most famous such case, though as Paul Rozin points out, the French don't regard the matter as paradoxical at all. We Americans resort to that term because the French experience— a population of wine-swilling cheese eaters with lower rates of heart disease and obesity—confounds our orthodoxy about food. That orthodoxy regards certain tasty foods as poisons (carbs now, fats then), failing to appreciate that how we eat, and even how we feel about eating, may in the end be just as important as what we eat. The French eat all sorts of supposedly unhealthy foods, but they do it according to a strict and stable set of rules: They eat small portions and don't go back for seconds; they don't snack; they seldom eat alone; and communal meals are long, leisurely affairs. In other words, the French culture of food successfully negotiates the omnivore's dilemma, allowing the French to enjoy their meals without ruining their health.

Perhaps because we have no such culture of food in America almost every question about eating is up for grabs. Fats or carbs? Three squares or continuous grazing? Raw or cooked? Organic or industrial? Veg or vegan? Meat or mock meat? Foods of astounding novelty fill the shelves of our supermarket, and the line between a food and a "nutritional supplement" has fogged to the point where people make meals of protein bars and shakes. Consuming these neo-pseudo-foods alone in our cars, we have become a nation of antinomian eaters, each of us struggling to work out our dietary salvation on our own. Is it any wonder Americans suffer from so many eating disorders? In the absence of any lasting consensus about what and how and where and when to eat, the omnivore's dilemma has returned to America with an almost atavistic force.

This situation suits the food industry just fine, of course. The more anxious we are about eating, the more vulnerable we are to the seductions of the marketer and the expert's advice. Food marketing in particular thrives on dietary instability and so tends to exacerbate it. Since it's difficult to sell more food to such a well-fed population (though not, as we're discovering, impossible), food companies put their efforts into grabbing market share by introducing new kinds of highly processed foods, which have the virtue of being both highly profitable and infinitely adaptable. Sold under the banner of "convenience," these processed foods are frequently designed to create whole new eating occasions, such as in the bus on the way to school (the protein bar or Pop-Tart) or in the car on the way to work (Campbell's recently introduced a one-handed microwaveable microchunked soup in a container designed to fit a car's cup holder).

The success of food marketers in exploiting shifting eating patterns and nutritional fashions has a steep cost. Getting us to change how we eat over and over again tends to undermine the various social structures that surround and steady our eating, institutions like the family dinner, for example, or taboos on snacking between meals and eating alone. In their relentless pursuit of new markets, food companies (with some crucial help from the microwave oven, which made "cooking" something even small children could do) have broken Mom's hold over the American menu by marketing to every conceivable demographic—and especially to children.

A vice president of marketing at General Mills once painted for me a picture of the state of the American family dinner, courtesy of video cameras that the company's consulting anthropologists paid families to let them install in the ceiling above the kitchen and dining room tables. Mom, perhaps feeling sentimental about the dinners of her childhood, still prepares a dish and a salad that she usually winds up eating by herself. Meanwhile, the kids, and Dad, too, if he's around, each fix something different for themselves, because Dad's on a low-carb diet, the teenager's become a vegetarian, and the eight-year-old is on a strict ration of pizza that the shrink says it's best to indulge (lest she develop eating disorders later on in life). So over the course of a half hour or so each family member roams into the kitchen, removes a single-portion entrée from the freezer, and zaps it in the microwave. (Many of these entrées have been helpfully designed to be safely "cooked" by an eight-year-old.) After the sound of the beep each diner brings his microwaveable dish to the dining room table, where he or she may or may not cross paths with another family member at the table for a few minutes. Families who eat this way are among the 47 percent of Americans who report to pollsters that they still sit down to a family meal every night.

Several years ago, in a book called The Cultural Contradictions of Capitalism, sociologist Daniel Bell called attention to the tendency of capitalism, in its single-minded pursuit of profit, to erode the various cultural underpinnings that steady a society but often impede the march of commercialization. The family dinner, and more generally a cultural consensus on the subject of eating, appears to be the latest such casualty of capitalism. These rules and rituals stood in the way of the food industry's need to sell a well-fed population more food through ingenious new ways of processing, packaging, and marketing it. Whether a stronger set of traditions would have stood up better to this relentless economic imperative is hard to say; today America's fast-food habits are increasingly gaining traction even in places like France.

So we find ourselves as a species almost back where we started: anxious omnivores struggling once again to figure out what it is wise to eat. Instead of relying on the accumulated wisdom of a cuisine, or even on the wisdom of our senses, we rely on expert opinion, advertising, government food pyramids, and diet books, and we place our faith in science to sort out for us what culture once did with rather more success. Such has been the genius of capitalism, to re-create something akin to a state of nature in the modern supermarket or fast-food outlet, throwing us back on a perplexing, nutritionally perilous landscape deeply shadowed again by the omnivore's dilemma.

SEVENTEEN

THE ETHICS OF
EATING ANIMALS

1. THE STEAKHOUSE DIALOGUES

The first time I opened Peter Singer's *Animal Liberation* I was dining alone at the Palm, trying to enjoy a rib-eye steak cooked medium rare. If that sounds like a recipe for cognitive dissonance, if not indigestion, well, that was sort of the idea. It had been a long time since this particular omnivore had felt any dilemma about eating meat, but then I had never before involved myself so directly in the processes of turning animals into food: owning a steak-bound steer, working the killing cones in Joel Salatin's processing shed, and now preparing to hunt a wild animal. The steak dinner in question took place on the evening before steer number 534's slaughter, the one event in his life I was not allowed to witness or even learn anything about, save its likely date. This didn't exactly surprise me: The meat industry understands that the more people know about what happens on the kill floor, the less meat they're likely to eat. That's not because slaughter is necessarily inhumane, but because most of us would simply rather not be reminded of exactly what meat is or what it takes to bring it to our plates. My steak dinner, eaten in the company of the world's leading philosopher of animal rights, represented my somewhat tortured attempt to mark the occasion, and to try—a bit belatedly, I know—to see if I could defend what I had done already and what I was about to do.

Eating meat has become morally problematic, at least for people who take the trouble to think about it. Vegetarianism is more popular than it has ever been, and animal rights, the fringiest of fringe movements until just a few years ago, is rapidly finding its way into the cultural mainstream. I'm not completely sure why this should be happening now, given that humans have been eating animals for tens of thousands of years without too much ethical heartburn. Certainly there have been dissenters over the years—Ovid, St. Francis, Tolstoy, and Gandhi come to mind. But the general consensus has always been that humans were indeed omnivores and, whatever spiritual or moral dilemmas the killing and eating of animals posed, our various cultural traditions (everything from the rituals governing slaughter to saying grace before the meal) resolved them for us well enough. For the most part our culture has been telling us for millennia that animals were both good to eat and good to think.

In recent years medical researchers have raised questions about the good to eat part, while philosophers like Singer and organizations like People for the Ethical Treatment of Animals (PETA) have given us new reasons to doubt meat is good to think—that is, good for our souls or our moral self-regard. Hunting is in particularly bad odor these days, even among people who still eat meat; apparently it's the fact of killing that these people most object to (as if a steak could be gotten any other way), or perhaps it's the taking pleasure in killing an animal that is the trouble. It may be that as a civilization we're groping toward a higher plane of consciousness. It may be that our moral enlightenment has advanced to the point where the practice of eating animals—like our former practices of keeping slaves or treating women as inferior beings—can now be seen for the barbarity it is, a relic of an ignorant past that very soon will fill us with shame.

That at least is the animal philosophers' wager. But it could also be that the cultural norms and rituals that used to allow people to eat meat without agonizing about it have broken down for other reasons. Perhaps as the sway of tradition in our eating decisions weakens, habits we once took for granted are thrown up in the air, where they're more easily buffeted by the force of a strong idea or the breeze of fashion.

Whatever the cause, the effect is an unusual amount of cultural confusion on the subject of animals. For at the same time many of us seem eager to extend the circle of our moral consideration to other species, in our factory farms we're inflicting more suffering on more animals than at any time in history. One by one science is dismantling our claims to uniqueness as a species, discovering that such things as culture, tool making, language, and even possibly self-consciousness are not, as we used to think, the exclusive properties of Homo sapiens. And yet most of the animals we eat lead lives organized very much in the spirit of Descartes, who famously claimed that animals were mere machines, incapable of thought or feeling. There's a schizoid quality to our relationship with animals today in which sentiment and brutality exist side by side. Half the dogs in America will receive Christmas presents this year, yet few of us ever pause to consider the life of the pig—an animal easily as intelligent as a dog—that becomes the Christmas ham.

We tolerate this schizophrenia because the life of the pig has moved out of view; when's the last time you saw a pig in person? Meat comes from the grocery store, where it is cut and packaged to look as little like parts of animals as possible. (When was the last time you saw a butcher at work?) The disappearance of animals from our lives has opened a space in which there's no reality check on the sentiment or the brutality; it is a space in which the Peter Singers and the Frank Perdues of the world fare equally well.

A few years ago the English writer John Berger wrote an essay called "Why Look at Animals?" in which he suggested that the loss of everyday contact between ourselves and animals—and specifically the loss of eye contact—has left us deeply confused about the terms of our relationship to other species. That eye contact, always slightly uncanny, had brought the vivid daily reminder that animals were both crucially like and unlike us; in their eyes we glimpsed something unmistakably familiar (pain, fear, courage) but also something irretrievably other (?!). Upon this paradox people built a relationship in which they felt they could both honor and eat animals without looking away. But that accommodation has pretty much broken down; nowadays it seems we either look away or become vegetarians. For my own part, neither option seemed especially appetizing; certainly looking away was now completely off the table. Which might explain how it was I found myself attempting to read Peter Singer in a steakhouse.

THIS IS NOT something I'd recommend if you're determined to continue eating meat. *Animal Liberation*, comprised of equal parts philosophical argument and journalistic description, is one of those rare books that demands you either defend the way you live or change it. Because Singer is so skilled in argument, for many readers it is easier to change. *Animal Liberation* has converted countless thousands to vegetarianism, and it didn't take me long to see why: within a few pages he had succeeded in throwing me and my meat eating, not to mention my hunting plans, on the defensive.

Singer's argument is disarmingly simple and, provided you accept its premises, difficult to refute. Take the premise of equality among people, which most of us readily accept. Yet what do we really mean by it? After all, people are not, as a matter of fact, equal at all—some are smarter than others, handsomer, more gifted, whatever. "Equality is a moral idea," Singer points out, "not an assertion of fact." The moral idea is that everyone's interests ought to receive equal consideration, regardless of "what they are like or what abilities they have." Fair enough; many philosophers have gone this far. But few have then taken the next logical step. "If possessing a higher degree of intelligence does not entitle one human to use another for his or her own ends, how can it entitle humans to exploit non-humans for the same purpose?"

This is the nub of Singer's argument, and right away, here on page six, I began scribbling objections in the margin. *But humans differ from animals in morally significant ways.* Yes they do, Singer readily acknowledges, which is why we shouldn't treat pigs and children alike. Equal consideration of interests is not the same as equal treatment, he points out; children have an interest in being educated, pigs in rooting around in the dirt. But where their interests are the same, the principle of equality demands they receive the same consideration. And the one all-important interest humans share with pigs, as with all sentient creatures, is an interest in avoiding pain.

Here Singer quotes a famous passage from Jeremy Bentham, the eighteenth-century utilitarian philosopher. Bentham is writing in 1789, after the French had freed their black slaves and granted them fundamental rights, but before the British or Americans had acted. "The day may come," Bentham wrote, "when the rest of the animal creation may acquire those rights." Bentham then asks what characteristics entitle any being to moral consideration. "Is it the faculty of reason, or perhaps the faculty of discourse?" Bentham asks. "But a full-grown horse or dog is beyond comparison a more rational, as well as a more conversational animal, than an infant."

"The question is not Can they reason? Or Can they *talk*? But Can they *suffer*?"

Bentham here is playing a powerful card philosophers call the "argument from marginal cases," or AMC for short. It goes like this: There are humans—infants, the severely retarded, the demented—whose mental function does not rise to the level of a chimpanzee. Even though these people cannot reciprocate our moral attentions (obey the golden rule, etc.) we nevertheless include them in the circle of our moral consideration. So on what basis do we exclude the chimpanzee?

Because he's a chimp, I furiously scribble in the margin, *and they're human beings!* For Singer that's not good enough. To exclude the chimp from moral consideration simply because he's not human is no different than excluding the slave simply because he's not white. In the same way we'd call that exclusion "racist" the animal rightist contends it is "speciesist" to discriminate against the chimpanzee solely because he's not human. But *the differences between blacks and whites are trivial compared to the differences between my son and the chimp.* Singer asks us to imagine a hypothetical society that discriminates on the basis of something nontrivial—intelligence, say. If that scheme offends our sense of equality, as it surely does, then why is the fact that animals lack this or that human characteristic any more just as a basis for discrimination? Either we do not owe any justice to the severely retarded, he concludes, or we do owe it to animals with higher capabilities.

This is where I put down my fork. If I believe in equality, and equality is based on interests rather than characteristics, then either I have to take the steer's interest into account or accept that I'm a speciesist.

For the time being, I decided, I'll plead guilty as charged. I finished my steak.

But Singer had planted a troubling notion, and in the days afterward it grew and grew, watered by the other animal rights thinkers I began reading: the philosophers Tom Regan and James Rachels, the legal theorist Steven M. Wise, writers like Joy Williams and Matthew Scully. I didn't think I minded being called a speciesist, but could it be, as these writers suggest, we will someday come to regard speciesism as an evil comparable to that of racism? Is it possible that history will someday judge us as harshly as it judges the Germans who went about their lives in the shadow of Treblinka? The South African novelist J. M. Coetzee posed precisely that question in a lecture at Princeton not long ago; he answered it in the affirmative. If the animal rightists are right, then "a crime of stupendous proportions" (in Coetzee's words) is going on all around us every day, just beneath our notice.

THE IDEA is almost impossible to seriously entertain, much less to accept, and in the months after the restaurant face-off between Singer and my steak at the Palm I found myself marshalling whatever mental power I could command to try to refute it. Yet one by one Singer and his colleagues managed to trump nearly every objection I could muster.

The meat eaters' first line of defense is obvious: *Why should we treat an-*

imals any more ethically than they treat one another? Ben Franklin actually tried this tack long before me. He tells in his autobiography of one day watching friends catch fish and wondering, "If you eat one another, I don't see why we may not eat you." He admits, however, that this rationale didn't occur to him until the fish were in the frying pan, beginning to smell "admirably well." The great advantage of being a "reasonable creature," Franklin remarks, is that you can find a reason for whatever you want to do.

To the "they do it, too" argument the animal rightist has a simple, devastating reply: Do you really want to base your moral code on the natural order? Murder and rape are natural, too. Besides, we can choose: Humans don't need to kill other creatures in order to survive; carnivorous animals do. (Though if my cat Otis is any guide, animals sometimes kill for the sheer pleasure of it.)

Which brings up another objection for the case of domestic animals: *Wouldn't life in the wild be worse for these creatures?* "Defenders of slavery imposed on black Africans often made a similar point," Singer retorts. "[T]he life of freedom is preferred."

But most domesticated animals can't survive in the wild; in fact, *without us eating them they wouldn't exist at all!* Or as one nineteenth-century political philosopher put it, "The pig has a stronger interest than anyone in the demand for bacon. If all the world were Jewish, there would be no pigs at all." Which as it turns out would be just fine by the animal rightist: If chickens no longer exist, they can no longer be wronged.

Animals on factory farms have never known any other life. The rightist rightly points out that "animals feel a need to exercise, stretch their limbs or wings, groom themselves and turn around, whether or not they have ever lived in conditions that permit this." The proper measure of their suffering, in other words, is not their prior experiences but the unremitting daily frustration of their instincts.

Okay, granted the suffering of animals at our hands is a legitimate problem, but *the world is full of problems, and surely solving human problems must come first.* Sounds high-minded . . . and yet all the animal people are ask-

ing me to do is to stop eating meat. There's no reason I can't devote myself to solving humankind's problems as a vegetarian.

But doesn't the very fact that we could choose to forego meat for moral reasons point
to a crucial difference between animals and humans, one that justifies our speciesism? The
very indeterminacy of our appetites, and the ethical prospects that
opens up, marks us as a fundamentally different kind of creature. We
alone are (as Kant pointed out) the moral animal, the only one capable
of even entertaining a notion of "rights." Hell, we invented the damned
things—for us. So what's wrong with reserving moral consideration for
those able to understand it?

Well, right here is where you run smack into the AMC: the moral
status of the retarded and the insane, the two-day-old infant and the advanced Alzheimer's patient. These people ("marginal cases," in the de-
testable language of modern moral philosophy) cannot participate in
ethical decision making any more than a monkey can, yet we nevertheless grant them rights. Yes, I respond, for the obvious reason: They're one
of us. Isn't it natural to give special consideration to one's kind?

Only if you're a speciesist, the animal rightist replies. Not so long
ago many white people said the same thing about being white: We look
out for our kind. Still, I would argue that there is a nonarbitrary reason
we protect the rights of human "marginal" cases: We're willing to make
them part of our moral community because we all have been and will
probably once again be marginal cases ourselves. What's more, these
people have fathers and mothers, daughters and sons, which makes our
interest in their welfare deeper than our interest in the welfare of even
the most intelligent ape.

A utilitarian like Singer would agree that the feelings of relatives
should count for something in our moral calculus, but the principle of
equal consideration of interests demands that given the choice between
performing a painful medical experiment on a severely retarded or-
phaned child and a normal ape, we must sacrifice the child. Why? Because the ape has a greater capacity for pain.

Here in a nutshell is the practical problem with the philosopher's
argument from marginal cases: It can be used to help the animals, but

just as often it ends up hurting the marginal cases. Giving up our speciesism can bring us to an ethical cliff from which we may not be prepared to jump, even when logic is pushing us to the edge.

And yet this isn't the moral choice I'm being asked to make here. (*Too bad!* It would be so much easier.) In everyday life the choice is not between the baby and the chimp but between the pig and the tofu. Even if we reject the hard utilitarianism of a Peter Singer, there remains the question of whether we owe animals that can feel pain any moral consideration, and this seems impossible to deny. And if we owe them moral consideration, then how do we justify killing and eating them?

This is why meat eating is the most difficult animal rights case. In the case of laboratory testing of animals, all but the most radical animal people are willing to balance the human benefit against the cost to the animals. That's because the unique qualities of human consciousness carry weight in the utilitarian calculus of pleasure and pain: Human pain counts for more than that of a mouse, since our pain is amplified by emotions like dread; similarly, our deaths are worse than an animal's because we understand what death is in a way that they don't. So the argument around animal testing is in the details: Is that particular animal experiment really necessary to save human lives? (Very often it's not.) But if humans no longer *need* to eat meat to survive, then what exactly are we putting on the human side of the scale to outweigh the interests of the animal?

I suspect this is finally why the animal people managed to throw me on the defensive. It's one thing to choose between the chimp and the retarded child, or to accept the sacrifice of all those pigs surgeons practiced on to develop heart bypass surgery. But what happens when the choice is, as Singer writes, between "a lifetime of suffering for a non-human animal and the gastronomic preferences of a human being?" You look away—or you stop eating animals. And if you don't want to do either? I guess you have to try to determine if the animals you're eating have really endured a lifetime of suffering.

According to Peter Singer I can't hope to answer that question objectively as long as I'm still eating meat. "We have a strong interest in convincing ourselves that our concern for other animals does not require us to stop eating them." I can sort of see his point: I mean, why *am* I working so hard to justify a dinner menu? "No one in the habit of eating an animal can be completely without bias in judging whether the conditions in which that animal is reared cause suffering." In other words, I'm going to have to stop eating meat before I can in good conscience decide if I can continue eating meat, much less go hunting for meat. This struck me as a challenge I had no choice but to accept. So on a September Sunday, after dining on a delicious barbecued tenderloin of pork, I became a reluctant and, I fervently hoped, temporary vegetarian.

2. THE VEGETARIAN'S DILEMMA

Like any self-respecting vegetarian (and we are nothing if not self-respecting) I will now burden you with my obligatory compromises and ethical distinctions. I'm not a vegan (I will eat eggs and dairy), because eggs and milk can be coaxed from animals without hurting or killing them—or so at least I thought. I'm also willing to eat animals without faces, such as mollusks, on the theory that they're not sufficiently sentient to suffer. No, this isn't "facist" of me: Many scientists and animal rights philosophers (Peter Singer included) draw the line of sentience at a point just north of scallop. No one knows for absolute certain if this is right, but I'm joining many dedicated animal people in giving myself the benefit of the doubt.

A month or so into the experiment I'm still feeling reluctant about it. I find making a satisfying vegetarian dinner takes a lot more thought and work (chopping work in particular); eating meat is simply more convenient. It's also more sociable, at least in a society where vegetarians still represent a relatively tiny minority. (*Time* magazine recently estimated there are 10 million of us in America.) What troubles me most about my vegetarianism is the subtle way it alienates me from other people and, odd as this might sound, from a whole dimension of human experience.

Other people now have to accommodate me, and I find this uncomfortable: My new dietary restrictions throw a big wrench into the basic host-guest relationship. As a guest, if I neglect to tell my host in advance that I don't eat meat, she feels bad, and if I *do* tell her, she'll make something special for me, in which case I'll feel bad. On this matter I'm inclined to agree with the French, who gaze upon any personal dietary prohibition as bad manners.

Even if the vegetarian is a more highly evolved human being, it seems to me he has lost something along the way, something I'm not prepared to dismiss as trivial. Healthy and virtuous as I may feel these days, I also feel alienated from traditions I value: cultural traditions like the Thanksgiving turkey, or even franks at the ballpark, and family traditions like my mother's beef brisket at Passover. These ritual meals link us to our history along multiple lines—family, religion, landscape, nation, and, if you want to go back much further, biology. For although humans no longer need meat in order to survive (now that we can get our B-12 from fermented foods or supplements), we have been meat eaters for most of our time on earth. This fact of evolutionary history is reflected in the design of our teeth, the structure of our digestion, and, quite possibly, in the way my mouth still waters at the sight of a steak cooked medium rare. Meat eating helped make us what we are in a physical as well as a social sense. Under the pressure of the hunt, anthropologists tell us, the human brain grew in size and complexity, and around the hearth where the spoils of the hunt were cooked and then apportioned, human culture first flourished.

This isn't to say we can't or shouldn't transcend our inheritance, only that it *is* our inheritance; whatever else may be gained by giving up meat, this much at least is lost. The notion of granting rights to animals may lift us up from the brutal, amoral world of eater and eaten—of predation—but along the way it will entail the sacrifice, or sublimation, of part of our identity—of our own animality. (This is one of the odder ironies of animal rights: It asks us to acknowledge all we share with animals, and then to act toward them in a most unanimalistic way.) Not that the sacrifice of our animality is necessarily regrettable; no one regrets our giving up raping and pillaging, also part of our inheritance. But we should at least acknowledge that the human desire to eat meat is not, as the animal rightists would have it, a trivial matter, a mere gastronomic preference. By the same token we might call sex—also now technically unnecessary for reproduction—a mere recreational preference. Rather, our meat eating is something very deep indeed.

4. ANIMAL SUFFERING

Whether our interest in eating animals outweighs their interest in not being eaten (assuming for a moment that is their interest) ultimately turns on the vexed question of animal suffering. Vexed, because in a certain sense it is impossible to know what goes on in the mind of a cow or pig or ape. Of course, you could say the same about other humans too, but since all humans are wired in more or less the same way, we have good reason to assume other people's experience of pain feels much like our own. Can we say the same thing about animals? Yes—and no.

I have yet to find any serious writer on the subject who still subscribes to Descartes's belief that animals cannot feel pain because they lack a soul. The general consensus among both scientists and philosophers is that when it comes to pain, the higher animals are wired much like we are for the same evolutionary reasons, so we would do well to take the writhing of the kicked dog at face value.

That animals feel pain does not seem in doubt. The animal people claim, however, that there are neo-Cartesian scientists and thinkers about who argue that animals are incapable of suffering because they lack language. Yet if you take the trouble to actually read the writers in question (Daniel Dennett and Stephen Budiansky are two of the ones often cited), you quickly realize they're being unfairly caricatured.

The offending argument, which does not seem unreasonable to me, is that human pain differs from animal pain by an order of magnitude. This qualitative difference is largely the result of our possession of language and, by virtue of language, our ability to have thoughts about thoughts and to imagine what is not. The philosopher Daniel Dennett suggests we can draw a distinction between pain, which a great many animals obviously experience, and suffering, which depends on a degree of self-consciousness only a handful of animals appear to command. Suffering in this view is not just lots of pain but pain amplified by distinctly human emotions such as regret, self-pity, shame, humiliation, and dread.

Consider castration, an experience endured by most of the male mammals we eat. No one would deny the procedure is painful to animals, yet very shortly afterward the animals appear fully recovered. (Some rhesus monkeys competing for mates will bite off a rival's testicle; the very next day the victim may be observed mating, seemingly little the worse for wear.) Surely the suffering of a man able to comprehend the full implications of castration, to anticipate the event and contemplate its aftermath, represents an agony of a different order.

By the same token, however, language and all that comes with it can also make some kinds of pain more bearable. A trip to the dentist would be an agony for an ape that couldn't be made to understand the purpose and duration of the procedure.

As humans contemplating the suffering or pain of animals we do need to guard against projecting onto them what the same experience would feel like to us. Watching a steer force-marched up the ramp to the kill-floor door, as I have done, I have to forcibly remind myself this is not Sean Penn in *Dead Man Walking*, that the scene is playing very differently in a bovine brain, from which the concept of nonexistence is thankfully absent. The same is true of the deer staring down the barrel of the hunter's rifle. "If we fail to find suffering in the [animal] lives we can see," Daniel Dennett writes in *Kinds of Minds*, "we can rest assured there is no invisible suffering somewhere in their brains. If we find suffering, we will recognize it without difficulty."

Which brings us—reluctantly, necessarily—to the American factory farm, the place where all such distinctions promptly turn to dust. It's not easy to draw lines between pain and suffering in a modern egg or hog operation. These are places where the subtleties of moral philosophy and animal cognition mean less than nothing, indeed where everything we've learned about animals at least since Darwin has been simply . . . put aside. To visit a modern Concentrated Animal Feeding Operation (CAFO) is to enter a world that for all its technological sophistication is still designed on seventeenth-century Cartesian principles: Animals are treated as machines—"production units"—incapable of feeling pain. Since no thinking person can possibly believe this anymore, industrial animal agriculture depends on a suspension of disbelief on the part of the people who operate it and a willingness to avert one's eyes on the part of everyone else.

Egg operations are the worst, from everything I've read; I haven't managed to actually get into one of these places since journalists are unwelcome there. Beef cattle in America at least still live outdoors, albeit standing ankle-deep in their own waste eating a diet that makes them sick. And broiler chickens, although they get their beaks snipped off with a hot knife to keep them from cannibalizing one another under the stress of their confinement, at least don't spend their lives in cages too small to ever stretch a wing.

That fate is reserved for the American laying hen, who spends her brief span of days piled together with a half-dozen other hens in a wire cage the floor of which four pages of this book could carpet wall to wall. Every natural instinct of this hen is thwarted, leading to a range of behavioral "vices" that can include cannibalizing her cage mates and rubbing her breast against the wire mesh until it is completely bald and bleeding. (This is the chief reason broilers get a pass on caged life; to scar so much high-value breast meat would be bad business.) Pain? Suffering? Madness? The operative suspension of disbelief depends on the acceptance of more neutral descriptors, such as "vices" and "stereo-types" and "stress." But whatever you want to call what goes on in those cages, the 10 percent or so of hens that can't endure it and sim-ply die is built into the cost of production. And when the output of the survivors begins to ebb, the hens will be "force-molted"—starved of food and water and light for several days in order to stimulate a final bout of egg laying before their life's work is done.

I know, simply reciting these facts, most of which are drawn from poultry trade magazines, makes me sound like one of the animal peo-ple, doesn't it? I don't mean to (remember, I got into this vegetarian deal assuming I could go on eating eggs), but this is what can happen to you when . . . you look. And what you see when you look is the cruelty—and the blindness to cruelty—required to produce eggs that can be sold for seventy-nine cents a dozen.

A tension has always existed between the capitalist imperative to maximize efficiency at any cost and the moral imperatives of culture, which historically have served as a counterweight to the moral blind-ness of the market. This is another example of the cultural contradic-tions of capitalism—the tendency over time for the economic impulse to erode the moral underpinnings of society. Mercy toward the animals in our care is one such casualty.

The industrial animal factory offers a nightmarish glimpse of what capitalism is capable of in the absence of any moral or regulatory con-straint whatsoever. (It is no accident that the nonunion workers in these factories receive little more consideration than the animals in their care.) Here in these wretched places life itself is redefined—as "protein production"—and with it "suffering." That venerable word becomes "stress," an economic problem in search of a cost-effective solution such as clipping the beaks of chickens or docking the tails of pigs or, in the industry's latest initiative, simply engineering the "stress gene" out of pigs and chickens. It all sounds very much like our worst nightmares of confinement and torture, and it is that, but it is also real life for the billions of animals unlucky enough to have been born beneath those grim sheet-metal roofs into the brief, pitiless life of a production unit in the days before the suffering gene was found.

5. ANIMAL HAPPINESS

Vegetarianism doesn't seem an unreasonable response to the existence of such an evil. Who would want to be complicit in the misery of these animals by eating them? You want to throw *something* against the walls of those infernal sheds, whether it's the Bible, with its call for mercy to the animals we keep, or a new constitutional right, or a whole platoon of animal people in chicken suits bent on breaking in and liberating the inmates. In the shadow of these factory farms Coetzee's notion of a "stupendous crime" doesn't seem far-fetched at all.

And yet there are other images of animals on other kinds of farms that contradict the nightmare ones. I'm thinking of the hens I saw at Polyface Farm, fanning out over the cow pasture on a June morning, pecking at the cowpats and the grass, gratifying their every chicken instinct. Or the image of pig happiness I witnessed in that cattle barn in March, watching the hogs, all upturned pink hams and corkscrew tails, nosing their way through that deep cake of compost in search of alcoholic morsels of corn. It is true that farms like this are but a speck on the monolith of modern animal agriculture, yet their very existence, and the possibility that implies, throws the whole argument for animal rights into a different light.

To many animal people even Polyface Farm is a "death camp"—a way station for doomed creatures awaiting their date with the executioner. But to look at the lives of these animals is to see this holocaust analogy for the sentimental conceit it really is. In the same way we can probably recognize animal suffering when we see it, animal happiness is unmistakable, too, and during my week on the farm I saw it in abundance.

For any animal, happiness seems to consist in the opportunity to express its creaturely character—its essential pigness or wolfness or chickenness. Aristotle talked about each creature's "characteristic form of life." At least for the domestic animal (the wild animal is a different case) the good life, if we can call it that, simply doesn't exist, cannot be achieved, apart from humans—apart from our farms and therefore from our meat eating. This, it seems to me, is where the animal rightists betray a deep ignorance about the workings of nature. To think of domestication as a form of slavery or even exploitation is to misconstrue that whole relationship—to project a human idea of power onto what is in fact an example of mutualism or symbiosis between species.

Domestication is an evolutionary, rather than a political, development. It is certainly not a regime humans somehow imposed on animals some ten thousand years ago. Rather, domestication took place when a handful of especially opportunistic species discovered, through Darwinian trial and error, that they were more likely to survive and prosper in an alliance with humans than on their own. Humans provided the animals with food and protection in exchange for which the animals provided the humans their milk, eggs, and—yes—their flesh. Both parties were transformed by the new relationship: The animals grew tame and lost their ability to fend for themselves in the wild (natural selection tends to dispense with unneeded traits) and the humans traded their hunter-gatherer ways for the settled lives of agriculturists. (Humans changed biologically, too, evolving such new traits as the ability to digest lactose as adults.)

From the animals' point of view the bargain with humanity turned out to be a tremendous success, at least until our own time. Cows, pigs, dogs, cats, and chickens have thrived, while their wild ancestors have languished. (There are ten thousand wolves left in North America and fifty million dogs.) Nor does the loss of autonomy seem to trouble these creatures. It is wrong, the rightists say, to treat animals as means rather than ends, yet the happiness of a working animal like the dog consists precisely in serving as a means to human ends. Liberation is the last thing such a creature wants. (Which might explain the contempt many animal people display toward domesticated species.) To say of one of

Joel Salatin's caged broilers that "the life of freedom is to be preferred" betrays an ignorance about chicken preferences that, around his place at least, revolve around not getting one's head bitten off by a weasel.

It is probably safe to say, however, that chicken preferences do not include living one's entire life six to a battery cage indoors. The crucial moral difference between a CAFO and a good farm is that the CAFO systematically deprives the animals in it of their "characteristic form of life."

But haven't Salatin's chickens simply traded one predator for another—weasels for humans? True enough, and for the chickens this is probably not a bad deal, either. It is precisely the evolutionary reason the species entered into its relationship with humans in the first place. For, brief as it is, the life expectancy of a farm animal would be considerably briefer in the world beyond the pasture fence or chicken coop. (Pigs, which often can survive in the wild, are the exception that proves the rule.) It's brutal out there. A bear will eat a lactating ewe alive, starting with her udders. As a rule, animals in the wild don't get good deaths surrounded by their loved ones.

Which brings us to the case of animals in the wild. The very existence of predation in nature, of animals eating animals, is the cause of much anguished hand-wringing in the animal rights literature. "It must be admitted," Peter Singer writes, "that the existence of carnivorous animals does pose one problem for the ethics of Animal Liberation, and that is whether we should do anything about it." (Talk about the need for peacekeeping forces!) Some animal people train their dogs and cats to become vegetarians. (Note: The cats will require nutritional supplements to survive.) Matthew Scully, in *Dominion*, a Christian-conservative treatment of animal rights, calls predation "the intrinsic evil in nature's design . . . among the hardest of all things to fathom." *Really?* Elsewhere, acknowledging the gratuitous suffering inflicted by certain predators (like cats), Scully condemns "the level of moral degradation of which [animals] are capable." *Moral degradation?*

A deep current of Puritanism runs through the writings of the animal philosophers, an abiding discomfort not just with our animality, but with the animals' animality, too. They would like nothing better than to airlift us out from nature's "intrinsic evil"—and then take the animals with us. You begin to wonder if their quarrel isn't really with nature itself.

But however it may appear to those of us living at such a remove from the natural world, predation is not a matter of morality or of politics; it, too, is a matter of symbiosis. Brutal as the wolf may be to the individual deer, the herd depends on him for its well-being. Without predators to cull the herd deer overrun their habitat and starve—all suffer, and not only the deer but the plants they browse and every other species that depends on those plants. In a sense, the "good life" for deer, and even their creaturely character, which has been forged in the crucible of predation, depends on the existence of the wolf. In a similar way chickens depend for their well-being on the existence of their human predators. Not the individual chicken, perhaps, but Chicken—the species. The surest way to achieve the extinction of the species would be to grant chickens a right to life.

Long before human predation was domesticated (along with the select group of animals we keep) it operated on another set of species in the wild. The fact of human hunting is, from the point of view of a great many creatures in a great many habitats, simply a fact of nature. We are to them as wolves. And in the same way the deer evolved a specific set of characteristics under the pressure of hunting by wolves (fleetness, sensory acuity, coloration, etc.), so have the animals that humans have hunted. Human hunting, for example, literally helped form the American Plains bison, which the fossil record suggests changed both physically and behaviorally after the arrival of the Indians. Before then the bison did not live in big herds and had much larger, more outstretched horns. For an animal living in a wide-open environment like the Great Plains and facing a sophisticated predator armed with spears, mobbing in big groups is the best defense, since it affords the vigilance of many eyes; yet big, outstretched horns pose a problem for creatures living in such close proximity. It was human hunting that selected for herd behavior and the new upright arrangement of bison horns, which appears in the fossil record not long after the arrival of human hunters.

"While a symbol of the 'wild west,'" Tim Flannery writes in *The Eternal Frontier*, an ecological history of North America, "the bison is a human artifact, for it was shaped by Indians."

Until the advent of the rifle and a global market in bison hides, horns, and tongues, Indian hunters and bison lived in a symbiotic relationship, the bison feeding and clothing the hunters while the hunters, by culling the herds and forcing them to move frequently, helped keep the grasslands in good health. Predation is deeply woven into the fabric of nature, and that fabric would quickly unravel if it somehow ended, if humans somehow managed "to do something about it." From the point of view of the individual prey animal predation is a horror, but from the point of view of the group—and of its gene pool—it is indispensable. So whose point of view shall we favor? That of the individual bison or Bison? The pig or Pig? Much depends on how you choose to answer that question.

Ancient man regarded animals much more as a modern ecologist would than an animal philosopher—as a species, that is, rather than a collection of individuals. In the ancient view "they were mortal and immortal," John Berger writes in "Looking at Animals." "An animal's blood flowed like human blood, but its species was undying and each lion was Lion, each ox was Ox." Which, when you think about it, is probably pretty much how any species in nature regards another.

Until now. For the animal rightist concerns himself only with individuals. Tom Regan, the author of *The Case for Animal Rights*, bluntly asserts that because "species are not individuals . . . the rights view does not recognize the moral rights of species to anything, including survival." Singer concurs, insisting that only sentient individuals can have interests. But surely a species has interests—in its survival, say, or the health of its habitat—just as a nation or a community or a corporation can. Animal rights' exclusive concern with the individual might make sense given its roots in a culture of liberal individualism, but how much sense does it make in nature? Is the individual animal the proper focus of our moral concern when we are trying to save an endangered species or restore a habitat?

As I write, a team of sharpshooters in the employ of the National Park Service and the Nature Conservancy is at work killing thousands of feral pigs on Santa Cruz Island, eighteen miles off the coast of Southern California. The slaughter is part of an ambitious plan to restore the island's habitat and save the island fox, an endangered species found on a handful of Southern California islands and nowhere else. To save the fox the Park Service and Nature Conservancy must first undo a complicated chain of ecological changes wrought by humans beginning more than a century ago.

That's when the pigs first arrived on Santa Cruz, imported by ranchers. Though pig farming ended on the island in the 1980s, by then enough pigs had escaped to establish a wild population that has done grave damage to the island ecosystem. The rooting of the pigs disturbs the soil, creating ideal conditions for the establishment of invasive exotic species like fennel, now rampant. The pigs also eat so many acorns that the island's native oaks have trouble reproducing. But the most serious damage the pigs have done has been to feed golden eagles with their piglets, sparking an explosion in the eagle population. That's when the island fox's troubles began.

Golden eagles are not native to the island; they've taken over a niche formerly occupied by the bald eagle, which lost its place on the island after a chemical maker dumped large quantities of DDT into the surrounding waters in the 1950s and 1960s. (Settlement money from the company is underwriting the habitat restoration project.) The DDT damaged the eggshells of the bald eagles, crashing their population and creating an opening for the more aggressive golden eagles. Unlike bald eagles, which dine mostly on seafood, golden eagles feed on small terrestrial mammals. But while the golden eagles have a taste for pig, piglets are harder to catch than the cubs of island fox, which the eagles have now hunted to the edge of extinction. To save the fox, the plan is to kill every last pig, trap and remove the golden eagles, and then reintroduce the bald eagles—essentially, rebuild the island's food chain from the ground up.

The wholesale slaughter of thousands of pigs has predictably drawn the protests of animal welfare and rights groups. The Channel Islands Animal Protection Association has been flying banners from small planes imploring the public to "Save the Pigs" and friends of the animal have sued to stop the hunt. A spokesman for the Humane Society of the United States claimed in an op-ed article that "wounded pigs and or-phaned piglets will be chased with dogs and finished off with knives and bludgeons." Note the rhetorical shift in focus from the Pig, which is how the Park Service ecologists would have us see the matter, to im-ages of individual pigs, wounded and orphaned, being hunted down by dogs and men wielding bludgeons. Same story, viewed through two entirely different lenses.

The fight over the pigs at Santa Cruz Island suggests at the very least that a human morality based on individual rights makes for an awk-ward fit when applied to the natural world. This should come as no sur-prise: Morality is an artifact of human culture devised to help humans negotiate human social relations. It's very good for that. But just as we recognize that nature doesn't provide a very good guide for human so-cial conduct, isn't it anthropocentric of us to assume that our moral system offers an adequate guide for what should happen in nature? Is the individual the crucial moral entity in nature as we've decided it should be in human society? We simply may require a different set of ethics to guide our dealings with the natural world, one as well suited to the particular needs of plants and animals and habitats (where sen-tience counts for little) as rights seem to suit us and serve our purposes today.

6. THE VEGAN UTOPIA

To contemplate such questions from the vantage of a farm, or even a garden, is to appreciate just how parochial, and urban, an ideology an-imal rights really is. It could only thrive in a world where people have lost contact with the natural world, where animals no longer pose any threat to us (a fairly recent development), and our mastery of nature seems unchallenged. "In our normal life," Singer writes, "there is no serious clash of interests between human and nonhuman animals." Such a statement assumes a decidedly citified version of "normal life," certainly one no farmer—indeed, no gardener—would recognize.

The farmer would point out to the vegan that even she has a "serious clash of interests" with other animals. The grain that the vegan eats is harvested with a combine that shreds field mice, while the farmer's tractor wheel crushes woodchucks in their burrows and his pesticides drop songbirds from the sky; after harvest whatever animals that would eat our crops we exterminate. Killing animals is probably unavoidable no matter what we choose to eat. If America was suddenly to adopt a strictly vegetarian diet, it isn't at all clear that the total number of animals killed each year would necessarily decline, since to feed everyone animal pasture and rangeland would have to give way to more intensively cultivated row crops. If our goal is to kill as few animals as possible people should probably try to eat the largest possible animal that can live on the least cultivated land: grass-finished steaks for everyone.

The vegan utopia would also condemn people in many parts of the country to importing all their food from distant places. In New England, for example, the hilliness of the land and rockiness of the soil has dictated an agriculture based on grass and animals since the time of the Puritans. Indeed, the New England landscape, with its rolling patchwork of forest and fields outlined by fieldstone walls, is in some sense a creation of the domestic animals that have lived there (and so in turn of their eaters). The world is full of places where the best, if not the only, way to obtain food from the land is by grazing (and hunting) animals on it—especially ruminants, which alone can transform grass into protein.

To give up eating animals is to give up on these places as human habitat, unless of course we are willing to make complete our dependence on a highly industrialized national food chain. That food chain would be in turn even more dependent than it already is on fossil fuels and chemical fertilizer, since food would need to travel even farther and fertility—in the form of manures—would be in short supply. Indeed, it is doubtful you can build a genuinely sustainable agriculture without animals to cycle nutrients and support local food production. If our concern is for the health of nature—rather than, say, the internal consistency of our moral code or the condition of our souls—then eating animals may sometimes be the most ethical thing to do.

ARE THESE good enough reasons to give up my vegetarianism? Can I in good conscience eat a happy and sustainably raised chicken? I'm mindful of Ben Franklin's definition of a reasonable creature as one who can come up with reasons for whatever he wants to do. So I decided I would track down Peter Singer and ask him what he thought. I hatched a scheme to drive him down from Princeton to meet Joel Salatin and his animals, but Singer was out of the country, so I had to settle for an exchange of e-mail. I asked him about the implications for his position of the "good farm"—one where animals got to live according to their natures and to all appearances do not suffer.

"I agree with you that it is better for these animals to have lived and died than not to have lived at all . . . ," Singer wrote back. Since the utilitarian is concerned exclusively with the sum of happiness and suffering, and the slaughter of an animal with no comprehension of death need not entail suffering, the Good Farm adds to the total of animal happiness, provided you replace the slaughtered animal with a new one. However, this line of thinking does not obviate the wrongness of killing an animal that "has a sense of its own existence over time, and can have preferences about its own future." In other words, it might be okay to eat the chicken or the cow, but perhaps not the (more intelligent) pig. Yet, he continued, "I would not be sufficiently confident of my argument to condemn someone who purchased meat from one of these farms."

Singer went on to express doubts that such farms could be practical on a large scale, since the pressures of the marketplace will lead their owners to cut costs and corners at the expense of the animals. Also, since humanely raised food is more expensive, only the well-to-do can afford morally defensible animal protein. These are important considerations, but they don't alter what seems to me the essential concession: What's wrong with eating animals is the practice, not the principle.

What this suggests to me is that people who care about animals should be working to ensure that the ones they eat don't suffer, and that their deaths are swift and painless—for animal welfare, in others words, rather than rights. In fact, the "happy life and merciful death" line is how Jeremy Bentham justified his own meat eating. Yes, the philosophical father of animal rights was himself a carnivore. In a passage seldom quoted by animal rightists Bentham defended meat eating on the grounds that "we are the better for it, and they are never the worse. . . . The death they suffer in our hands commonly is, and always may be, a speedier and, by that means a less painful one, than that which would await them in the inevitable course of nature."

My guess is that Bentham never looked too closely at what actually happens in a slaughterhouse, but the argument suggests that in theory at least a utilitarian can justify eating humanely raised and slaughtered animals. Eating a wild animal that had been cleanly shot presumably would fall under the same dispensation. Singer himself suggests as much in *Animal Liberation*, when he asks, "Why . . . is the hunter who shoots a deer for venison subject to more criticism than the person who buys a ham at the supermarket? Overall it is probably the intensively reared pig who has suffered more."

All of which was making me feel pretty good about eating meat again and going hunting—until I recalled that these utilitarians can also justify killing retarded orphans. Killing just isn't the problem for them that it is for other people, including me.

7. A CLEAN KILL

The day after my steak-and-Singer dinner at the Palm I found myself on a plane flying from Atlanta to Denver. A couple of hours into the flight the pilot, who hadn't uttered word one until now, came on the public address system to announce, apropos of nothing, that we were passing over Liberal, Kansas. This was the first, last, and only landmark on our flight path that the pilot deigned to mention, which seemed very odd, given its obscurity to everyone on the plane but me. For Liberal, Kansas, happens to be the town where my steer, very possibly that very day, was being slaughtered. I'm not a superstitious person, but this struck me as a most eerie coincidence. I could only wonder what was going on just then, thirty thousand feet below me, on the kill floor of the National Beef Plant, where steer number 534 had his date with the stunner.

I could only wonder because the company had refused to let me see. When I'd visited the plant earlier that spring I was shown everything but the kill floor. I watched steers being unloaded from trailers into corrals and then led up a ramp and through a blue door. What happened on the other side of the blue door I had to reconstruct from the accounts of others who had been allowed to go there. I was fortunate to have the account of Temple Grandin, the animal-handling expert, who had designed the ramp and killing machinery at the National Beef Plant, and who audits the slaughter there for McDonald's. Stories about cattle "waking up" after stunning only to be skinned alive—stories documented by animal rights groups—had prompted the company to hire Grandin to audit its suppliers. Grandin told me that in cattle slaughter, "there is the pre-McDonald's era and the post-McDonald's era—it's night and day." We can only imagine what night must have been like.

Here's how Grandin described what steer 534 experienced after passing through the blue door:

"The animal goes into the chute single file. The sides are high enough so all he sees is the butt of the animal in front of him. As he walks through the chute, he passes over a metal bar, with his feet on either side. While he's straddling the bar, the ramp begins to decline at a twenty-degree angle, and before he knows it, his feet are off the ground, and he's being carried along on a conveyor belt. We put in a false floor so he can't look down and see he's off the ground. That would panic him."

I had been wondering what 534 would be feeling as he neared his end. Would he have any inkling—a scent of blood, a sound of terror from up the line—that this was no ordinary day? Would he, in other words, suffer? Grandin anticipated my question.

"Does the animal know it's going to get slaughtered? I used to wonder that. So I watched them going into the squeeze chutes on the feedlot, getting their shots, and going up the ramp at a slaughter plant. No difference. If they knew they were going to die you'd see much more agitated behavior.

"Anyway, the conveyor is moving along at roughly the speed of a moving sidewalk. On a catwalk above stands the stunner. The stunner has a pneumatic-powered 'gun' that fires a steel bolt about seven inches long and the diameter of a fat pencil. He leans over and puts it smack in the middle of the forehead. When it's done correctly it will kill the animal on the first shot.

"After the animal is shot while he's riding along a worker wraps one of his feet and hooks it to an overhead trolley. Hanging upside down by one leg, he's carried by the trolley into the bleeding area, where the bleeder cuts his throat. Animal rights people say they're cutting live animals, but that's because there's a lot of reflex kicking. What I look for is, is the head dead? It should be flopping like a rag, with the tongue hanging out. He'd better not be trying to hold it up—then you've got a live one on the rail. Just in case, they have another stunner in the bleed area."

I found Temple Grandin's account both reassuring and troubling. Reassuring, because the system sounds humane, and yet I realize I'm relying on the account of its designer. Troubling, because I can't help dwelling on all those times "you've got a live one on the rail." Mistakes are inevitable on an assembly line that is slaughtering four hundred head of cattle every hour. (McDonald's tolerates a 5 percent "error rate.") So is it possible to slaughter animals on an industrial scale without causing them to suffer? In the end each of us has to decide for himself whether eating animals that have died in this manner is okay. For my part, I can't be sure, because I haven't been able to see for myself.

This, I realize, is why Joel Salatin's open-air abattoir is such a morally powerful idea. Any customer who so desires can see how his chicken meets its end—can look and then decide. Few will take up such an offer; many of us would prefer to delegate the job of looking to a government bureaucrat or a journalist, but the very option of look-ing—that transparency—is probably the best way to ensure that ani-mals are killed in a manner we can abide. No doubt some of us will decide there is no killing of animals we can countenance, and they probably shouldn't eat meat.

When I was at the farm I asked Joel how he could bring himself to kill a chicken. "That's an easy one. People have a soul, animals don't. It's a bedrock belief of mine. Animals are not created in God's image, so when they die, they just die."

The idea that it is only in modern times that people have grown queasy about killing animals is of course a flattering myth. Taking a life is momentous, and people have been working to justify the slaughter of animals to themselves for thousands of years, struggling to come to terms with the shame they feel even when the killing is necessary to their survival. Religion, and ritual, has played a crucial part in this process. Native Americans and other hunter-gatherers give thanks to the animal for giving up its life so the eater might live. The practice sounds a little like saying grace, a ceremony hardly anyone bothers with any-more. In biblical times the rules governing ritual slaughter stipulated a rotation, so that no individual would have to kill animals every day, lest he become dulled to the gravity of the act. Many cultures have offered sacrificial animals to the gods, perhaps as a way to convince themselves it was the gods' appetite that demanded the slaughter, and not their own. In ancient Greece, the priests responsible for the slaughter (Priests! Now we give the job to migrant workers paid the minimum wage) would sprin-kle holy water on the sacrificial animal's head. The beast would promptly shake its head, and this was taken as a necessary sign of assent.

For all these people it was the ritual—the cultural rules and norms—that allowed them to look, and then to eat. We no longer have any ritu-als governing either the slaughter or eating of animals, which perhaps helps explain why we find ourselves in this dilemma, in a place where we feel our only choice is either to look away or give up meat. National Beef is happy to serve the first customer, Peter Singer the second.

My own wager is that there might still be another way open to us, and that finding it will begin with looking once again—at the animals we eat, and at their deaths. People will see very different things when they look into the eyes of a pig or a chicken or a steer: a being without a soul, a "subject of a life" entitled to rights, a receptacle of pleasure and pain, an unambiguously tasty lunch.

We certainly won't philosophize our way to a single answer. I remember a story Joel told me about a man who showed up at the farm one Saturday morning to have a look. When Joel noticed a PETA bumper sticker on the man's car he figured he was in for some unpleasantness. But the man had a different agenda. He explained that after being a vegetarian for sixteen years he had decided that the only way he could ever eat meat again was if he killed the animal himself. So Joel grabbed a chicken and took the man into the processing shed.

"He slit the bird's throat and watched it die," Joel recalled. "He saw that the animal did not look at him accusingly, did not do a Disney double take. He saw that the animal had been treated with respect while it was alive and that it could have a respectful death—that it wasn't being treated like a pile of protoplasm." I realized I'd seen this, too, which perhaps explains why I was able to kill a chicken one day and eat it the next. Though the story did make me wish I had killed and eaten mine with as much consciousness and attention as that man; perhaps hunting would give me a second chance.

Sometimes I think that all it would take to clarify our feelings about eating meat, and in the process begin to redeem animal agriculture, would be to simply pass a law requiring all the sheet-metal walls of all the CAFOs, and even the concrete walls of the slaughterhouses, to be replaced with glass. If there's any new right we need to establish, maybe this is the one: the right, I mean, to look. No doubt the sight of some of these places would turn many people into vegetarians. Many others would look elsewhere for their meat, to farmers willing to raise and kill their animals transparently. Such farms exist; so do a handful of small processing plants willing to let customers onto the kill floor, including one—Lorentz Meats, in Cannon Falls, Minnesota—that is so confident of their treatment of animals that they have walled their abattoir in glass.

The industrialization—and brutalization—of animals in America is a relatively new, evitable, and local phenomenon: No other country raises and slaughters its food animals quite as intensively or as brutally as we do. No other people in history has lived at quite so great a remove from the animals they eat. Were the walls of our meat industry to become transparent, literally or even figuratively, we would not long continue to raise, kill, and eat animals the way we do. Tail docking and sow crates and beak clipping would disappear overnight, and the days of slaughtering four hundred head of cattle an hour would promptly come to an end—for who could stand the sight? Yes, meat would get more expensive. We'd probably eat a lot less of it, too, but maybe when we did eat animals we'd eat them with the consciousness, ceremony, and respect they deserve.

EIGHTEEN

HUNTING

The Meat

1. A WALK IN THE WOODS

Walking with a loaded rifle in an unfamiliar forest bristling with the signs of your prey is thrilling. It embarrasses me to write that, but it is true. I am not by nature much of a noticer, yet here, now, my attention to everything around me, and deafness to everything else, is complete. Nothing in my experience (with the possible exception of certain intoxicants) has prepared me for the quality of this attention. I notice how the day's first breezes comb the needles in the pines, producing a sotto voce whistle and an undulation in the pattern of light and shadow tattooing the tree trunks and the ground. I notice the specific density of the air. But this is not a passive or aesthetic attention; it is a hungry attention, reaching out into its surroundings like fingers, like nerves. My eyes venture deep into thickets my body could never penetrate, picking their way among the tangled branches, sliding over rocks and around stumps to bring back the slenderest hint of movement. In the places too deeply shadowed to admit my eyes my ears roam at will, returning with the report of a branch cracking at the bottom of a ravine, or the snuffling of a . . . wait: *What was* that? Just a bird. Everything is amplified. Even my skin is alert, so that when the shadow launched by the sudden ascent of a turkey vulture passes overhead I swear I can feel the temperature momentarily fall. I am the alert man.

Hunting powerfully inflects a place. The ordinary prose of the ground, the literally down-to-earth, becomes as layered and springy as verse. Angelo, my Virgil in this world, has taught me how to read the ground for signs of pig. Notice the freshly rototilled soil at the base of that oak tree? Look how the earth has not yet been crisped by the midday sun; this means pigs have been rooting here since yesterday afternoon, either overnight or earlier this morning. See that smoothly scooped-out puddle of water? That's a wallow, but notice how the water is perfectly clear: Pigs haven't disturbed it yet today. We could wait here for them. Angelo says that the pigs, who travel in groups of a half dozen or so, follow a more or less fixed daily routine, moving from place to place, feeding, sleeping, cooling off. This grove of oaks is where they root for acorns, tubers, and grubs. In the afternoon heat they snooze in oval nests scooped out of the dusty dirt beneath that protective tangle of manzanita. They cool off in these muddy wallows, the shores of which are letterpressed with dainty hoof prints. They scrape the mud from their backs on that pine tree there, the one where the lower bark is rubbed smooth and tan. And they move from one such pig place to another along narrow lanes that temporarily part the thick pelts of rattlesnake grass clothing the hillsides; since the grass springs back to erase their lanes after a few hours of sunshine, you can form a pretty good idea of how recently they've passed through here. On their appointed rounds the pigs can cover forty square miles in a day.

After hunting here for years Angelo has come to the conclusion there are three distinct groups sharing the oak forest and the grassy ridge above it like three overlapping nations, each with a slightly different map of good pig places. The hunter maintains his own mental map of the same ground, marked with auspicious spots, the places he has encountered pigs before and the connecting routes he can navigate, which of course are far fewer than those available to the pigs. Unlike that of the pigs, the hunter's map also contains legal things like property lines and rights of way.

The hunter's aim is to have his map collide with the pig's map, which, should it happen, will do so at a moment of no one's choosing. For although there's much the hunter can know about pigs and about their places, in the end he knows nothing about what is going to happen here today, whether the longed-for and dreaded encounter will actually take place and, if it does, how it will end.

Since there's nothing he can do to make the encounter happen, the hunter's energy goes into readying himself for it, and attempting, by the sheer force of his attention, to summon the animal into his presence. The drama of the hunt links the actors in it, predator and prey, long before they actually meet. Approaching his prey, the hunter instinctively becomes more like the animal, straining to make himself less visible, less audible, more exquisitely alert. Predator and prey alike move according to their own maps of this ground, their own forms of attention, and their own systems of instinct, systems that evolved expressly to hasten or avert precisely this encounter . . .

WAIT A MINUTE. Did I really write that last paragraph? Without irony? That's embarrassing. I'm actually writing about the hunter's "instinct," suggesting that the hunt represents some sort of primordial union between two kinds of animals, one of which is me? This seems a bit much. I recognize this kind of prose: hunter porn. And whenever I've read it in the past, in Ortega y Gasset and Hemingway and all those hardbitten, big-bearded American wilderness writers who still pine for the Pleistocene, it never failed to roll my eyes. I never could stomach the straight-faced reveling in primitivism, the barely concealed bloodlust, the whole macho conceit that the most authentic encounter with nature is the one that comes through the sight of a gun and ends with a large mammal dead on the ground—a killing that we are given to believe constitutes a gesture of respect. So it is for Ortega y Gasset, the

Spanish philosopher, who writes in his *Meditations on Hunting* that "the greatest and most moral homage we can pay to certain animals on certain occasions is to kill them." Please.

And yet here I find myself sliding into the hunter's ecstatic purple, channeling Ortega y Gasset. It may be that we have no better language in which to describe the experience of hunting, so that all of us who would try sooner or later slide into this overheated prose ignorant of irony. Or it could be that hunting is one of those experiences that appear utterly different from the inside than from the outside. That this might indeed be the case was forcibly impressed upon me after my second trip hunting with Angelo when, after a long and gratifying day in the woods, we stopped in at a convenience store for a bottle of water. The two of us were exhausted and filthy, the fronts of our jeans stained dark with blood. We couldn't have smelled terribly fragrant. And under the bright fluorescence of the 7-Eleven, in the mirror behind the cigarette rack behind the cashier, I caught a glimpse of this grungy pair of self-satisfied animal killers and noted the wide berth the other customers in line were only too happy to grant them. Us. It is a wonder that the cashier didn't preemptively throw up his hands and offer us the contents of the cash register.

Irony—the outside perspective—easily withers everything about hunting, shrinks it to the proportions of boy's play or atavism. And yet at the same time I found that there is something about the experience of hunting that puts irony itself to rout. In general, experiences that banish irony are much better for living than for writing. But there it is: I enjoyed shooting a pig a whole lot more than I ever thought I should have.

2. A CANNABINOID MOMENT

Part of me did not want to go. The night before I had anxiety dreams about hunting. In one I was on a bobbing boat trying to aim a rifle at a destroyer that was firing its cannons at me; in the other the woods were crawling with Angelo's Sicilian relatives, and I couldn't for the life of me remember how my gun worked, whether the safety was on when the little button popped up on the left side of the trigger or the right.

I had tried out my rifle only once before taking it to the woods, at a firing range in the Oakland Hills, and by the end of the morning my paper target had sustained considerably less damage than my left shoulder, which ached for a week. I wasn't ready to buy a gun of my own, so Angelo had borrowed a fairly basic pump-action rifle, a .270 Winchester with an old-fashioned sight that I had trouble getting used to. After my session at the range, the first-order worry that I wouldn't have whatever it takes to fire a rifle aimed at an animal was overtaken by a second-order worry that, assuming I did manage to pull the trigger, nothing of consequence would happen to the animal.

The plan was to hunt boar in the sparsely populated northern reaches of Sonoma County, on a thousand-acre property owned by a friend of Angelo's named Richard. Angelo hunts deer and turkey and duck, too, but for a number of reasons I felt more comfortable going for wild pig. The animal is regarded as a pest in many parts of California and it seemed to me much easier to justify killing a pest than a wild native species that, like many waterfowl these days, is threatened by loss of habitat or overhunting. The pigs have been here a long time, but they are not native or even exactly wild; feral would be more accurate. They are also, by reputation, vicious; one of the nicknames the California pig has earned is "dog ripper."

Columbus brought pigs to the New World on his second voyage, in 1493. By the end of the following century the Spanish had introduced domestic swine into the American South and California; it was their practice to release the animals into the woods, let them fatten on acorns and grasses, and then hunt them as needed. In the 1840s, Russian settlers brought domestic pigs with them to Northern California and, some years later, landowners introduced an unknown number of wild Eurasian boar, probably as a big game species. The wild boar and the feral pigs have long since interbred in California and their hardy, intelligent offspring have flourished in California's oak forests and chaparral. (People commonly refer to the animals as boar, but from the looks of them the genes of domestic swine predominate; that said, California's wild pigs do have longer snouts, straighter tails, and much thicker bristles than their domestic forbears.) In the absence of serious predators, the population of wild pigs has overrun many habitats, threatening farmland and vineyards and forest; they rip up great swaths of land with their rooting, exposing it to erosion and invasive weeds.

So there was a story I could tell myself about the environmental rationale for hunting wild pig in California. But I also wanted to eat wild pig, more than I wanted to eat venison or duck or all the more diminutive birds Angelo likes to hunt. I like pork, and since coming to California I'd heard how much tastier wild pigs are than either domestic hogs or the more full-blooded boar hunted in the South. (I'd tasted that once, in a stew, and found it a little too musky.) When I asked Angelo why he hunted wild pig he didn't hesitate (or utter a word about the environment): rather, he just kissed the tips of his fingers and said, "Because it is the most delicious meat. And there is nothing that tastes so good as boar prosciutto. You'll see. You shoot a big one and we'll make some."

In a sense, that's what Angelo was really hunting, not pigs so much as prosciutti. On one of our drives up to Sonoma he'd talked a little about his philosophy of hunting and fishing. "For me it is all about the eating. Not the 'sport.' I am not what you call a trophy hunter. I take what I need, enough to make a nice dinner for me and my friends, maybe some salami, a prosciutto, but then: That's it, I go home. My friend Xavier and I have this argument every time we go hunting or fishing. He keeps on fishing even after he's caught the limit, throwing the fish back then catching it again. You know, 'catch and release.' I tell him he's catching the same fish over and over. To me that's playing with your food. You shouldn't play with your food."

On this, my first outing, we were joined by Richard, the property's owner (whom Angelo had introduced to pig hunting), and Angelo's friend Jean-Pierre, a Frenchman who works as a chef at Chez Panisse in Berkeley. Jean-Pierre hadn't hunted in years, though he had grown up hunting boar with his relatives in Brittany. He had on one of those green felt Alpine fedoras with the feather (a hat he managed to wear without so much as a trace of irony) and a pair of tall black riding boots. We didn't look much the part of an American hunting party (Angelo had on a pair of flouncy Euro-style black pants), though Richard did have on the full international orange regalia and I was wearing my brightest orange sweater. We divided into pairs, me with Angelo, and went our separate ways, with a plan to meet back at the cars for lunch around noon. Jean-Pierre and Richard set off on the logging road that descended into the lower forest, while Angelo and I planned to reconnoiter the grassy ridge in Angelo's four-wheel-drive ATV—what he calls his "bike." The bike made a racket, but Angelo claimed it didn't bother the pigs and would allow us to cover a lot more ground than we could on foot. So we put our loaded rifles on mounts on the vehicle's hood, I arranged as much of my butt as I could fit onto the narrow plywood platform behind the driver's seat, and we set off in quest of Pig, bouncing noisily down the dirt road.

"You are going to kill your first pig today," Angelo shouted over the roar of the engine. Given the nature of hunting, not to mention me, I understood this as less a prediction than a prayer. Around every curve of road we came upon another "really good spot" or "very prevalent area," and each such place had a hunting story attached to it. Indeed, the whole landscape soon became an epic of pig death and pig death narrowly averted. There was the one about the sow Angelo couldn't bring himself to shoot because her piglets were trailing behind. ("But since then I learned that another pig would have adopted the babies, that's what they do, so maybe next time. . . .") There was the spot where he'd fired into a knot of pigs and hit two with one bullet. And then there was the place where he'd taken a long shot at a boar that must have been easily three, four hundred pounds, but missed. A story about the big one that got away is all-important, of course, since it imbues the hunting ground with mythical possibility. The big one was still out there, somewhere.

After a while we parked the bike and set out on foot on our own. Angelo gave me a route and a destination—a wallow in a grassy opening at the bottom of a ravine—and told me to find a tree with a good view of it and wait there, perfectly still, for twenty minutes until I heard him whistle. He would make his way toward the same spot from another direction, in the hopes of driving some pigs into my field of vision.

When I could hear Angelo's footsteps no more my ears and eyes started tuning in—everything. It was as if I'd dialed up the gain on all my senses or quieted myself to such an extent that the world itself grew louder and brighter. I quickly learned to filter out the static of birdsong, of which there was plenty at that early hour, and to listen for the frequency of specific sounds—the crack of branches or the snuffling of animals. I found I could see farther into the woods than I ever had before, picking out the tiniest changes in my visual field at an almost inconceivable distance, just so long as those changes involved movement or blackness. The sharpness of focus and depth of field was uncanny, though, being nearsighted, I knew it well from the experience of putting on glasses with a strong new prescription for the first time. "Hunter's eye," Angelo said later when I described the phenomenon; he knew all about it.

I found a shaded spot overlooking the wallow and crouched down in the leaves, steadying my back against the smooth trunk of a madrone. I rested my gun across my thighs and got quiet. The whoosh of air through my nostrils suddenly sounded calamitous, so I began inhaling and exhaling through my mouth, silencing my breath. So much sensory information was coming into my head that it seemed to push out the normal buzz of consciousness. The state felt very much like meditation, though it took no mental effort or exercise to achieve that kind of head-emptying presence. The simple act of looking and listening, tuning my senses to the forest frequencies of Pig, occupied every quadrant of mental space and anchored me to the present. I must have lost track of time because the twenty minutes flashed by. Ordinarily my body would have rebelled at being asked to hold a crouch that long, but I felt no need to change position or even to shift my weight.

Later it occurred to me that this mental state, which I quite liked, in many ways resembled the one induced by smoking marijuana: the way one's senses feel especially acute and the mind seems to forget every-thing outside the scope of its present focus, including physical discom-fort and the passing of time. One of the more interesting areas of research in the neurosciences today is the study of the brain's "cannabi-noid network," a set of receptors in the nervous system that are acti-vated by a group of unusual compounds called cannabinoids. One of these compounds is THC, the active ingredient in marijuana; another is anandamide, a recently discovered neurotransmitter manufactured in the brain (and named by its discoverer for the Sanskrit word for inner bliss). Whether made by the plant or the brain, cannabinoids have the effect of intensifying sensory experience, disabling short-term mem-ory, and stimulating appetite. Scientists still aren't certain what the evolutionary utility of such a system might be. Some researchers hypothesize that the cannabinoids, like the opiates, play a role in the brain's pain relief and reward system; others that they help regulate ap-petite, or emotion.

The experience of hunting suggests another theory. Could it be that the cannabinoid network is precisely the sort of adaptation that natural selection would favor in the evolution of a creature who survives by hunting? A brain chemical that sharpens the senses, narrows your men-tal focus, allows you to forget everything extraneous to the task at hand (including physical discomfort and the passage of time), and makes you hungry would seem to be the perfect pharmacological tool for man the hunter. All at once it provides the motive, the reward, and the optimal mind-set for hunting. I would not be the least bit surprised to discover that what I was feeling in the woods that morning, crouching against a tree avidly surveying that forest grove, was a tide of anandamide wash-ing over my brain.

But whether I was actually having a cannabinoid moment or not, in the moments before Angelo's whistle pierced my vigil I did feel as though I had somehow entered nature through a new door. For once I was not a spectator but a full participant in the life of the forest. Later, when I reread Ortega y Gasset's description of the experience, I decided that maybe he wasn't so crazy after all, not even when he asserted that hunting offers us our last best chance to escape history and return to the state of nature, if only for a time—for what he called a "vacation from the human condition."

> When one is hunting, the air has another, more exquisite feel as it glides over the skin or enters the lungs, the rocks acquire a more expressive physiognomy, and the vegetation becomes loaded with meaning. But all this is due to the fact that the hunter, while he advances or waits crouching, feels tied through the earth to the animal he pursues, whether the animal is in view, hidden, or absent.

The tourist in nature achieves no such immersion or connection; all he sees is a landscape, which is something made by history (and rather recently at that). His gaze conditioned by art and expectation, the tourist remains a spectator to a scene, unable to get outside himself or history, since the landscape he beholds is as much the product of his civilization as of nature.

> The tourist sees broadly the great spaces, but his gaze glides, it seizes nothing, it does not perceive the role of each ingredient in the dynamic architecture of the countryside. Only the hunter, imitating the perpetual alertness of the wild animal, for whom everything is danger, sees everything and sees each thing functioning as facility or difficulty, as risk or protection.

Ortega believed that in hunting we returned to nature because "hunting is the generic way of being a man" and because the animal we are stalking summons the animal still in us. This is atavism pure and simple—the recovery of an earlier mode of being human—and that for Ortega is the supreme, and the exclusive, value of hunting. For perhaps his most outrageous claim is that the hunt is the only such return available to us—we can't ever, as he points out, go back to being Christian in the manner of St. Augustine, say, because once history begins it is irreversible. So how is it we can still go back to being Paleolithic? Because our identity as hunters is literally prehistoric—is in fact inscribed by evolution in the architecture of our bodies and brains. (Of course, the same might be said about gathering, too, which Ortega doesn't address; my guess is that that way of being in nature is insufficiently dramatic or masculine for the Spaniard's taste.) Much that surrounds hunting nowadays is completely artificial, Ortega freely admits, yet the experience itself, the encounter of predator and prey, is no fiction. (Just ask the animals.) Even though the hunt takes place during a brief "vacation" from modern life, what occurs in the space of that electrifying parenthesis will ever and always be, in a word Ortega never shrinks from using, "authentic."

3. READY. OR NOT.

As I said, all this seemed much less crazy to me after I'd been in the woods that first morning with my gun, long before I even had occasion to fire it. I'm chagrined to report that that occasion never presented itself during that first hunting trip—or rather, when it did present itself I was in no position to do anything about it. I know, I've been talking here like Mister Big Game Hunter, comparing notes on the experience with the likes of Ortega y Gasset, but I returned from the woods that day not only empty-handed, which in hunting is entirely forgivable, but, what is not, having failed as a hunter because I was not *ready*.

I blame this, at least partly, on lunch.

By the end of the morning one animal had been shot, a small boar taken by Jean-Pierre. He and Richard had spotted two pigs in the lower forest, a big one and a little one, but by the time they could agree on whose shot it was (Richard politely deferring to his guest, Jean-Pierre to his host) the bigger one had bolted. On our way back up to the ridge in the ATV Angelo and I picked up Jean-Pierre's animal; it wasn't a whole lot bigger than a poodle, with a florid red blotch erupting from the side of its bristly black head. Angelo hung it by its ankles from the limb of a tree near the cars; he planned to dress it after lunch.

Being Europeans, as well as accomplished cooks, Angelo and Jean-Pierre take lunch very seriously, even when out in the woods some distance from civilization. "So I brought with me a few little things to nibble on," Jean-Pierre mumbled. "Me, too," chimed Angelo. And out of their packs came course after course of the most astonishing picnic, which they proceeded to lay out on the hood of Angelo's SUV: a terrine of lobster and halibut *en gelée*, artisanal salami and prosciutto and mortadella, Angelo's homemade pâté of boar and home-cured olives, cornichons, chicken salad, a generous selection of cheeses and breads, fresh strawberries and pastries, silverware and napkins, and, naturally, a bottle each of red and white wine.

It was a delicious lunch, but arguably it took off some of my hunter's edge. One of the easier questions on my hunter education exam went something like this: "Hunting while intoxicated is an acceptable practice, true or false." Not that I was intoxicated, but I *was* feeling notably relaxed and loquacious when Richard and I set off to look for another pig, while Angelo dressed the little one and Jean-Pierre, having already shot a pig, had a postprandial snooze in the grass. Our rifles slung over our shoulders, we strolled down a shady trail toward a spot where Richard had once had some luck, all the while getting acquainted and chatting about one thing or another. We soon discovered we'd both once worked for the same newspaper, so there was fresh gossip to be traded, scandals to dissect. Thoroughly absorbed in conversation, our attention gradually floated off from these woods all the way to a building in midtown Manhattan. Until, that is, I happened to glance up ahead and saw directly in front of us, not thirty yards away, three or four large black shapes swimming in the shadows. The path ahead was deeply shaded by a steep embankment and a large oak, but the sight of these pigs, my first, was incontrovertible, and their sudden appearance violently wrenched my attention back to the forest's present. There they were, four large pigs milling beneath the oak, their attention fixed on the acorns littering the path that connected us. In-credibly, they gave no sign that they'd spotted us or heard our yammering.

I grabbed Richard by the shoulder, put my finger to my lips, and pointed ahead. He stopped. "It's your shot," he whispered. "Go ahead. Take it." It is the custom when hunting with companions that the first shot belongs to the person who spotted the animal, perhaps in recog-nition of the fact that skill in hunting is as much about finding the game as killing it. In fact, in many hunter-gatherer societies, the first portion of meat belongs not to the hunter who kills the animal but to the one to whom it first appeared. These pigs were mine.

One little problem. I had neglected to pump my rifle before we set out on the trail. There was no bullet in the chamber, and to cock my gun now would almost surely alert the pigs to our presence. I could take the chance, but to do so probably meant the pigs would be on the run by the time I was ready to shoot. I explained all this in a whisper to Richard, whose own gun, a fancy new Finnish bolt-action job, could be loaded with little more than a click of the little bolt. I gave him my shot.

Richard got down on one knee and slowly raised his rifle to his shoulder. I braced for the explosion, preparing to pump my gun the moment it came; perhaps I could still get off a shot at one of the oth-ers. Richard took his time, aiming carefully, waiting for one of the ani-mals to turn and offer its flank. The pigs had their heads down, eating acorns, utterly oblivious to our presence. Then the woods exploded. I saw a pig stagger and fall back against the embankment, then struggle drunkenly to its feet. I pumped my rifle but it was already too late: The other pigs were gone. Richard fired again at the wounded pig and it crumpled.

The other pigs had run down the path away from us, and we gave chase for a few minutes, but they took off around a bend and we lost them. By the time we returned to the scene Richard's pig was dead. It was considerably bigger than Jean-Pierre's poodle, and appeared to have taken a bullet in the butt. I felt a rush of adrenaline; maybe it had surged earlier, but only now did I feel all light-headed and shaky. It hadn't been my shot, yet I felt I'd been party to something momentous, something that felt like a collision of worlds. The shadowy realm of the pigs had crashed into our brightly lit world, and this emissary of that other nation had crossed over from wild life, become "meat."

The pig, a sow weighing perhaps a hundred pounds, was too heavy to carry, so we took turns dragging it by its rear leg up the path back toward the cars; I now understood in a way I never had before the expression "dead weight." Holding the pig by the ankle just above its delicate hoof, I could still feel its warmth beneath the bristly skin, some fading remnant of its formidable energy. It felt wrong to be dragging the body over the rocky ground, and I had to remind myself that the pig, though still warm, felt nothing. By the time we had dragged the carcass all the way back to the cars its skin felt cool to the touch.

Angelo trotted over to see the animal, excited and impressed and eager to hear our story. It's curious how the hunting story takes shape in the minutes after the shot, as you work through the chaotic simultaneity of that lightning, elusive moment, trying to tease out of the adrenaline fog something linear and comprehensible. Even though we'd both witnessed the event together, Richard and I had taken turns carefully telling each other the story on the long march back, rehearsing our lack of readiness, reviewing the reasons Richard had taken the shot instead of me, trying to nail down the precise distance and number of pigs involved, carefully unpacking the moment and turning our shaky recollections into a consensus of fact—a hunting story. As I watched Angelo drink in our hunting story I could see the disappointment bloom on his face. It had been my shot, my pig, but I hadn't taken it.

"You weren't ready," Angelo said, levelly. "In hunting you always need to be ready. So, okay, you learned something today. Next time you will be ready and you will take your shot." He was trying hard not to sound like the disappointed father; even so, I couldn't help feeling like the disappointing son.

So what had really happened? I hadn't been ready to shoot. But why? The practical reasons seemed clear enough; surely it had made more sense to give my shot to Richard than to risk losing the animal. It was because of my unselfish decision that we now had this pig. Yet maybe there was some deeper sense in which I hadn't been ready; maybe my failure to have a bullet in the chamber reflected some unconscious reluctance about doing what I was asking myself to do. The fact is I'd blown it, and I wasn't sure how deep I should go in search of an explanation. And yet I had been, and still was, determined to shoot a pig: I had a meal to cook, for one thing, but I was also genuinely hungry for the experience, to learn whatever it had to teach me. So I spent the rest of the afternoon hunting intently alone, walking the ridge, raking the shadows for signs of pig, looking and listening as hard as I could to will another animal out of the woods. When Angelo announced it was time to go home, I felt deflated.

Jean-Pierre generously offered to give me some cuts of his pig. Since I needed the meat for my meal I was grateful for his offer, yet I understood that to accept it underscored my inferior status in our little society of hunters. To the successful hunter goes the privilege of giving away the spoils, and I'd read a lot in the anthropological literature suggesting just how important that privilege was. The sheer nutritional density of meat has always made it a precious form of social currency among hunter-gatherers. Since the successful hunter often ends up with more meat than he or his family can eat before it spoils, it makes good sense for him to, in effect, bank the surplus in the bodies of other people, trading meat for obligations and future favors. Chimps will do the same thing. Not to say that Jean-Pierre was lording it over me or demanding anything in return; he wasn't. But that didn't change the fact that here I stood, on the vaguely pathetic receiving end of the alpha hunter's meat gift. (I briefly considered trying to educate Richard about the traditional meat rights of the game spotter, but thought better of it.) I thanked Jean-Pierre for the gift.

IN THE DAYS after I wasn't sure whether I needed to go hunting again. I had my meat. And I had *been* hunting: I felt like I had a good idea of what it was all about, or nearly all about—the hunter's way of being in nature and the way of the pigs. I'd spotted the prey and witnessed the kill. I had a pretty good story, too. And yet everyone to whom I told it managed to remind me how unsatisfactory the ending was. *You mean you never even fired your gun?!* I'd violated the Chekhovian dramatic rule: Having introduced a loaded gun in Act One, the curtain can't come down until it is fired. I might miss, but the gun had to be fired. That at least seemed to be the narrative imperative.

And then of course there was Señor Ortega y Gasset, who, as you might expect, was not about to accept me into the fellowship of hunters until I'd actually killed an animal. Mere spectatorship, or "platonic" analogues of hunting such as photography or bird watching, doesn't cut it for him. ("Platonism," he writes, "represents the maximum tradition of affected piety.")

"One can refuse to hunt," he allows, "but if one hunts one has to accept certain ultimate requirements without which the reality 'hunting' evaporates." Killing is one of those requirements. And although Ortega says one does not hunt in order to kill, he also says that one must kill in order to have hunted. Why? For authenticity's sake. If for me this venture was about taking ultimate responsibility for the animals I eat, their deaths included, well, I hadn't done that yet, had I?

I e-mailed Angelo and asked him to let me know the next time he planned to go hunting. He wrote back saying he would give me forty-eight hours notice, to get ready.

4. MY PIG

Word came about a month later, on a May Friday, that we were to meet at a gas station in Sonoma the following Monday morning, 6:00 A.M. sharp. This time it would be just the two of us.

We drove the last few miles together in Angelo's SUV, following a deserted road north of Healdsburg that curved extravagantly through deeply creased hills in the process of turning from winter green to summery gold. To me that morning all the hills looked like the backs and shoulders of great beasts, the thick grasses covering them like pelts.

Coming around the final bend before we reached Richard's gate I spotted on my side a large group of pigs, big ones and babies together, right there on the hillside that sloped down to meet the road. Angelo pulled over onto the shoulder; the pigs were on Richard's land, he said. I remembered from hunter ed that you weren't allowed to shoot from a public roadway. So we decided we'd try to spook the pigs, force them up over the crest of the hill and down the other side, which would bring them into Richard's forest. We honked, we hollered, we got out of the truck and waved our arms like lunatics, and eventually the pigs started to move up the hill.

"This gives me a very good feeling," Angelo said as we climbed back into the truck. And then he offered the prediction/prayer: "You are going to shoot your pig today. A big pig." I had my doubts, yet seeing those pigs did seem like a promising sign: They were up and about, feeding and on the move.

We spent the first part of the morning doing the circuit of Angelo's customary spots, patrolling first the ridge in the ATV and then moving down into the lower forest on foot. The entire day I kept a round in my chamber. It was hotter than last time, so Angelo felt the pigs would be keeping to the shadier parts of the property. We staked out a wallow deep in the woods, and then a trampled clearing of ferns on the near side of the hill that abuts the road, but saw no signs of the group we'd tried to herd this way.

A little after nine in the morning we were walking together down a logging road cut into a steep hillside when we were stopped in our tracks by a grunt so loud and deep and guttural that it seemed to be coming from the bowels of the earth. A very big pig was very close by. But where? What direction to look? The sound had no address; this was the grunt of the ground itself, omnipresent, more audible to my torso than my ears. We crouched down low, making ourselves as inconspicuous as possible, and listened as hard as I've ever listened for anything before, listened the way you listen when you hear a strange sound in the night.

I needn't have strained so, because the next sound we heard was nearly as loud as the first: the sharp clean crack of a branch coming from above us to our right, where the thickly oaked hillside climbed steeply to a crest. A stream ran down the hillside and crossed the path in front of us about thirty yards ahead. With my eyes I followed the silvery line of the stream up through the woods to the crest, and that's when I saw it: a rounded black form, a negative of sunrise, coming over the top of the hill. Then another black sun, and another, a total of five or six, I couldn't be sure, popping over the crest in a line like a string of huge black pearls.

I touched Angelo on the shoulder and pointed toward the pigs. *What should I do?* This time my gun was cocked, of course, and now, for the first time, I took off the safety. *Should I shoot? No, you wait,* Angelo said. *See—they're coming down the hill now.* I followed the pigs with the barrel of my gun, trying to get one of them in my sight. My finger rested lightly on the trigger, and it took all the self-restraint I could summon not to squeeze, but I didn't have a clear shot—too many trees stood in the way. *Take your time,* Angelo whispered. *They will come to us.* And so they did, following the streambed down to the road directly in front of us, moving toward us in an excruciatingly slow parade. I have no idea how long it took the pigs to pick their way down the steep hill, whether it was minutes or just seconds. At last the first animal, a big black one, stepped out into the clearing of the dirt road, followed by another that was just as big but much lighter in color. The second pig presented its flank. *Now!* Angelo whispered. *This is your shot!*

I could sense Angelo a step or two behind me, preparing to take his shot the second I took mine. We were both down on one knee. I braced the rifle against my shoulder and lined up my sight. I felt calmer and clearer than I expected to; at least, when I looked down the barrel of the rifle it didn't appear to be wagging uncontrollably. I took aim at the shoulder of the grayish pig, aligning the site's U and I with the top of the animal's front leg, and then inched down a shade, hoping to correct for the fact that at the rifle range my shots had all landed several inches high. I held my breath, resisted a sudden urge to clamp my eyes shut, and gently squeezed.

The crystal stillness of the scene and the moment in time now exploded into a thousand shards of sense. The pigs erupted in panic, moving every which way at once like black bumper cars, and then the *blam!* of Angelo's shot directly behind made me jump. One pig was down; another seemed to stagger. I pumped my gun to fire again but the adrenaline was surging now and I was shaking so violently that my finger accidentally pressed the trigger before I could lower my gun; the shot went wild, skying far over the heads of the rioting pigs. Something like the fog of war now descended on the scene, and I'm uncertain exactly what happened next, but I believe Angelo fired a second time. I collected myself just enough to pump and fire one more poorly aimed round before the pigs dispersed, most of them tumbling down the steep embankment to our left.

We ran forward to the downed animal, a very large grayish sow beached on her side across the dirt road; a glossy marble of blood bubbled directly beneath her ear. The pig thrashed briefly, attempting to lift her head, then gave it up. Death was quickly overtaking her, and I was relieved she wouldn't need a second shot. We ran past her, looking for the others. Angelo said he thought he had grazed another one, and I climbed down the embankment looking for it, but very quickly the going got treacherously steep and Angelo called me back up to the road to see what I'd done.

Angelo clapped me on the back and congratulated me extravagantly. "Your first pig! Look at the size of it. And with a perfect shot, right in the head. You did it!" *Did* I do it? Was that really my shot? I *thought* my first shot had dropped the pig but already that moment was blurred irretrievably, and when I saw what a clean shot it was I suddenly had my doubts. Yet Angelo was adamant—he had fired at a different pig, a black one. "No, this is your pig, Michael, you killed it, there's no doubt in my mind." Our hunting story was taking form, the fluid confusion of the moment rapidly hardening into something sturdier and sharper than it really was. "What a great shot," Angelo continued. "You got yourself a big one. That's some very nice prosciutti!"

Meat I was not yet quite ready to see. What I saw was a dead wild animal, its head lying on the dirt in a widening circle of blood. I kneeled down and pressed the palm of my hand against the pig's belly above the nipples and felt beneath the dusty, bristly skin her warmth, but no heartbeat. My emotions were as surging and confused as the knot of panicked pigs had been on this spot just a moment before. The first to surface was this powerful upwelling of pride: I had actually done this thing I'd set out to do, had successfully shot a pig. I felt a flood of relief, too, that the deed was done, thank God, and didn't need to be done again. And then there was this wholly unexpected feeling of gratitude. But for what exactly, or to whom? For my good fortune, I guess, and to Angelo, of course, but also to this animal, for stepping unbidden over the crest of that hill, out of the wild and into my sight, to become what Angelo kept calling her: *your pig*. More than the product of any labor of mine (save receptiveness) the animal was a gift—from whom or what I couldn't say—but gratitude seemed in order, and gratitude is what I felt.

The one emotion I expected to feel but did not, inexplicably, was remorse, or even ambivalence. All that would come later, but now, I'm slightly embarrassed to admit, I felt absolutely terrific—unambiguously *happy*. Angelo wanted to take my picture, so he posed me behind my pig, one hand cradling the rifle across my chest, the other resting on the animal. I couldn't decide whether to smile or to compose a more somber expression. I opted for the latter, but I couldn't quite manage to untie the knot of my smile. "Every good hunter is uneasy in the depths of his conscience when faced with the death he is about to inflict on the enchanting animal," I'd read in Ortega y Gasset's *Meditations*, but I was unable to locate this feeling, either immediately before or after the fateful shot. Nor did I register, yet anyway, the slightest disgust at the creeping stain of the animal's blood on the ground, the stain that I remembered Ortega calling a "degradation." I was still too excited, too *interested* in this most improbable drama in which I had somehow found myself playing the hero's part.

5. MAKING MEAT

The sense of elation didn't last. Less than an hour later I found myself in a much less heroic role, embracing the pig's hanging carcass from behind to steady it so Angelo could reach in and pull out its viscera. I was playing the nurse now, passing him tools and holding the patient still. Using a block and tackle and a stainless steel hanger with two hooks that Angelo had forged expressly for this purpose, we'd managed to raise and hang the pig by its rear ankles from the stout limb of an oak tree. A scale attached to the rig gave the weight of the animal: 190 pounds. The pig weighed exactly as much as I did.

Dead bodies are awkward, among other things, and negotiating one this big proved a difficult, clumsy, and oddly intimate operation. It took us a while, but we managed to hoist the pig up onto the hood of the ATV, get it up the hill without falling off and then into this tree. I kept finding myself in awkward embraces with my pig, as when I had to press with all my weight against the carcass when it threatened to slide down off the hood, or when Angelo needed me to wrap my arms around the pig to keep it from swinging while he cut into it. Dressing the pig was further complicated by the fact that we planned on making prosciutto, which requires that the hide covering the hams be left intact. So instead of skinning the hindquarters we had to shave them, painstakingly scraping the animal's dust-caked thighs with our knife blades to remove all her bristles.

Next Angelo made a shallow incision along an equator circling the pig's belly and began to gently work the hide loose. I held down a narrow flap of skin while he cut into the fat behind it, leaving as much of the creamy white adipose layer on the carcass as possible. "This is really good fat," Angelo explained, "for the salami." The flap of skin grew larger as we worked our way down the body and then slowly pulled it down over the pig's shoulders, until the inside-out skin looked like a discarded sweater caught in the instant it comes over the head. What hunters call dressing an animal is really an undressing.

As we drew the skin down over the rib cage it exposed the bullet, or what remained of the bullet. It had torn a ragged slot in the last rib and come to rest there, just beneath the hide. "Here's a souvenir for you," Angelo said, extracting the bloody, mangled chunk of metal from the bone like a tooth and handing it to me. The bullet was too smashed up to easily identify its caliber, though it occurred to me a forensics expert could probably determine whether it had really come from my rifle, settle once and for all—the words "Warren Commission!" popped into my head—whether or not there had been a second gunman.

Angelo worked with a small cigar clamped between his teeth; the smoke discouraged the flies and yellow jackets, which had taken an avid interest in the dead animal. There were also a pair of turkey vultures circling high overhead, patiently waiting for us to finish. Whatever parts of this pig we didn't take the local fauna were preparing to set upon and consume, weaving this bonanza of fat and protein back into the fabric of the land. Using a short knife, Angelo made another shallow incision the length of the animal's belly, moving very slowly so as not to pierce any internal organs. A punctured bladder would give the meat a nasty, ineradicable taint, he explained, and cutting into the colon risked contaminating it with intestinal bacteria.

Angelo talked while he worked, mostly, if you can believe it, about food. As he cut into the thin visceral membrane that held all the organs in the body cavity in a translucent bag, he told me all about ventricina, a dish made in Abruzzi that involves stuffing the visceral membrane with various "noble cuts" of the pig and then hanging it to cure like a salami. "It's tricky to keep the bag from tearing, but one of these days I'll make some."

I could not believe Angelo was still talking about food. The pig was splayed open now, all its internal organs glistening in their place like one of those cutaway anatomy dolls from biology: the bluish links of intestine coiled beneath the stout muscle of the heart, beribboned with

its map of veins; the spongy pink pair of lungs like outspread wings behind; and below, the sleek chocolate slab of liver. I'd handled plenty of
viscera in the chickens I'd gutted on Joel's farm, but this was different
and more disturbing, probably because the pig's internal organs, in
their proportions and arrangement and colors, looked exactly like human organs. Which is why, I recalled, surgeons hone their skills by operating on pigs.

I held the cavity open while Angelo reached in to pull out the mass
of organs, hoping to save the liver, which had a jagged tear across it. The
bullet had apparently crossed the rib cage diagonally from upper left to
lower right, tearing through a lobe of the liver. But Angelo thought the
liver was salvageable ("for a nice pâté"), so he cut it free and dropped
it into a Ziploc bag. Then he reached in and pulled gently and the rest
of the viscera tumbled out onto the ground in a heap, up from which
rose a stench so awful it made me gag. This was not just the stink of pig
shit or piss but those comparatively benign smells compounded by an
odor so wretched and ancient that death alone could release it. I felt a
wave of nausea begin to build in my gut. The clinical disinterest with
which I had approached the whole process of cleaning my pig collapsed all at once: This was disgusting.

I still had my arms wrapped around the pig from behind, holding it
steady and open, but I needed, badly, to break away for a moment to locate an uncontaminated breath. So I told Angelo I wanted to take a picture of him working on the pig. This was not a picture I particularly
wanted (to the contrary), but the time and distance that snapping it
now promised suddenly seemed precious beyond reason. I turned away
and gulped a breath of clean air, then went off—blessed errand!—in
search of Angelo's camera.

Since it was my plan to cook, serve, and eat this animal, the revulsion at its sight and smell that now consumed me was discouraging, to
say the least. That plan was no longer just a narrative conceit either,
since the moment I killed this pig I felt it descend on me with the
weight of a moral obligation. And yet now the prospect of sitting down

to a meal of this animal was unthinkable. Pâté? Prosciutto? *Ventricina?* Just then I could have made myself vomit simply by picturing myself putting a fork to a bite of this pig. How was I ever going to get past this? And what was this spasm of revulsion all about, anyway?

Disgust, I understood, is one of the tools humans have evolved to navigate the omnivore's dilemma. The emotion alerts us to things we should not ingest, like rotten meat or feces. And surely that protective reflex figured in what I was feeling as I beheld these viscera, which no doubt did contain things that could sicken me. The stink in my nostrils was probably the contents of the pig's intestines, which had split open on the ground, in various stages of digestion and decomposition. So here, I guess, was the "intuitive microbiology" of disgust at work.

But there had to be more to it than that, and later, when I went back and reread Paul Rozin on disgust, I had a better idea of what else might underlie my revulsion. As Rozin has written, most of the things that disgust people universally do come from animals—bodily fluids and secretions, decaying flesh, corpses. This makes meat eating especially problematic, which might explain why cultures have more rules and taboos governing the eating of meat than of any other food, rules that specify not only which animals are okay to eat and not, but which parts of those animals and how they must be killed.

Beyond the sanitary reasons for avoiding certain parts and products of animals, these things disgust us, Rozin suggests, because they confront us with the reality of our own animal nature. So much of the human project is concerned with distinguishing ourselves from beasts that we seem strenuously to avoid things that remind us that we are beasts too—animals that urinate, defecate, copulate, bleed, die, stink, and decompose. Rozin tells a story about Cotton Mather, who confided to his journal the powerful revulsion he felt at finding himself urinating alongside a dog. Mather turned his self-disgust into a resolution of self-transcendence: "Yet I will be a more noble creature; at the very time when my natural necessities debase me into the condition of the beast, my spirit shall (I say at that very time!) rise and soar."

Exactly why we would strive so hard to distance ourselves from our animality is a large question, but surely the human fear of death figures in the answer. What we see animals do an awful lot of is die, very often at our hands. Animals resist dying, but, having no conception of death, they don't give it nearly as much thought as we do. And one of the main thoughts about it we think is, will my own death be like this animal's or not? The belief, or hope, that human death is somehow different from animal death is precious to us—but unprovable. Whether it is or is not is one of the questions I suspect we're trying to answer whenever we look into the eyes of an animal.

From the moment I laid eyes on my animal straight through to the moment Angelo sawed off her head her eyes remained tightly shut beneath her disconcerting eyelashes, yet everything else about the episode asked me to confront these kinds of questions. What disgusted me about "cleaning" the animal was just how messy—in every sense of the word—the process really was, how it forced me to look at and smell and touch and even to taste the death, at my hands, of a creature my size that, on the inside at least, had all the same parts and probably looked an awful lot like I did. Such an encounter is no doubt more disturbing to someone who, like me, lacks the religious certainty that humans have souls and animals don't, period. The line between human and animal that I could discern here was nowhere near that sharp. Cannibalism is one of the things that most deeply disgusts us, and while this isn't by any reasonable definition that, you could forgive the mind for being fooled into reacting as if it were—in disgust.

Here, I decided, was one of the signal virtues of hunting: It puts large questions about who we and the animals are, and the nature of our respective deaths, squarely before the hunter, and while I'm sure there are many hunters who manage to avoid their gaze, that must take some doing. As Ortega writes in his *Meditations*, hunting plunges us into the intertwined enigmas of death and animals, enigmas that admit of no easy answers or resolution. This for him is the wellspring of the hunter's uneasiness: "He does not have the final and firm conviction that his conduct is correct. But neither, it should be understood, is he certain of the opposite."

Ambivalence and ambiguity are the hunter's lot, and according to Ortega it has probably ever been thus. Like John Berger he believes that the mystery of animals—how they can be at once so like and unlike us—has always been one of the central mysteries of human life: "Humanity sees itself as something emerging from animality, but it cannot be sure of having transcended that state completely. The animal remains too close for us not to feel mysterious communication with it." Those moderns who have had the clearest idea about the animals, and therefore the least uneasiness about killing them, were the Cartesians, who decided animals were, in effect, mineral—insensible machines. Unfortunately for us, they were wrong.

So we are left standing there in the woods with our uneasiness and our disgust, and disgust's boon companion, shame. I mentioned earlier that I had not registered any such emotion in the moments after shooting my pig, but eventually it dawned, or fell, on me, like a great and unexpected weight. It happened late that evening, when back at home I opened my e-mail and saw that Angelo had sent me some digital pictures under the subject heading Look the great hunter! I was eager to open them, excited to show my family my pig, since it hadn't come home with me, but was hanging in Angelo's walk-in cooler.

The image that appeared on my computer screen hit me like an unexpected blow to the body. A hunter in an orange sweater was kneeling on the ground behind a pig the side of whose head has erupted in blood that is spreading like a river delta toward the bottom of the frame. The hunter's rifle is angled just so across his chest; clearly he is observing some hoary convention of the hunter's trophy portrait. One proprietary hand rests on the dead animal's broad flank. The man is looking into the camera with an expression of unbounded pride, wearing a big shit-eating grin that might have been winning, if perhaps incomprehensible, had the bloodied carcass sprawled beneath him been cropped out of the frame. But the bloodied carcass was right there, front and center, and it rendered that grin—there's no other word for it—obscene. I felt as though I had stumbled on some stranger's pornography. I hurried my mouse to the corner of the image and clicked, closing it as quickly as I could. No one should ever see this.

What could I possibly have been thinking? What was the man in that picture feeling? I couldn't for the life of me explain what could have inspired such a mad grin, it seemed so distant and alien from me now. If I didn't know better I would have said that the man in the picture was drunk. And perhaps he was, captured in the throes of some sort of Dionysian intoxication, the "bloodlust" that Ortega says will sometimes overtake the successful hunter. And what was I so damned proud of, anyway? I'd killed a pig with a gun, big deal.

Like the mirror in the convenience store earlier that afternoon, Angelo's digital photo had shown me the hunt, and the hunter, from the outside, subjecting it to a merciless gaze that hunting can't withstand, at least not in the twenty-first century. Yet I'm not prepared to say that that gaze offers the more truthful view of the matter. The picture is a jolting dispatch from the deep interior of an experience that does not easily travel across the borders of modern life. Angelo's pictures—there were more, and eventually I looked at them all—resemble in certain respects the trophy photos sent home by soldiers, who shock their brides and mothers with images of themselves grinning astride the corpses of the enemy dead. They are entitled to their pride—killing is precisely what we've asked them to do—and yet do we really have to look at the pictures?

I've looked at Angelo's pictures again, trying to figure out why they should have shamed me so. I realize it isn't the killing they record that I felt ashamed of, not exactly, but the manifest joy I seemed to be feeling about what I'd done. This for many people is what is most offensive about hunting—to some, disgusting: that it encourages, or allows, us not only to kill but to take a certain pleasure in killing. It's not as though the rest of us don't countenance the killing of tens of millions of animals every year. Yet for some reason we feel more comfortable with the mechanical killing practiced, out of view and without emotion, by industrial agriculture.

Perhaps there is a more generous light in which to regard the hunter's joy. Perhaps it is the joy of a creature succeeding at something he has discovered his nature has superbly equipped him to do, an action that is less a perversion of that nature, his "creaturely character," than a fulfillment of it. But what of the animal in the picture? Well, the animal too has had the chance to fulfill its wild nature, has lived, and arguably even died, in a manner consistent with its creaturely character. Hers is, by the standards of animal death, a good one. But could I really say that yet? What if it turned out I couldn't eat this meat? I realized that the drama of the hunt doesn't end until the animal arrives at the table.

"For one creature to mourn the death of another is a new thing under the sun," wrote Aldo Leopold, himself a deeply conflicted hunter. It is a very good thing indeed, he suggests, but we would do well to recognize just how new it is, what a departure from the usual order of nature this mourning represents. What shames at least some of us about hunting is the same thing that shames us about every other reminder of our origins: that is, the incompleteness of our transcendence of our animal nature.

So which view of me the hunter is the right one, the shame at the photograph or the joy of the man in it, the outside gaze or the inside one? The moralist is eager to decide this question once and for all, to join Cotton Mather in his noble quest for a more complete transcendence. The hunter—or at least the grown-up hunter, the uneasy hunter—recognizes the truths disclosed in both views, which is why his joy is tempered by shame, his appetite shadowed by disgust.

The fact that you cannot come out of hunting feeling unambiguously good about it is perhaps what should commend the practice to us. You certainly don't come out of it eager to protest your innocence. If I've learned anything about hunting and eating meat it's that it's even messier than the moralist thinks. Having killed a pig and looked at myself in that picture and now looking forward (if that's the word) to eating that pig, I have to say there is a part of me that envies the moral clarity of the vegetarian, the blamelessness of the tofu eater. Yet part of me pities him, too. Dreams of innocence are just that; they usually depend on a denial of reality that can be its own form of hubris. Ortega suggests that there is an immorality in failing to look clearly at reality, or in believing that the sheer force of human will can somehow overcome it. "The preoccupation with what should be is estimable only when the respect for what is has been exhausted."

"What is." I suppose that this as much as anything else, as much as a pig or a meal, is what I was really hunting for, and what I returned from my hunt with a slightly clearer sense of. "What is" is not an answer to anything, exactly; it doesn't tell you what to do or even what to think. Yet respect for what is does point us in a direction. That direction just happens to be the direction from which we came—to that place and time, I mean, where humans *looked* at the animals they killed, regarded them with reverence, and never ate them except with gratitude.

THERE WAS one other picture in Angelo's e-mail that I didn't look at very closely until some time later, no doubt because it hadn't hit me over the head the way the trophy portrait had. This was the picture I took of Angelo cleaning my pig when I needed to break away. It's a straightforward snapshot of the pig hanging from the tree but taken from a sufficient distance that you can see in that one frame the animal and the butcher and the oak tree against the sun-filled sky and the pig-plowed earth, sloping down to a brook below. You can't make out the buzzing yellow jackets or the turkey vultures doing lazy circles overhead or the acorns littering the ground, but I realized that here in this single picture you could actually observe this food chain in its totality, the entire circuit of energy and matter that had created the pig we were turning into meat for our meal. For there was the oak tree standing in the sun, light which it had transformed into the acorns that littered the ground and fed the pig that the man in the picture was turning into food. The man had done nothing to create this food chain, only stepped into a role prepared long ago for the Predator. And whatever of this Prey the man left behind the other animals here, the Scavengers, would in due course fold back into the earth, nourishing the oak so that it might in turn nourish another pig. Sun-soil-oak-pig-human: There it was, one of the food chains that have sustained life on earth for a million years made visible in a single frame, one uncluttered and most beautiful example of what is.

NINETEEN

GATHERING

The Fungi

Isn't it curious how in so many of our pastimes and hobbies we play at supplying one or another of our fundamental creaturely needs—for food, shelter, even clothing? So some people knit, others build things or chop wood, and a great many of us "work" at feeding ourselves—by gardening or hunting, fishing or foraging. An economy organized around a complex division of labor can usually get these jobs done for a fraction of the cost, in time or money, that it takes us to do them ourselves, yet something in us apparently seeks confirmation that we still have the skills needed to provide for ourselves. You know, just in case. Evidently we want to be reminded how the fundamental processes that sustain us, by now hidden behind a globe-spanning scrim of economic complexity, actually work. It may be little more than a conceit at this point, but we like to think of ourselves as self-reliant, even if only for a few hours on the weekend, even when growing the stuff yourself winds up costing twice as much as it would to buy it at the store.

Playing at self-reliance takes different forms in different people, and you can probably tell a lot about a person by his choice of atavism:

whether he's drawn to the patient and solitary attentiveness of fishing, the strict mathematical syntax of building, the emotional drama of the hunt, or the mostly comic dialogue with other species that unfolds in the garden. Most of us have a pretty good idea which of these jobs we'd try for if somehow a time machine were to plunk us down in the Pleistocene or Neolithic.

At least until my adventures in hunting and gathering I'd always thought of myself as a Neolithic kind of guy. Growing food has been my atavism of choice since I was ten years old, when I planted a "farm" in my parents' suburban yard and set up a farm stand patronized, pretty much exclusively, by my mother. The mysteries of germination and flowering and fruiting engaged me from an early age, and the fact that by planting and working an ordinary patch of dirt you could in a few months' time harvest things of taste and value was, for me, nature's most enduring astonishment. It still is.

Gardening is a way of being in nature steeped in assumptions of which the gardener is seldom more than vaguely aware—if at all. To work exclusively with domesticated species, for example, is bound to color your view of nature as being a fairly benign place, one that answers to human desires (for beauty, for tastiness). In the garden you will also, understandably, come to think of whatever grows there as belonging to you, since it is more or less the product of your labors performed on your land. And you will regard the wilder, less tractable residents of your garden, the ones you didn't invite, as "pests"—the Other. The gardener is a confirmed dualist, dividing his world into crisp categories: cultivated land and wilderness, domestic and wild species, mine and theirs, home and away. The gardener, like the farmer, lives in a well-marked and most legible world.

I hadn't actually thought about the gardener's worldview in this light till I'd spent some time mushroom hunting, which proposes a whole other way of being in nature. Hunting for mushrooms is an operation that superficially resembles harvesting—you're looking around in nature for the ready-to-eat—yet you quickly discover that the two activities could hardly be more different. For starters, mushrooms are usually hunted in an unfamiliar place where you stand a very good chance of getting lost, particularly since you are looking down at the ground so determinedly the whole time. Getting lost just isn't much of a problem in the garden. (Which is why gardeners looking to create that experience plant mazes.) And whereas in your garden the ready-to-eat tomato beckons to you, flashing red from out of the undifferentiated green, mushrooms definitely hide. Picking and eating the wrong ones could get you killed, too, something not easily done in the garden. No, gratifying human needs and desires is just not what mushrooms are about. Mushrooms, you soon discover, are wild things in every way, beings pursuing their own agenda quite apart from ours. Which is why "hunting," rather than harvesting, is the mycophile's preferred term of art.

1. FIVE CHANTERELLES

It was a Sunday morning in late January when I got the call from Angelo.

"The chanterelles are up," he announced.

"How do you know? Have you been out looking?"

"No, not yet. But it's been three weeks since the big rains." We'd had a torrential week between the holidays. "They're up now, I'm sure of that. We should go tomorrow."

At the time I barely knew Angelo (we had yet to go pig hunting), which made his invitation to come mushrooming with him all the more generous. Mushroom hunters are famously protective of their "spots," and a good chanterelle spot is a precious personal possession (though not quite as precious as a good porcini spot). Before Angelo agreed to take me I'd asked a slew of acquaintances I knew to be mycophiles if I might accompany them. (The Bay Area is home to many such people, probably because mushroom hunting marries the region's two guiding obsessions: eating and the outdoors.) I was always careful to solemnly swear to protect the location of their spots. For some people you could see at once that this was an entirely outrageous request, tantamount to asking if I might borrow their credit card for the afternoon. Others reacted more calmly, yet always cagily. Angelo's friend Jean-Pierre is reputed to have good chanterelle spots right within the Berkeley city limits, but he repeatedly found polite ways to deflect my entreaties into the distant future. Several mushroom hunters responded to my request with the same joke: "Sure, you can come mushroom hunting with me, but I must tell you that immediately afterward I will have to kill you." What you fully expect to follow such a jokey warning (a warning I always parried with an offer to wear a blindfold coming and going) is some sort of conditional invitation, but it never arrives. Without ever exactly saying no, the mushroom hunter will deftly beg off or change the subject. I thought maybe the problem was that I was a writer, somebody who might do something as crazy as publish the location of a favorite spot, so I emphasized that a journalist would sooner go to jail than reveal a secret from a confidential source. This swayed precisely no one. I was beginning to think it was hopeless, that I was going to have to learn to hunt mushrooms from books—a dubious, not to mention dangerous, proposition. And then Angelo called.

Though I probably shouldn't overstate Angelo's generosity. The place he took me mushrooming was on private and gated land owned by an old friend of his, so it wasn't as though he was giving away the family jewels. The property was a vineyard outside of Glen Ellen, with several hundred untended acres of oak chaparral stretching to the northeast toward St. Helena. As soon as you stepped out of the manicured vineyard the land relaxed into gently rolling savanna, with broad sloping passages of grass, verdant after the winter rains, punctuated by shady groves of live oak and bay laurel.

The chanterelle is a mycorrhizal species, which means it lives in association with the roots of plants—oak trees, in the chanterelle's case, and usually oak trees of a venerable age. Though there must have been hundreds of promisingly ancient oaks here, Angelo, who had been hunting chanterelles on the property for years, seemed to be on a first name basis with every one of them. "That one there is a producer," he'd tell me, pointing across the meadow with his forked walking stick to an unremarkable tree. "But the one next to it, I never once found a mushroom there."

I cut my own walking stick from an oak branch and set off across the meadow to hunt beneath the tree Angelo had declared a good producer. He had instructed me to use the stick to turn over the leaf litter wherever it seemed uplifted. The stick also would carry spores from one tree to another, Angelo explained; evidently he regarded himself as something of a bumblebee to the chanterelles, transporting their genes from tree to tree. (In general mushroom hunters view their role in nature as benign.) I looked around my tree for a few minutes, walking a stooped circle under its drip line, flicking the leaf litter here and there with my stick, but I saw nothing. Eventually Angelo came over and pointed to a spot no more than a yard from where I stood. I looked, I stared, but still saw nothing but a chaotic field of tan leaves and tangled branches. Angelo got down on his knees and brushed the leaves and soil away to reveal a bright squash-colored trumpet the size of his fist. He cut it at the base with a knife and handed it to me; the mushroom was unexpectedly heavy, and cool to the touch.

How in the world had he spotted it? The mushroom hadn't even peeked up from the leaf litter yet. Apparently you had to study the leaves for subtle signs of hydraulic lift from below, and then look at the ground sideways, because the fat gold shafts of the chanterelles often reveal themselves before their tops break through the leaves. Yet when Angelo pointed to another spot under the same tree, a spot where he had obviously seen another mushroom, I was still blind. Not until he had shuffled the leaves with the tip of his stick did the golden nugget of fungus flash at me. I became convinced that Angelo had some other sense working for him besides sight, that he must be smelling the chanterelles before looking down to see them.

But that's apparently how it goes with hunting mushrooms: You have to *get your eyes on*, as hunters will sometimes put it. And after following Angelo around for a while, I did begin to get my eyes on, a little, though at first, oddly enough, this would only happen when I was in

Angelo's presence, working the same oak tree. Other novices talk about this phenomenon, and I suspect it's a little like the trick of the counting horse, who is not really doing arithmetic, as it appears, but is merely picking up subtle clues in the body language of its trainer. Wherever Angelo lingered, wherever the beams of his gaze raked the ground with particular intensity, I would look and occasionally would see. I was the horse who could count, the man who could find a chanterelle using someone else's eyes.

But before the morning was out I'd begun to find a few chanterelles on my own. I began to understand what it meant to have my eyes on, and the chanterelles started to pop out of the landscape, one and then another, almost as though they were beckoning to me. So had I stumbled on a particularly good spot or had I learned at last how to see them? Nature or nurture? There was no way of telling, though I did have the eerie experience of resurveying the very same patch of ground and finding a Siamese pair of chanterelles, bright as double egg yolks, in a spot where a moment before I could swear there had been nothing but the tan carpet of leaves. Either they had just popped up or visual perception is a lot more variable, and psychological, than we think. It is certainly ruled by expectation, because whenever I was convinced I was in a good spot the mushrooms were more likely to appear. "Seeing is believing" has it backward when it comes to hunting mushrooms; in this case, believing is seeing. My ability to see mushrooms seemed to function less like a window than a tool, a constructed and wielded thing.

After spotting a couple of nice ones I developed a measure of confidence that ultimately proved to be unfounded. Based on my still modest scores I worked out a snap theory of the Good Spot, which involved the optimal springiness of the soil and the distance from the trunk, but the theory didn't hold up. After a brief run of luck I promptly went blind again—and failed to find another mushroom all day. I would say there were no more mushrooms left to find, except that Angelo was still finding them under canopies I had supposedly exhausted; not a lot— we were a few days early, he decided—but enough to fill a grocery bag.

I had managed to find a total of five, which doesn't sound like much except that several of them weighed close to a pound each. My five chanterelles were tremendous, beautiful things I couldn't wait to taste.

And that night I did. I washed off the dirt, patted them dry, and then sliced the chanterelles into creamy white slabs. They smelled faintly of apricots, and I knew at once that this was the same mushroom I had found near my house, the one I had been afraid to taste. The squashy hue matched, and these had the same shallow gills, ridges really, running up the stalk, which flared out to meet the gently in-folded cap like a stout golden vase. I sautéed the chanterelles as Angelo had recommended, first in a dry frying pan to sweat out their water, which was copious, and then with butter and shallots. The mushrooms were delicious in a subtle way that could easily be overwhelmed or overlooked. They had a delicate flavor, fruity with a hint of pepper, and a firm but silky texture.

You might reasonably ask if, eating my wild mushrooms, I felt the least bit concerned about waking up dead. Did I harbor any lingering doubts that these mushrooms were really chanterelles—edible delicacies and not some deadly poison Angelo had mistaken for chanterelles? An understandable question, yet oddly enough, in view of my mycophobic predilections, it was no longer an issue. Oh, maybe I felt the vaguest shadow of a doubt as I lifted the first forkful, but it was easily brushed aside. I trusted Angelo implicitly, and besides, these mushrooms smelled and tasted right.

At dinner that night we joked about mushroom poisoning, recalling the time Judith had stumbled upon a prodigious patch of morels while biking with her friend Christopher in Connecticut. She came home with a trash bag half full of them, an astounding haul. But I could not bring myself to serve the mushrooms until we could get some kind of confirmation that these were indeed morels and not, say, the "false morels" that the field guides warned against. But how to be sure? I couldn't quite trust the books, or at least my reading of them. The solution to the dilemma seemed obvious, if perhaps a little heartless. I proposed to Judith we put the morels in the refrigerator overnight, and then give Christopher a call in the morning. Assuming he was sufficiently alive to answer his phone, he would undoubtedly mention whether he'd eaten the morels the previous night, and we would then know ours were safe to eat. I saw no reason to mention his role as an experimental human subject.

Well, that's one way of dealing with the omnivore's dilemma. Wild mushrooms in general throw that dilemma into particularly sharp relief, since they confront us simultaneously with some of the edible world's greatest rewards and gravest risks. Arguably, mushroom eating poses the starkest case of the omnivore's dilemma, which could explain why people hold such strong feelings, pro or con, on the subject of wild mushrooms. As mycologists are fond of pointing out, you can divide most people, and even whole cultures, into mycophiles and mycophobes. Anglo-Americans are notoriously mycophobic, while Europeans and Russians tend to be passionate mycophiles, or so mushroomers will tell you. But I suspect most of us harbor both impulses in varying proportions, approaching the wild mushroom with a heightened sense of the omnivore's basic tension as we struggle to balance our adventurousness in eating against a protective fear, our neophilia against our neophobia.

As the case of mushrooms suggests the omnivore's dilemma often comes down to a question of identification—to knowing exactly what it is you are preparing to eat. From the moment Angelo handed me that first mushroom, what is and is not a chanterelle suddenly seemed as plain to me as sunshine. I knew right then that the next time I found a chanterelle, anywhere, I would recognize it and not hesitate to eat it. Which is peculiar, when you consider that in the case of the chanterelle I found in my neighborhood, a half dozen authoritative field guides by credentialed mycologists had failed to convince me beyond a reasonable doubt of something I now was willing to bet my life on, based on the say-so of one Sicilian guy with no mycological training whatsoever. How could that be?

In deciding whether or not to ingest a new food, the omnivore will happily follow the lead of a fellow omnivore who has eaten the same food and lived to talk about it. This is one advantage we have over the rat, which has no way of sharing with other rats the results of his digestive experiments with novel foodstuffs. For the individual human, his community and culture successfully mediate the omnivore's dilemma, telling him what other people have safely eaten in the past as well as how they ate it. Just imagine if we had to decide every such edibility question on our own; only the bravest or most foolish of us would ever eat a mushroom. The social contract is a great boon to omnivores in general, and to mushroom eaters in particular.

The field guides contain our culture's accumulated wisdom on the subject of mushrooms. Curiously, though, the process of imparting and absorbing this life-and-death information works much better in person than it does on paper, whether through writing or even photography. Andrew Weil discusses this phenomenon in a wonderful series of essays on mushrooms he's collected in a volume called *The Marriage of the Sun and Moon*. "One learns most mushrooms in only one way: through people who know them. It is terribly difficult to do it from books, pictures, or written descriptions."

I wonder if books fail us here because the teaching transaction— *This one is good to eat, that one not*—is so fundamental, even primordial, that we're instinctively reluctant to trust it to any communication medium save the oldest: that is, direct personal testimony from, to put it bluntly, survivors. After all, precisely what is meant by "this one," the myriad qualities embedded in that modest little pronoun, can be conveyed only imperfectly in words and pictures. Our ability to identify plants and fungi with confidence, which after all is one of the most critical tools of our survival, involves far more sensory information than can ever be printed on a page; it is, truly, a form of "body knowledge" not easily reduced or conveyed over a distance. But now that I have held a freshly picked chanterelle in my hands, smelled its apricoty scent, registered its specific heft and the precise quality of its cool dampness (and absorbed who knows how many other qualities beneath the threshold of conscious notice), I'll recognize the next one without a moment's hesitation. At least in the case of this one species, my mycophobic instinct has been stilled, allowing me to enjoy. It's not every day you acquire such a sturdy piece of knowledge.

2. MUSHROOMS ARE MYSTERIOUS

I put that knowledge to good use the following week, when I returned to the oak tree near my house and found beneath it a gold rush of chanterelles. I hadn't thought to bring a bag, and there were more chanterelles than I could carry, so I made a carrier of my T-shirt, folding it up in front of me like a basket, and then filled it with the big, mud-encrusted mushrooms. I drew looks from passers-by—looks of envy, I decided, though at the time I was so excited I may have gotten that wrong. So now I have a spot and, just like Jean-Pierre's, it's right here in town. (Please don't ask me where it is; I don't want to have to kill you.)

Once the rains stopped in April the chanterelles were done for the year, and there wouldn't be another important mushroom to hunt until the morels came up in May. I used the time before then to read about mushrooms and talk to mycologists, hoping to answer some of the questions I had collected about fungi, a life form I was beginning to regard as deeply mysterious. What made mushrooms mushroom when and where they did? Why do chanterelles associate with oaks and morels with pines? Why under this tree and not that one? How long do they live? Why do some mushrooms manufacture deadly toxins, not to mention powerful hallucinogens and a range of delicious flavors? I brought the gardener's perspective to these plantlike objects, but of course they're not plants, and plant knowledge is all but useless in understanding fungi, which are in fact more closely related to animals than they are to plants.

As it happens the answers to most of my questions about mushrooms, even the most straightforward ones, are elusive. Indeed, it is humbling to realize just how little we know about this, the third kingdom of life on earth. The books I consulted brimmed with confessions of their ignorance: "it is not known why this should be" . . . "the number of genders among fungi is as yet undetermined" . . . "the exact mechanisms by which this phenomenon occurs are not entirely understood at this time" . . . "the fundamental chemistry responsible for the vivid hallucinations was a mystery then, and remains so today" . . . "it is not certain whether the morel is a saprophytic or a mycorrhizal species, or perhaps it is both, a changeling" . . . and so on, through thousands of pages of the mycological literature. When I went to visit David Arora, the renowned mycologist whose doorstop of a field guide, Mushrooms Demystified, is the West Coast mushroomer's bible, I asked him what he considered the big open questions in his field. Without a moment's hesitation he named two: "Why here and not there? Why now and not then?"

In other words, we don't know the most basic things about mushrooms.

Part of the problem is simply that fungi are very difficult to observe. What we call a mushroom is only the tip of the iceberg of a much bigger and essentially invisible organism that lives most of its life underground. The mushroom is the "fruiting body" of a subterranean network of microscopic hyphae, improbably long rootlike cells that thread themselves through the soil like neurons. Bunched like cables, the hyphae form webs of (still microscopic) mycelium. Mycologists can't dig up a mushroom like a plant to study its structure because its mycelia are too tiny and delicate to tease from the soil without disintegrating. Hard as it may be to see a mushroom—the most visible and tangible part!—to see the whole organism of which it is merely a component may simply be impossible. Fungi also lack the comprehensible syntax of plants, the orderly and visible chronology of seed and vegetative growth, flower, fruit, and seed again. The fungi surely have a syntax of their own, but we don't know all its rules, especially the ones that govern the creation of a mushroom, which can take three years or thirty, depending. On what? We don't really know. All of which makes mushrooms seem autochthonous, arising seemingly from nowhere, seemingly without cause.

Fungi, lacking chlorophyll, differ from plants in that they can't manufacture food energy from the sun. Like animals, they feed on organic matter made by plants, or by plant eaters. Most of the fungi we eat obtain their energy by one of two means: saprophytically, by decomposing dead vegetable matter, and mycorrhizally, by associating with the roots of living plants. Among the saprophytes, many of which can be cultivated by inoculating a suitable mass of dead organic matter (logs, manure, grain) with their spores, are the common white button mushrooms, shiitakes, cremini, Portobellos, and oyster mushrooms. Most of the choicest wild mushrooms are impossible to cultivate, or nearly so, since they need living and often very old trees in order to grow, and can take several decades to fruit. The mycelium can grow more or less indefinitely, in some cases for centuries, without necessarily fruiting. A single fungus recently found in Michigan covers an area of forty acres underground and is thought to be a few centuries old. So inoculating old oaks or pines is no guarantee of harvesting future mushrooms, at least not on a human time scale. Presumably, these fungi live and die on an arboreal time scale.

Mycorrhizal fungi have coevolved with trees, with whom they've worked out a mutually beneficial relationship in which they trade the products of their very different metabolisms. If the special genius of plants is photosynthesis, the ability of chlorophyll to transform sunlight and water and soil minerals into carbohydrates, the special genius of fungi is the ability to break down organic molecules and minerals into simple molecules and atoms through the action of their powerful enzymes. The hyphae surround or penetrate the plant's roots, providing them with a steady diet of elements in exchange for a drop of simple sugars that the plant synthesizes in its leaves. The network of hyphae vastly extends the effective reach and surface area of a plant's root system, and while trees can survive without their fungal associates, they seldom thrive. It is thought that the fungi may also protect their plant hosts from bacterial and fungal diseases.

The talent of fungi for decomposing and recycling organic matter is what makes them indispensable, not only to trees but to all life on earth. If the soil is the earth's stomach, fungi supply its digestive enzymes—literally. Without fungi to break things down, the earth would long ago have suffocated beneath a blanket of organic matter created by plants; the dead would pile up without end, the carbon cycle would cease to function, and living things would run out of things to eat. We tend to train our attention and science on life and growth, but of course death and decomposition are no less important to nature's operations, and the fungi are the undisputed rulers of this realm.

That the fungi are so steeped in death might account for much of their mystery and our mycophobia. They stand on the threshold between the living and the dead, breaking the dead down into food for the living, a process on which no one likes to dwell. Cemeteries are usually good places to hunt for mushrooms. (Mexicans call mushrooms *carne de los muertos*—"flesh of the dead.") The fact that mushrooms can themselves be direct agents of death doesn't exactly shine their reputation, either. Just why they should produce such potent toxins isn't well understood; many mycologists assume the toxins are defenses, but others point out that if poisoning the animals that eat you is such a good survival strategy, then why aren't all mushrooms poisonous by now? Some of their toxins may simply be fungal tools for doing what fungi do: breaking down complicated organic compounds. What the deadly amanita does to a human liver is, in effect, to digest it from within.

The evolutionary reason many mushrooms produce powerful hallucinogens is even more mysterious, though it probably has nothing to do with creating hallucinations in human brains. As the word intoxication implies, substances that poison the body sometimes can change consciousness, too. This might explain why mycophiles think civilians make far too much of the dangers of mushrooms, which they see as occupying a continuum from the deadly to the really interesting. The dose makes the poison, as they say, and the same mushroom toxins that can kill can also, in smaller doses, produce astonishing mental effects, ranging from the ecstatic to the horrific. No doubt the mind-altering properties of many common mushrooms, known to people for thousands of years, have nourished the cult of mystery surrounding the fungal kingdom, in this case feeding both mycophobia and mycophilia alike.

Andrew Weil points out an interesting paradox about mushrooms: It's difficult to reconcile the extraordinary energies of these organisms with the fact that they contain relatively little of the kind of energy that scientists usually measure: calories. Because they don't supply many calories, nutritionists don't regard mushrooms as an important source of nutrition. (They do provide some minerals and vitamins, as well as a few essential amino acids, which are what give some species their meaty flavor.) But calories are simply units of solar energy that have been captured and stored by green plants and, as Weil points out, "mushrooms have little to do with the sun." They emerge at night and wither in the light of day. Their energies are of an entirely different order from those of plants, and their energies are prodigious and strange. Consider:

There are fungi like the shaggy mane (*Coprinus comatus*) that can push their soft fleshy tissue through asphalt. Inky caps (*Coprinus atramentarius*) can mushroom in a matter of hours and then, over the course of a day, dissolve themselves into a puddle of blackish ink. Oyster mushrooms (*Pleurotus ostreatus*) can digest a pile of petrochemical sludge in a fortnight, transforming the toxic waste into edible protein. (This alchemy makes more sense when you recall that what saprophytic mushrooms have evolved to do is break down complex organic molecules, which is precisely what petrochemicals are.) Jack o'lanterns (*Omphalotus olivascens*) can glow in the dark, emitting an eerie blue bioluminescence for reasons unknown. The psilocybes can alter the texture of human consciousness and inspire visions; *Amanita muscaria* can derange the mind. And of course there are the handful of fungi that can kill.

We don't have the scientific tools to measure or even account for these fungi's unusual powers. Weil speculates that their energies derive from the moon rather than the sun, that mushrooms contain, instead of calories of solar origin, prodigious amounts of lunar energy.

Okay, it is hard, I agree, to avoid the conclusion that some of the people who write about mushrooms have themselves partaken, perhaps immoderately, of the mind-altering kinds. Their reverence for their subject runs so deep that they will pursue it wherever it leads, even if that means occasionally leaping the fence of current scientific understanding. In the case of mushrooms, that's not a very tall or sturdy fence. A powerful and compelling strain of mysticism runs like branching mycelium through the mycological literature, where I encountered one incredible speculation after another: that the mycelia of fungi are literally neurons, together comprising an organ of terrestrial intelligence and communication (Paul Stamets); that the ingestion of hallucinogenic mushrooms by the higher primates spurred the rapid evolution of the human brain (Terence McKenna); that the hallucinogenic mushrooms ingested by early man inspired the shamanic visions that led to the birth of religion (Gordon Wasson); that the ritual ingestion of a hallucinogenic fungus—called ergot—by Greek thinkers (including Plato) at Eleusis is responsible for some of the greatest achievements of Greek culture, including Platonic philosophy (Wasson again); that wild mushrooms in the diet, by nourishing the human unconscious with lunar energy, "stimulate imagination and intuition" (Andrew Weil).

I'm not prepared to discount any of these speculations just because they're not provable by our science. Mushrooms are mysterious. Who's to say the day won't come when science will be able to measure the fungi's exotic energies, perhaps even calculate our minimum daily requirement of lunar calories?

3. WORKING THE BURN

After the first pig hunt Jean-Pierre had driven me home, and I used the captive car time to probe him once again on the subject of mushrooms. He yielded no ground, but did mention a mushroom hunter by the name of Anthony Tassinello, who had shown up at his restaurant earlier in the week with several pounds of morels. Jean-Pierre offered to put me in touch with Anthony. (It is amazing the lengths to which some people will go to deflect attention from their own mushroom spots.)

True to his word, Jean-Pierre sent an e-mail to Anthony, who expressed a willingness to take me morel hunting. I was surprised he'd let a complete stranger tag along, but after some back and forth by e-mail, it began to make more sense. The morels were "on," and Anthony could use an extra pair of hands, especially ones that were asking for nothing in return. As for concerns I might compromise his spots (I swore my usual oaths), the secrecy issue is not nearly so touchy in the case of the "burn morels" we would be hunting. These are morels that fruit in profusion the spring following a pine forest fire. Even if I were to disclose classified information, it would have little value beyond this spring— indeed, beyond the next couple of weeks, since he fully expected the entire California mycological community to descend on this burn the moment the word got out.

Anthony e-mailed that I should meet him in front of his house Friday morning at 6:00 A.M. sharp, warning me to come prepared for a harsh and unpredictable environment. "We'll go rain, snow, or shine. Don't laugh: It has snowed once already this spring, and we managed to find morels poking through the accumulation. It wasn't fun, but memorable.

"The weather where we'll hunt can be very different from here, and even from the valley. We'll be hiking nearly a mile above sea level, and it can be hot, cold, or wet, all in a matter of hours. Bring light layers and rain gear just in case. A solid pair of hiking boots with ankle support is a must: It is very steep, rocky terrain with huge, burned fallen trees and ground that is thoroughly soaked. Bring a hat, the sun is stronger at this elevation, plus it keeps cedar needles and spiderwebs out of your face and can double as a mushroom sack when your basket is full." Anthony also advised me to bring sunscreen and bug spray (for mosquitoes), at least a gallon of water, ChapStick, and, if I owned one, a walkie-talkie.

Morel hunting didn't sound like much fun, more like survival training than a walk in the woods. I crossed my fingers that Anthony was just trying to scare me, and set my alarm for 4:30 A.M., wondering why it is all these hunting-gathering expeditions had to begin at such ungodly hours in the morning. In the case of the pigs, I understood the need to be ready when the animals were still active early in the day, but it's not as though these morels were going anywhere after lunch. Perhaps when you're foraging you just want as much daylight as possible. Or maybe we wanted the early start to beat competing foragers to the best spots.

I pulled up to Anthony's curb a little before six to find two thirtyish-looking men in rain slickers loading an SUV with enough matériel to provision a weeklong campaign in hostile territory. Anthony was a rail-thin, angular six-footer with a goatee in the style of Frank Zappa; his friend Ben Baily was a somewhat rounder and softer man with an easy laugh. I learned on the long ride across the Central Valley that Anthony and Ben were childhood friends from Piscataway, New Jersey; after college they'd both made the pilgrimage to the Bay Area to become chefs. Anthony had been working as a pastry chef at Chez Panisse when one afternoon a backwoodsy-looking fellow dressed in camouflage showed up at the kitchen door with crates of wild mushrooms.

"I love to eat mushrooms, so I put it in his ear that I wanted to go out with him some time, and eventually it happened. He took me up to Sonoma and we found boletes and chanterelles. We just went outside and found dinner! It was such a feeling of empowerment, to feed yourself by figuring out the puzzle of nature." Anthony still works as a chef, mostly doing private dinners, which leaves many of his days free for mushroom hunting, usually with Ben (who also works as a chef). Anthony mentioned that we were going to be joined today by someone they'd met at the burn the week before, a young guy known to them only by his mycological handle: Paulie Porcini.

I gathered that Paulie Porcini was part of the subculture of mushroom hunters who travel up and down the West Coast, following the seasonal fruiting of the fungi: porcinis in the fall, chanterelles in winter, morels in the spring. "These are people living out of vans," Ben explained, "not the types who ever watch the five o'clock news." They cobble together a living selling their mushrooms to brokers who set up shop in motel rooms near the forests, post signs, and pay the hunters in cash. Anthony and Ben aren't really a part of this world; they hold jobs, live in houses, and sell their mushrooms directly to restaurants. "We don't think of ourselves as professionals yet," Anthony said.

We drove for several hours through the valley and then gradually ascended into the Sierra to Eldorado National Forest, a twelve-hundred-square-mile swath of pine and cedar stretched between Lake Tahoe and Yosemite. As we climbed into the mountains the temperature dropped down into the thirties and a frozen rain began to pelt the windshield. Along the roadside patches of old, dirty snow grew steadily larger and fresher until they expanded to cover everything. It was early May, but we had driven back into winter.

The morels come up on pine fire lands just as the snow cover retreats and the soil begins to warm, so after entering the burn area at about five thousand feet we descended along a logging road, looking for the frontier of white snow and blackened earth. At about forty-five hundred feet we found it: a forbidding moonscape in black and white. We knew our altitude because Anthony and Ben, like many mushroom hunters these days, carry portable Global Positioning System (GPS) locators—to mark good spots, calculate their altitude, and keep from getting lost.

We parked the SUV and had an initial look around. Soon after, Paulie Porcini appeared, a bearded, self-contained fellow in his twenties who carried a walking stick and had a bandanna wrapped around his head. Paulie, a man of few words, seemed like someone who was very comfortable in the woods.

The forest was gorgeous, and the forest was ghastly. Ghastly because it was, for as far as you could see, a graveyard of vertically soaring trunks that had been shorn of every horizontal, every branch, by the fire. For five days the previous October the "power fire," as it was called (it began near a power station), had roared across these mountains, consuming seventeen thousand acres of pine and cedar before a change in the wind direction allowed firefighters to contain it. The fire had been so fierce in places that it had vaporized whole trees. The only reason you knew this was because the flames, still ravenous for wood, had followed the trunks all the way down beneath the forest floor to consume the tree's roots, creating voids that reached deep into the earth. These blackened craters resembled molds that were you to fill them with plaster would yield a ghostly model of a pine tree's entire root system, accurate to the last detail. Not much lived in this desolate landscape: a handful of raptors (we heard owls), the occasional dazed squirrel, and here and there a fresh green spray of miner's lettuce that shocked the black ground.

And yet if you achieved a slightly more aestheticized view of the scene, the same landscape exhibited a tranquil, almost modernist abstraction that was just beautiful. The dead-straight black verticals ordered the hillsides as evenly as bristles on a brush, their steady rhythm varied every so often by a heavy black slash angled weirdly across the grid. The underlying shapes of the land, which was deeply creased into ravines gushing with snowmelt, had the explicitness of a line drawing, everything in view reduced to its formal essentials.

But this was virtually the last time all day I lifted my gaze to take in the panorama: As soon as Ben announced he'd spotted his first morel, I began, exclusively and determinedly, looking down. There I found a thick carpet of pine needles amid the charred carcasses of pine. A morel resembles a tanned finger wearing a dark and deeply honeycombed dunce cap. They're a decidedly comic-looking mushroom, resembling leprechauns or little penises. The morel's distinctive form and patterning would make it easy to spot if not for its color, which ranges from dun to black and could not blend in more completely with a charred landscape. From a distance, the tiny stumps of burned saplings are easily mistaken for morels; so are the blackened pinecones, many of which stuck straight out of the ground like chubby thumbs and fooled you with their patterning, rhythmic like the morels. This was hard looking, and for the first hour or so, every auspicious sighting turned out on closer inspection to be one or the other of these morel imposters.

To help me get my eyes on Ben—who by common consent had the best eyes on our team—began leaving in place patches of morels he'd found, so I could study them in situ, approaching a patch from various angles until I'd settled on the proper focal length and angle. The trigonometry of the gaze was everything, and I found that if I actually got down on the ground—which just below the layer of pine duff formed a mattress of sliding mud—I could see the little hats popping up here and there, morels that a moment before had been utterly invisible. When Ben spotted me hunting in a prone position, he approved. "We say, 'Stop, drop, and roll,' because you can see things at ground level you'll never see from above."

Ben and Anthony had a slew of these mushroom-hunting adages and I collected them over the course of the day. "Seeing is boleting" means you never see any mushrooms until someone else has demonstrated their presence by finding one. "Mushroom frustration" is what you feel when everyone around you is seeing them and you're still blind—until, that is, you find your first, thereby breaking your "mushroom virginity." Then there's the "cluster fuck," when your eyes are on and other hunters crowd you, hoping your good fortune will rub off. Cluster fucking, I was given to understand, was bad manners. And then there was the "screen saver"—the fact that after several hours interrogating the ground for little brown dunce caps, their images will be burned on your retinas. "You'll see. When you get into bed tonight," Ben said, "you'll shut your eyes and there they'll be again—wall-to-wall morels."

Anthony and Ben had dozens of theories about mushrooms—as well as a healthy appreciation for the limitations of all theories involving something as mysterious as mushrooms. They cataloged for me the "indicator species" for morels: other, more conspicuous plants and fungi that signaled their likely presence. Dogwood in bloom was a good sign that the soil had reached the proper temperature, as was, allegedly, the appearance of the ice plant, a big bright red phallus rising up from the otherwise lifeless forest floor; however, there were no morels in the vicinity of the one ice plant I spotted. A tiny brown cup fungus was another indicator species that proved somewhat more reliable. Anthony and Ben were convinced the morels would appear at the same altitude in any given week, so wherever we wandered we consulted the GPS to ascertain how high we were and tried to stay around forty-four hundred feet.

I could see why you would want theories to organize your hunting; I'd worked up my own while hunting chanterelles with Angelo. There was so much ground to cover, and the morels were so damned quiet, that theories helped divide the field on which we were playing this game of hide-and-seek into warmer and colder areas. The theories told you when to intensify your attention, scrupulously combing the forest floor with your eyes, and when you could safely rest it. For the hunter-gatherer, high-quality attentiveness is a precious but limited resource, and theories, by encapsulating past experience, help you to deploy it most efficiently.

"But you must never forget the final theory, the theory of all theories," Ben warned near the end of my morning tutorial. "We call it TPITP: The Proof Is in the Pudding." In other words, when hunting mushrooms you should be prepared to jettison all previous theories and go with whatever seems to be working in this particular place, at this particular time. Mushrooms behave unpredictably, and theories can go only so far in pushing back their mystery. "It's a lot like gambling," Ben said. "You're looking for the big score, the mother lode. The conditions might be perfect in every way, but you never know what you're going to find around the next bend—it could be a sea of mushrooms, or nothing at all."

The morning was spent wandering in more or less the same square mile or so, the four of us with our heads down, tracing utterly random patterns across the steep hillside, following trails of morels that went hot and cold. My gaze locked on a point about six steps in front of me, I'd completely lose track of my location in space and time. In this, mushroom hunting felt like a form of meditation, the morel serving as a kind of visual mantra shutting out almost every other thought. (Which was a good thing indeed, because my socks were soaked and icy.)

To regain my bearings I'd have to stop and reclaim the panoramic view, but because the day was foggy and the terrain was so heavily and deeply creased, I often had no idea in which direction the road was or the others had wandered. Every now and then a burst of static would shatter the meditative quiet, as my walkie-talkie erupted: "I've hit a mother lode down here by the creek" or "Where the hell are you guys?" (That's another kind of pleasure mushrooming affords: Boys in the woods with walkie-talkies hunting treasure.)

It was deeply satisfying when the morels appeared, a phenomenon you could swear was as much under their control as yours. I became, perforce, a student of the "pop-out effect," a term I'd first heard from mushroomers but subsequently learned is used by psychologists studying visual perception. To reliably distinguish a given object in a chaotic or monochromatic visual field is a daunting perceptual task, one so complex that researchers in artificial intelligence have struggled to teach it to computers. Yet when we fix in our mind some visual quality of the object we're hoping to spot—whether its color or pattern or shape—it will pop out of the visual field, almost as if on command. To get your eyes on is to have this narrow visual filter installed and functioning. That's why Ben had me practice on his finds, to fix in my mind's eye the pattern of morels as seen against the forest's layer of duff. To hunt for mushrooms makes you appreciate what a crucial evolutionary adaptation the pop-out effect is for a creature that forages for food in a forest—especially when that food doesn't want to be found.

Without the pop-out effect, finding one's dinner would depend on chance encounters with edible species and, of course, on fruit, the only important food source in nature that actually tries to pop out. Since the evolutionary strategy of fruiting plants is to recruit animals to transport their seeds, they've evolved to get themselves noticed, attracting us with their bright colors. In the case of the fruits and flowers, the pop-out effect is, in effect, collaborative. But just about everything else you might want to eat in the forest is hiding.

Wandering aimlessly and yet purposefully through the blackened forest, steadily turning blacker and blacker myself, I realized I had entered the existential opposite of a garden. In the garden almost every species you encounter engages with you. Nobody hides; nobody means you harm; your place in the local food chain is established and acknowledged. Everything you sense in the garden—the colors and patterns, the flavors and scents—is not only comprehensible but answers to your desires. Indeed, the plants have by now folded those desires into their genes, craftily exploiting them in order to expand their numbers and habitat. It is as much as anything else this mutualism that makes the garden the most hospitable of landscapes, for everything in it is, in some sense, an extension of ourselves, a kind of mirror. (And we are in some sense an extension of the garden's plants, unwitting means to their ends.) The domestic species in a garden (or farm) are figures in our world, live under the same roof. You can forage in the garden, in the way Adam and Eve presumably did, but there isn't much to it: no dilemmas, no hunting stories.

This forest proposes a completely different way of being in nature. The morels would just as soon I pass them by, and it will be a long time before the first berries return to this blasted landscape and declare their bright presence. It's a little like being in a foreign country: *No one knows me here!* In the forest you're encumbered by none of the agriculturist's obligations of citizenship; you feel some of the traveler's exquisite lightness of being in a place oblivious to his presence, as well as his hyperreal sense of first sight, first smell, first taste. That sense, too, of *something for nothing*, for all this is coming to you simply by dint of walking around and deploying your senses. Of course the rush of newness is usually shadowed by worry: *Am I getting lost? Should I pick that mushroom, too?*

And yet though the burned forest does not welcome us like the garden and exists completely outside the realm of our domestic arrangements, you nevertheless feel certain strands of affiliation with these wild species you're looking for: the affinities of the hunt. When it's working, the pop-out effect—this amazing perceptual tool we've developed to defeat the arts of camouflage—feels very much like the manifestation of such an affinity. Alone in the woods out of earshot of my fellow mushroom hunters, I found myself, idiotically, taunting the morels whenever a bunch of them suddenly popped out. "Gotcha!" I would cry, as if this were a game we were playing, the mushrooms and

I, and I'd just won a round. This is not something I can ever imagine saying to an apple in the garden; there, it just wouldn't be news.

I'd completely lost track of time and space when my walkie-talkie blurted, "Break for lunch—meet back at the car." I had wandered nearly a mile from the car, mostly downhill, and by the time I worked my way back up to the road, clambering up ankle-twisting hills that slid out from under my feet, the others were standing around on the roadside munching trail mix and admiring their impressive hauls. "You couldn't have picked a better day," Ben gushed when I wandered over with my own bag full of morels. "The mushrooms are so on today, I've never seen it like this—we're killing them!"

We sat on a charred log (by now we were well charred ourselves) and ate our lunch, talking about the mushrooms and the "mushroom trail" and this summer's upcoming big mycological happening. Apparently thousands of mushroom hunters were expected to descend on a vast burn deep in the Yukon, some by helicopter, to await what was expected to be a world-historical flush of morels. Paulie Porcini was thinking of going. "You get twenty-two hours of hunting up there," Paulie said, as if this were an unquestioned boon.

People have been gathering morels in burned forests forever; Ben mentioned that in Bavaria people would set forest fires for the express purpose of harvesting morels. I asked if mycologists had figured out what made the morels come up after forest fires. Were they saprophytes feeding on the roots of dead pines, suddenly plentiful, or mycorrhizal mushrooms that had suddenly lost their hosts? Nobody knew for sure, though one of Anthony's theories holds that "a bad year for the organism is a good year for us."

Mycologists I talked to later confirmed Anthony's hunch. The current thinking is that the morels found in pine forests are a mycorrhizal species for whom the death of their pine associates represents a crisis: Suddenly there are no more roots supplying them with food. So the fungus fruits, sending up morels to release trillions of spores that the wind will loft far from this blasted forest. In effect, the morels fruit in order to escape the burn, dispatching their genes to colonize new pine lands before the organism starves to death. Humans don't figure in their plans, though it may be that animals like us that eat morels do help them disperse their spores as we move them around on the way to our plates. Does hunting morels hurt the organism? Probably no more than picking apples hurts the tree, and because the morels do such a good job of hiding from us, there will always be plenty that escape our notice, each capable of releasing literally billions of spores.

Yet at the same time the morels are trying to escape the dying forest, they also play a role in its renewal. The slightly sulfurous, meaty odor of morels attracts flies, which lay eggs in the safety of the mushroom's hollow stalk. Larvae emerge and feed on the flesh of the morels; birds then return to the forest to feed on the larvae, in the process dropping seeds that sprout on the forest floor. Mushrooms are hinges in nature, now turning toward death, now toward new life.

After lunch we wandered off our separate ways again for a few more hours. I worked my way downhill, slip sliding in the mud along a steep embankment that followed a stream until it emptied into a creek. I had no idea where I was or where I was going: I was following the trail of mushrooms like a desultory train of thought, heedless of anything else. Including, as it turned out, property lines: I ran into a forester who told me I was on his company's land. But that was okay with him, just so long as I promised to tell people that logging companies aren't always evil. Logging companies aren't always evil. The forester, evidently thrilled to have someone to talk to, told me to keep an eye out along the creek—it was called Beaver Creek—for large boulders with blackened hollows scooped out of them like bowls. It seems the Washoe Indians would wash and mash acorns in these bowls and then bake them into a kind of flat bread.

I never did find one of the Indian bowls, but hearing about them made me realize that this forest had been part of a human food chain for hundreds, perhaps thousands of years. The Indians understood that you could work out relations with wild species that didn't necessarily involve bringing them under your roof. Oaks have always refused the

domestic bargain, clinging to their bitterness in the face of countless
human efforts to domesticate them. But the Indians found a way to live
off these trees even so, by devising a way to detoxify the acorns. (We
have to do something similar with these morels, which, uncooked,
would sicken us.) So much here has changed. The oaks have given way
to pine, obviously, and the forest food chain that once sustained the
Washoe along Beaver Creek now, attenuated and extended, reaches clear
to the coast, linking these woods to a pricey taste on tonight's menu at
Chez Panisse.

Along Beaver Creek that afternoon the morels were totally *on*, as Ben
would say; almost everywhere I looked the honeycombed dunce caps
appeared, and I filled a bag in less than an hour. My hands by now were
black with soot and stunk of smoke, but I could still smell the meaty
perfume of the morels, these fleshy buttons of protein popping out of
the dead earth, this seemingly spontaneous combustion of food. I was
talking to them, cheering on their every appearance, and they were
talking to me, or so it seemed. I exulted at their sudden ubiquity, which
I took, weirdly, as evidence of some new connection between us. It
sounds crazy, but there is something reciprocal about the transaction,
the looking and the appearing, as if we were each doing our part,
throwing a line of affiliation across the gulf of wildness. I've no idea
how deep into the woods I'd wandered, but I was more *outside* than I can
remember ever being, and more than a little lost, but not to the morels,
who weren't hiding from me any longer. Maybe I'd gotten good at this,
had my eyes on; or maybe it was them, revealing themselves at last because I had found a way out of my world and into theirs.

Whichever it was, here was the warm sun of fortune smiling on
me, this sudden shower of forest flesh, and I felt, again, the gratitude I'd
felt in that other forest, the moment that wild pig first appeared to me
on the top of that ridge. Oh, it can be hard work, hunting and gathering, but in the end it isn't really the work that produces the food you're
after, this effort for that result, for there's no sure correlation between effort and result. And no deserving of this: I felt none of the sense of
achievement you feel at the end of a season in the garden, when all

your work has paid off in the bounty of the harvest. No, this felt more like something for nothing, a wondrous and unaccountable gift.

BY THE END of the afternoon we'd all ended up down by Beaver Creek, and at around four we made our way back to the car. We changed our soaking socks on the tailgate and filled the entire cargo area of Anthony's SUV with morels, trying as best we could to hide them from view. No reason, really, but a big haul of mushrooms just isn't something you want to advertise. (Earlier that afternoon a couple of mushroom hunters in an old conversion van stopped to ask if I was having any luck. For no good reason I had lied through my teeth.) We'd found sixty pounds of morels, it turned out—a personal best for Anthony and Ben. Before we climbed into the car to head home, Paulie took a picture of the three of us holding a crate loaded with morels, an obscenely huge one propped up on top of the pile. We were filthy and exhausted, but felt rich as kings. It was a Friday, and as we drove out of the forest, we passed dozens of cars and vans and trucks driving in; the word on the Eldorado flush was apparently out on the Web, and the weekend morel hunters were arriving in force. That meant the price—now twenty dollars a pound—would probably collapse by Monday, so Anthony wasted no time. He started working the cell phone, calling his chefs in Berkeley and San Francisco, taking orders for delivery tonight, and by the time we hit traffic outside of Stockton, all the wild mushrooms had been sold.

TWENTY

THE PERFECT MEAL

Perfect?! A dangerous boast, you must be thinking. And, in truth, there was much about my personally hunted, gathered, and grown meal that tended more toward the ridiculous than the sublime. I burned, just slightly, the crust of the cherry galette, the morels were a little gritty, and the salt, which in keeping with the conceit of the meal I'd gathered myself in San Francisco Bay, tasted so toxic I didn't dare put it on the table. So I seriously doubt that any of my guests, assuming I was out of earshot, would declare this a "great meal." But for me it was the perfect meal, which is not quite the same thing.

I set the date for the dinner—Saturday, June 18—as soon as my animal was in the bag: Wild California pig would be the main course. Now I had a couple of weeks, while the pig hung in Angelo's walk-in, to coordinate the entrée with whatever else I could find to serve. In planning the menu the rules I imposed on myself were as follows (and the exceptions thereto follow what follows):

1. Everything on the menu must have been hunted, gathered, or grown by me.
2. The menu should feature at least one representative of each edible kingdom: animal, vegetable, and fungus, as well as an edible mineral (the salt).
3. Everything served must be in season and fresh. The meal would reflect not only the places that supplied its ingredients, but a particular moment in time.
4. No money may be spent on the meal, though already purchased items in the pantry could be deployed as needed.
5. The guest list is limited to those people who helped me in my foraging and their significant others. This included Angelo, Anthony, Richard, and a friend named Sue who took me on an unsuccessful chanterelle hunt on Mount Tamalpais. Plus, of course, Judith and Isaac. Unfortunately, Jean-Pierre was in France. There would be ten of us in all.
6. I would cook the meal myself.

As the rules suggest, the meal was a conceit—an ambitious, possibly foolhardy, and, I hoped, edible conceit. My aim in attempting it, as should be obvious, was not to propose hunting and gathering and growing one's own food as an answer to any question larger than the modest ones I started out with: Would it be possible to prepare such a meal, and would I learn anything of value—about the nature or culture of human eating—by doing so? I certainly don't mean to suggest that anyone else should try this at home, or that a return to finding and producing our own food is a practical solution to any of our culture's dilemmas surrounding eating and agriculture. No, little if anything about this meal was what anyone would call "realistic." And yet no meal I've ever prepared or eaten has been more real.

1. PLANNING THE MENU

I had better start by getting out of the way some of the exceptions to the foregoing rules and various compromises forced upon me by reality, personal limitation, and folly. This was a meal far richer in stories than calories, and some of those stories, like the one about the salt, did not end well.

Early in my menu planning I had learned that there are still a few salt ponds at the bottom of San Francisco Bay. You can see them flying into SFO, a sequence of arresting blocks of color—rust, yellow, orange, blood red—laid out below you as if in a Mondrian painting. The different colors, I learned, are created by different species of salt-tolerant algae and archaea; as the seawater evaporates from the ponds, the salinity rises, creating conditions suitable for one species of microorganism or another.

On the Saturday before my dinner an exceptionally game friend and I drove down to a desolate stretch of shoreline beneath the San Mateo Bridge. After an interminable trek through acrid and trash-strewn wetlands, we found the salt ponds: rectangular fields of shallow water outlined by grassy levees. The water was the color of strong tea and the levees were littered with garbage: soda cans and bottles, car parts and tires, and hundreds of tennis balls abandoned by dogs. Here, I realized, was the West Coast's answer to the Jersey Meadowlands, a no-man's-land where a visitor would not be wrong to worry about stumbling upon criminal activities or the washed-up corpse of a murder victim. This was definitely the sort of place where you could see too much . . .

. . . of anything, that is, except salt. This year the winter rains had persisted well into spring, making the ponds deeper and less saline than they would normally be in June. So instead of scraping snowy white crystals of sea salt off the rocks, as I'd anticipated, we ended up filling a couple of scavenged polyethylene soda bottles with the cloudy brown brine. That night I evaporated the liquid in a pan over a low flame; it filled the kitchen with a worrisome chemical steam, but after a few hours a promising layer of crystals the color of brown sugar formed in the bottom of the pan, and once it cooled I managed to scrape out a few tablespoons. Unfortunately this salt, which was a bit greasy to the touch, tasted so metallic and so much like chemicals that it actually made me gag, and required a chaser of mouthwash to clear from my tongue. I expect this was a case where the human disgust reflex probably saved lives. No doubt professional salt gatherers have sophisticated purification techniques, but I had no clue what these might be. So I abandoned plans to cook with and serve my own salt, and counted myself lucky not to have contracted hepatitis.

Perhaps the hardest rule to obey was the one about seasonality and freshness. Based on my experience, I would venture that the daily menus of real hunter-gatherers were limited to loads of whatever happens to be plentiful that day and very little of anything else. I had in mind a more varied and ambitious menu, but bringing to the table on a date certain freshly killed game, freshly foraged mushrooms, ripe local fruit, and just-picked garden vegetables turned out to be no mean feat, even in California. In the end I was forced to make an exception for the fungi, since there are no good mushrooms to hunt hereabouts in June. Luckily I had dried a pound of the morels that I'd gathered in the Sierra the previous month, and decided that, especially since dried morels are more intensely flavored than fresh ones, this could be the exception that proved the rule of freshness.

I also had to abandon my overly ambitious plans for a seafood appetizer: grilled abalone. Abalone is a large mollusk that grows on the undersides of underwater rocks along the Pacific coast. Since the abalone population is languishing in California, it can no longer be hunted or sold commercially, but individuals crazy enough to do so may still harvest a strictly limited number: three per day. When, a few days after I'd bagged my pig, a friend who lives on Point Reyes invited me to forage abalone with him the following week—during a once in a blue moon low tide occurring, as you've no doubt guessed, at 5:30 in the morning—I figured I had nailed down my appetizer. So I set my alarm and managed to straggle down to the designated beach at dawn, not quite believing I would have to get into the ocean.

Alas, after surviving the experience of finding an abalone, I learned that it must be eaten absolutely fresh, since freezing abalone utterly ruins its texture. Which is ironic, or something, because looking for abalone, at least on the Northern California coast, involves utterly and completely freezing yourself.

Abalone are gathered during unusually low tides by wading and diving among and beneath underwater boulders and feeling around blindly for their upside-down football-size shells with hands too numb to feel anything—except, that is, the barbed spines of sea urchins, which happen to occupy many of the same underwater crevices as abalone. And if you're lucky enough to avoid getting stuck by sea urchin spines, your probing fingers are liable to settle on the undulating slime of a sea anemone, recoiling abruptly therefrom in terror and disgust. All of this takes place beneath the bemused gaze of sea lions, the presence of whom I was informed is most welcome, since it indicates an absence of man-eating sharks. I might not have frozen myself quite so stiffly had I been wearing a wet suit that actually fit, but the only one available—my friend's grandfather's—was two sizes too small. This had the effect of cutting off circulation to my extremities at the very moment when they needed circulation more than they ever had before. I was out of the water for an hour before I regained enough sensation in my fingers to zip up my fly.

Gathering abalone was the most arduous foraging I did for my meal, and quite possibly the stupidest. I learned later that more Californians are killed gathering abalone each year—by getting dashed on the rocks, being attacked by sharks, or succumbing to hypothermia—than die in hunting accidents. Even if you're better at it than I was (my two hours in the water produced a single keeper), there's no question that you burn more calories looking for abalone than you can possibly collect, making this a perfectly absurd human enterprise. And yet one taste of fresh abalone supplies a fairly convincing explanation for the persistence of this folly.

We ate mine right on the beach, cleaning and pounding the big muscle on a rock, then slicing it and pounding it some more. We built a fire from some driftwood, and then cooked the abalone slices in a pan with butter, onions, and eggs. We ate our breakfast sitting on driftwood logs, watching the tide come in with the day, still fresh. The setting and the abalone, which has some of the chewiness of squid combined with the richer, sweeter flavor of a sea scallop, made this one of life's most memorable breakfasts, almost (though in honesty probably not quite) worth the trouble that went into procuring it. When I got home I made abalone another way, brushing thin, well-pounded slices with olive oil and quickly grilling them over wood. Delectable, but unfortunately for my dinner guests, I had to serve this appetizer several weeks before they'd been told to arrive, making it a purely notional item on their menu.

For the real menu's appetizer, I had to turn to the garden, where there were fava beans ready to pick. I'd planted them as a cover crop back in November, and by May had scores of fat glossy pods, which I held off harvesting in anticipation of the big meal. The fava, which is the only bean native to the Old World, is a broad, flat, bright green shelling bean that if picked young and quickly blanched has a starchy sweet taste that to me is as evocative of springtime as fresh peas or asparagus. But by June many of my beans were a bit long in the tooth, so I decided to make fava bean toasts: I'd mash the beans with roasted garlic and sage and serve them on toasted rounds of homemade sourdough bread. (The younger, sweeter beans I'd reserve for the pasta.) For a second appetizer, I asked Angelo to bring a block of the pâté he'd made from the liver of my pig.

So yes, okay, here was another exception to the rules: Angelo made the pâté. I also asked him to make the pasta for the first course: morels sautéed in butter with thyme and, for color, the tiny fava beans, over fresh egg fettuccine.

Wild California pig was the main course, but which cut and how to prepare it? Angelo recommended slowly braising the leg, in his opinion the most flavorful cut. I was curious to try the loin, and grilling outdoors over a fire seemed to me more in keeping with the season as well as the hunter-gatherer theme. Unable to choose between the two approaches, I decided to try both. I would braise the leg in red wine (Angelo's) and homemade stock, and serve it with a reduction of the cooking liquid. The loin I would brine overnight, to keep the lean meat from drying out on the grill, cover it with crushed peppercorns, and then roast it fairly quickly over olive wood. The stock I could make earlier in the week, and the olive wood I would forage not in an olive orchard but, with Jean-Pierre's blessing, in the woodshed behind Chez Panisse.

I wanted to make my own bread and decided it would be fitting to use wild yeast, thereby introducing a second species of foraged fungus into the proceedings. I found a recipe (in an excellent cookbook called *Bread Alone*) that gave instructions for gathering wild yeast, in a process that took several days but didn't sound too difficult. For the wine I had a couple bottles of Angelo's 2003 Syrah and he offered to bring a few more.

After the main course there would be a salad, which I had originally hoped to assemble from foraged wild greens. Earlier in the spring I had found a lush patch of miner's lettuce and wild rapini in the Berkeley Hills, but by June the greens had begun to yellow, so I decided to go instead with a simple salad of lettuces from the garden.

Which left dessert, and for a while that posed a problem. My plan was to forage fruit, for a tart, from one of the many fruit trees lining the streets in Berkeley. I see no reason why foraging for food should be restricted to the countryside, so in the weeks before the dinner I embarked on several urban scouting expeditions in quest of dessert. Actually these were just strolls around the neighborhood with a baggie. In the two years we've lived in Berkeley I've located a handful of excellent fruit trees—plum, apple, apricot, and fig—offering publicly accessible branches, but none of the usual suspects had quite ripened yet, with the exception of a Santa Rosa plum on Parker Street that was already past its peak.

So I started asking around, hoping somebody might point me in the direction of a promising neighborhood dessert tree. It was my sister-in-law, Dena, who saved my dessert. She reported that her neighbor's Bing cherry tree was so heavily laden with ripe fruit that several of its branches were at that very moment bending low over her backyard. I wasn't quite sure if picking cherries from a neighbor's tree was exactly kosher, either by my lights or the law. But isn't there some old legal principle that confers the right to pick fruit from trees overhanging your property? I did a little research and discovered that indeed there is. The Romans called it "usufruct," which the dictionary defines as "the right to enjoy the use and advantages of another's property short of the destruction or waste of its substance." Bingo! Here was a venerable legal principle that spoke to the very soul of foraging.*

With dessert I would serve a tisane, or herbal tea, made from wild chamomile I'd picked in the Berkeley Hills earlier in the spring and dried, mixed with mint and lemon balm from the garden. I also had a jar of honey made by a friend in town, the foraging in this case having been done in the Berkeley Hills by his bees.

Now I had my menu and I wrote it out on a card; this being Berkeley, I felt compelled to add a few pretentious restaurant menu flourishes:

> Fava Bean Toasts and Sonoma Boar Pâté
>
> Egg Fettuccine with Power Fire Morels
>
> Braised Leg and Grilled Loin of Wild Sonoma Pig
>
> Wild East Bay Yeast Levain
>
> Very Local Garden Salad
>
> Fulton Street Bing Cherry Galette
>
> Claremont Canyon Chamomile Tisane
>
> 2003 Angelo Garro Petite Syrah

It was still just a menu, okay, and admittedly it broke several of my own rules and leaned rather heavily on Angelo's generosity and talents, yet it promised an interesting meal and accomplished most of what I had set out to do.

---

* There's a Web site devoted to the principle of usufruct that offers maps to publicly accessible fruit trees in Los Angeles: fallenfruit.org.

As I looked over the menu, it occurred to me that besides representing several wild species and three edible kingdoms, not to mention the city and the country, this was a dinner drawn in large part from the forest. Here was the meal at the end of a woodland food chain, and that as much as anything else made it a little different. The pig and the morels came directly from the forest, obviously, but the cherry, too, is originally a woodland species that found its way to the orchard and then the city. (Cherry trees came originally from the forests of the Transcaucus, between the Black and Caspian seas. The Bing cherry is a chance seedling discovered in a Willamette Valley orchard in 1875 and named for a Mr. Ah Bing, the Chinese farmhand who tended it.) What this means is that the calories we'd be consuming represent energy captured by trees rather than, as is typical now, by annuals in farm fields or grasses in pastures. The sweetness of the dessert was made in the leaves of a cherry tree; the morels nourished themselves from sugars originally created in the needles of a pine tree and then absorbed from its roots by their mycelium; and the acorn-fed pig is a walking, snorting manifestation of the oak. Reversing the historical trajectory of human eating, for this meal the forest would be feeding us again.

2. IN THE KITCHEN

I started cooking Saturday's meal on Tuesday morning, when I made the stock and started the wild yeast culture for the bread. For the stock I used bones from both my pig and, because I'd never heard of a pure pork stock, from a grass-fed steer. A neighbor had recently bought a quarter of a beeve that arrived with a big bag of bones she didn't know what to do with, so I asked if I could forage them from her freezer. Similarly, I foraged from the depths of the produce bin in my refrigerator some past-due vegetables. After roasting the bones in the oven for an hour, I simmered them in a stockpot with the vegetables and some herbs for the rest of the day.

Gathering wild yeast turns out to be no big deal. The spores of various yeasts are floating in the air just about everywhere; collecting them is a matter of giving them a place to rest and something to eat. Some species of yeast taste better than others, however, and this is where geography and luck enter in. The Bay Area has a reputation for its sourdough bread, so I figured the air outside my house would be an excellent hunting ground for wild yeast. I made a thick soup of organic flour and spring water (the idea is to avoid any chemicals that might harm your yeast); then, after briefly exposing the mixture to the air on a windowsill, I sealed it in an airtight container and left it out on the kitchen counter overnight. By the following morning the surface of the chef, as it's called, was bubbling like pancake batter on a hot griddle, a good sign. Each day, you're supposed to feed fresh water and flour to the young colony of microbes, and sniff it. The chef should smell slightly alcoholic, sour, and yeasty—a bit like beer. The absence of bubbles is a bad sign. So is the presence of off odors or colorful scums, which indicate you've probably snared wilder and weirder microbes than you want; throw out the chef and start over. I counted myself lucky that by the second day my chef already smelled promisingly beery and breadlike.

Wednesday morning I drove into San Francisco to pick up the meat from Angelo at the forge. To get to his walk-in cooler you pass through a sequence of loftlike spaces of an almost Dickensian novelty and clutter, filled with metal scraps of every description, stacks of iron rods, ironworking tools and pieces of machinery, a small blast furnace raging heat and light, and, growing beneath an opening to the sky right in the middle of the forge, a fully grown fig tree. In the back there's a sunny kitchen with an industrial-duty espresso machine, a meat grinder, and a pasta machine and, to relieve the industrial clutter and clatter all around, big vases of fresh wildflowers. Industrial and domestic, hard and soft, metal and meat: The place was a lot like Angelo himself.

The carcass was hanging alongside a couple of others in the walk-in, amid racks holding prosciutto, pancetta, and salami in various stages of curing. Just outside the walk-in stood more racks holding oak barrels of wine and balsamic vinegar, hundreds of unlabeled bottles of wine, and fifty-pound bags of wheat, both durum and semolina. Angelo carried the stiff carcass out to the kitchen table and, with a cleaver, began expertly to disassemble my pig. We trimmed and salted the hams for prosciutto and, with a few well-placed blows of the cleaver, Angelo separated the rib cage from the spinal column and then the loins, one on either side of the spine like saddlebags of meat. Eyeing the mounting pile of trimmings—chunks of dark red meat and strips of snowy white fat—Angelo had an idea.

"Hey, you know, we should make a nice little ragout with all these scraps. For our lunch." And so we did, pushing the scraps through the grinder, stewing the ground meat with a can of tomatoes, and, while the ragout bubbled on the stove, making a batch of fresh pasta on which to serve it. Angelo showed me how to cut handfuls of the yellowy ribbons of fettuccine as they extruded themselves from the slots of his machine.

Ready or not, this would be my first taste of my pig, and I was a little taken aback at the speed with which it had just gone from hanging carcass to ground-up scraps of meat to lunch. But the ragout was delicious and, eating it at Angelo's kitchen table, even amid the raw cuts of meat arrayed on the counters around us, I suddenly felt perfectly okay about my pig—indeed, about the whole transaction between me and this animal that I'd killed two weeks earlier. Eating the pig, I understood, was the necessary closing act of that drama, and went some distance toward redeeming the whole play. Now it was all a matter of doing well by the animal, which meant making the best use of its meat by preparing it thoughtfully and feeding it to people who would appreciate it. Later, when I looked up the spelling of the word "ragout," I learned that it comes from the French verb *ragoûter*: "to restore the appetite." This one had done that, restoring my appetite for this meat after the disgust I'd felt cleaning the animal. I was reminded of what Paul Rozin had written about a traditional cuisine's power to obviate the omnivore's dilemma by clothing the exotic in familiar flavors. I left Angelo's with two gorgeous cuts of my pig neatly wrapped in butcher paper.

By the end of the week all the meal's raw ingredients were in place: I'd picked a gallon of cherries, harvested my fava beans, prepared the brine for the pig loin, made the stock and the chef, and soaked the dried morels in warm water to rehydrate them, a procedure that yielded an earthy black liquor that I decided would be good to add to the braising liquid. On Friday night, when I made a to-do list and schedule for Saturday, it hit me just how much I had to do, and, scarier still, how much of what I had to do I had never done before, including bake a wild yeast bread, pit a gallon of cherries, make a galette, and cook a wild pig two different ways. I also hadn't toted up until now how many total hours of oven time the meal would require, and since braising the pig leg at 250 degrees would take half the day, it wasn't clear how exactly I could fit in the bread and the galette. For some reason the very real potential for disaster hadn't dawned on me earlier, or the fact that I was cooking for a particularly discriminating group of eaters, several of them actual chefs. Now, dawn on me it did, and it left me feeling more than a little intimidated.

To give you a more comprehensive idea of exactly what I'd gotten myself into, here's the schedule I wrote out Friday evening on an index card:

8:00 brine the loin; shell and blanch and skin the fava beans. [Favas are one of nature's more labor-intensive legumes, requiring two separate peelings, with a blanching in between.]

9:00 make the bread dough. First rise.

10:00 brown the leg; prepare liquid for braise.

10:30 pit the cherries. Make pastry crust; refrigerate. Preheat oven for pig, 250°.

11:00 Pig in oven. Skin fava beans. Roast garlic, puree favas.

12:00 knead bread dough; second rise.

12:30 clean morels; harvest and chop herbs, sauté morels.

1:00 harvest and wash lettuce. Make vinaigrette.

2:00 knead dough again; proof loaves. Prepare grill, teapot, cut flowers, set table.

3:00 roll piecrust, make galette. Remove pig and heat oven for bread
   (450°). Score loaves and bake.

3:40 remove bread; bake galette (400°).

4:00 remove galette from oven; put pig back in (250°).

5:00 build fire. Crush peppercorns.

6:15 remove leg to rest; prepare loin (lard with garlic and herbs; roll
   in crushed pepper). Put loin on grill.

7:00 guests arrive. Remove loin to rest.

That was my Saturday in the kitchen, though of course the reality of
the day unfolded with none of the order or stateliness promised by the
schedule. No, in reality the day was a blizzard of harried labors, missing ingredients, unscheduled spills and dropped pots, unscheduled
trips to the store, unscheduled pangs of doubt, and throes of second-
guessing. There were moments when I sorely wished for another pair of
hands, but Judith and Isaac were away all day. Why, I asked myself when
I took a ten-minute break for lunch around 4:00, had I ever undertaken
to prepare such an elaborate meal by myself?

For a quick lunch I'd picked up a takeaway plastic tray of sushi—
Japanese fast food—and, you know, it tasted just great. So how much
better could I reasonably expect this dinner—this day-long (indeed,
months-long) extravaganza, this extremely slow food feast—to taste? Did I
really need to cook the pig two different ways? For dessert, why not just
serve the cherries in a bowl? Or open a can of beef stock for the braise?
Or a packet of fast-acting yeast?! Why in the world was I going to quite
this much trouble?

I thought of several answers while I wolfed my sushi, each of them
offering some sliver of a somewhat elusive larger truth. This meal was
my way of thanking these people, my patient and generous Virgils, for
all they'd contributed to my foraging education, and the precise
amount of thought and effort I put into the meal reflected the precise
depth of my gratitude. A bowl of fresh Bing cherries is nice, but to turn
them into a pastry is surely a more thoughtful gesture, at least provided
I managed not to blow the crust. It's the difference between a Hallmark

card and a handwritten letter. A cynical person might say that cooking like this—with ambition—is really just another way of showing off, a form of what might be called conspicuous production. It says, *I have the resources, sophistication, and leisure time to dazzle you with this meal.* No doubt there's often an element of truth to this, but cooking is many other things too, and one of them is a way to honor the group of people you have elected to call your guests.

Another thing cooking is, or can be, is a way to honor the things we're eating, the animals and plants and fungi that have been sacrificed to gratify our needs and desires, as well as the places and the people that produced them. Cooks have their ways of saying grace too. Maybe this explains why I wanted to prepare the pig two ways, and to serve Angelo's pig pâté. For me, doing right by my pig means wasting as little of it as possible and making the most of whatever it has to offer us. Cooking something thoughtfully is a way to celebrate both that species and our relation to it. By grilling one cut of my pig and braising the other, I was drawing on the two most elemental techniques people have devised for transforming raw meat into something not only more digestible but also more human: that is, cooking meat directly over a fire and, with liquid, in a pot. Both techniques promise to turn the flesh of animals into something good to eat and good to think, but each reflects a slightly different stance toward the animal. The second proposes a more "civilized" method of cooking meat, since it achieves a more complete transcendence or (take your pick) sublimation of the animal, and perhaps the animal in us, than the first. It leaves no trace of blood, which suits some meat eaters more than others, but it seemed to me I should give both approaches to the pig their due.

It was a long day of such transformations, as one after another the raw stuffs of nature—chunks of meat, piles of wild fungi, the leaves and pods of plants, and piles of pulverized grain—took on whole new forms, many of them wondrous. Bread dough magically rose and crisped; desiccated mushrooms came back to fleshy life; meat turned brown and caramelized; indigestible beans softened and sweetened; the leaves of herbs inflected whatever they touched; and all these unprepossessing parts of things combined into what promised to be greater and more delectable wholes.

The repetitive phases of cooking leave plenty of mental space for reflection, and as I chopped and minced and sliced I thought about the rhythms of cooking, one of which involves destroying the order of the things we bring from nature into our kitchens, only to then create from them a new order. We butcher, grind, chop, grate, mince, and liquefy raw ingredients, breaking down formerly living things so that we might recombine them in new, more cultivated forms. When you think about it, this is the same rhythm, once removed, that governs all eating in nature, which invariably entails the destruction of certain living things, by chewing and then digestion, in order to sustain other living things. In *The Hungry Soul* Leon Kass calls this the great paradox of eating: "that to preserve their life and form living things necessarily destroy life and form." If there is any shame in that destruction, only we humans seem to feel it, and then only on occasion. But cooking doesn't only distance us from our destructiveness, turning the pile of blood and guts into a savory salami, it also symbolically redeems it, making good our karmic debts: *Look what good, what beauty, can come of this!* Putting a great dish on the table is our way of celebrating the wonders of form we humans can create from this matter—this quantity of sacrificed life—just before the body takes its first destructive bite.

3. AT THE TABLE

It remained to be seen whether my own cooking would redeem any of these ingredients, but by the appointed hour everything was more or less ready, except me. I raced upstairs to change and, before I had my shoes tied, heard the doorbell ring. The guests were arriving. They came bearing feast-appropriate gifts: Angelo with his wine and pâté, Sue with a bouquet of lemon verbena picked from her garden, and An-

thony with a small carafe of homemade *nocino*, a jet-black Italian digestive he'd distilled from green walnuts—yet another gift of the forest to
our feast.

I'd been too busy worrying about the food to worry much about
the company, whether this somewhat haphazard assortment of people
would gel or not. A couple of paths had apparently crossed before, but
most of the guests were strangers to one another; what linked everyone
was foraging—and me. But as we settled into the living room with our
glasses of wine, it didn't take long for the stiffness of small talk to relax
into conversation and for the conversation, oiled by Angelo's excellent
Syrah, to steadily gain altitude. The fava bean toasts and boar pâté drew
appreciative murmurs and comments, launching a discussion about
boar hunting. Anthony was curious to go some time, but warned Angelo that he doubted he could bring himself to actually shoot anything.
"Maybe I could caddy for you," he proposed. When the mood in the
living room seemed to have attained a sustainable effervescence, I disappeared into the kitchen to ready the pasta course.

Within minutes Angelo appeared at my side, with an offer of help;
I think he was a little worried I was in over my head. While we waited
for the pasta water to come to a boil, I asked him to taste the morels.
"It's good, but maybe it needs a little more butter." I handed him a stick
and he dropped the whole thing in the pan. (So *that's* how the professionals do it!)

We plated the pasta, and I called everyone to the table for dinner. Vo-
tives were lit, wine was poured, the perfume of thyme and morels filled
the room, and I raised my glass for a toast. I'd actually meant to write
out something earlier in the day, because I'd wanted to organize my
thoughts on the meaning of the meal and everyone's contribution to it,
but the day had gotten away from me. So I kept it simple. I went around
the table and spoke of each person's contribution to my foraging education and to this meal that, though I had cooked most of it myself, was
in the deepest sense our collaboration. I talked about Sue's unprecedented generosity in sharing three of her choicest chanterelle spots
(one of them right in the front yard of an unsuspecting homeowner in

West Marin), and told the story of the afternoon we'd spent hunting mushrooms in a downpour—with nothing to show for it. I talked about Anthony's gameness in allowing a complete, and completely green, stranger to accompany him hunting morels in the Sierra. I talked about hunting with Richard in Sonoma during that first failed outing, how it had taught me the importance of preparedness, and temperance, in hunting. And lastly I talked about all the many things I'd learned from Angelo—things about mushrooms and pigs, about nature and the arts of cooking and eating well, and so much else besides. Then, worried I was in danger of melting down into sentiment, I raised my glass again and urged everyone to start.

I had actually wanted to say something more, to express a wider gratitude for the meal we were about to eat, but I was afraid that to offer words of thanks for the pig and the mushrooms and the forests and the garden would come off sounding corny and, worse, might ruin some appetites. The words I was reaching for, of course, were the words of grace. But as the conversation at the table unfurled like a sail amid the happy clatter of silver, tacking from stories of hunting to mother lodes of mushrooms to abalone adventures, I realized that in this particular case words of grace were unnecessary. Why? Because that's what the meal itself had become, for me certainly, but I suspect for some of the others, too: a wordless way of saying grace.

As you might expect from this crowd and occasion, the talk at the table was mainly about food. Yet this was not the usual food talk you hear nowadays; less about recipes and restaurants, it revolved around specific plants and animals and fungi, and the places where they lived. The stories told by this little band of foragers ventured a long way from the table, the words (the tastes, too) recalling us to an oak forest in Sonoma, to a pine burn in the Sierra Nevada, to the stinky salt flats of San Francisco Bay, to slippery boulders along the Pacific coast, and to a backyard in Berkeley. The stories, like the food that fed them, cast lines of relation to all these places and the creatures living (and dying) in them, drawing them all together on this table, on these plates, in what to me began to feel a little like a ceremony. And there's a sense in which the meal had become just that, a thanksgiving or a secular seder, for every item on our plates pointed somewhere else, almost sacramentally, telling a little story about nature or community or even the sacred, for mystery was very often the theme. Such storied food can feed us both body and soul, the threads of narrative knitting us together as a group, and knitting the group into the larger fabric of the given world.

I don't want to make too much of it; it was just a meal, after all. A very tasty meal, too, I don't mind saying, though I don't doubt that all the words and memories and stories in which the meal had marinated gave it much of its savor, and that a guest who spoke no English might not have enjoyed it half as much. The wild pig was delicious both ways, with a nutty sweetness to it that tasted nothing like store-bought pork, though I noticed that when the platter went around for seconds, the tender slices of braised leg went faster than the pink slabs of the roast. The sauce for the leg I'd reduced from the braising liquid was almost joltingly rich and earthy, powerfully reminiscent of the forest. So were the morels and butter (or perhaps I should say butter and morels), which had a deep, smoky, almost meaty flavor. My self-criticisms were that I could have done a better job cleaning the grit from the morels, and that the galette was a shade overcooked—though the cherries themselves detonated little bursts of summer on the tongue, and no one seemed to have any trouble polishing it off.

Angelo reserved his most enthusiastic praise for my bread, which I'll admit did have a perfect crust, an airy crumb, and a very distinctive (though not at all sour) flavor—the specific flavor, I guess, of the neighborhood yeasts. It occurred to me that the making of this meal, by acquainting me with these particular people, landscapes, and species, had succeeded in attaching me to Northern California, its nature and its culture both, as nothing I'd done before or since. Eating's not a bad way to get to know a place.

There comes a moment in the course of a dinner party when, with any luck, you realize everything's going to be okay. The food and the company having sailed past the shoals of awkwardness or disaster, and the host can allow himself at last to slip into the warm currents of the evening and actually begin to enjoy himself. For me that moment came just around the time that the platter of wild pig made its second circuit of the table and found so many eager takers. I was enjoying myself now, the words and the food in equal measure, and that's when I realized that this was, at least for me, the perfect meal, though it wasn't until some time later that I began to understand what that meant.

Was the perfect meal the one you made all by yourself? Not necessarily; certainly this one wasn't that. Though I had spent the day in the kitchen (a good part of the week as well), and I had made most everything from scratch and paid scarcely a dime for all the ingredients, it had taken many hands to bring this meal to the table. The fact that just about all of those hands were at the table was the more rare and important thing, as was the fact that every single story about the food on that table could be told in the first person.

I prized, too, the almost perfect transparency of this meal, the brevity and simplicity of the food chain that linked it to the wider world. Scarcely an ingredient in it had ever worn a label or bar code or price tag, and yet I knew almost everything there was to know about its provenance and its price. I knew and could picture the very oaks and pines that had nourished the pigs and the mushrooms that were nourishing us. And I knew the true cost of this food, the precise sacrifice of time and energy and life it had entailed. Some of that sacrifice had proven expensive to me, emotionally speaking, yet it was cheering to realize just how little this preindustrial and mostly preagricultural meal had diminished the world. My pig's place would soon be taken by another pig, and the life of these forests was little altered by our presence or what we had removed. Not just the Bing cherries but most of the meal owed its presence on our table to usufruct, which was a fact of nature long before it became an axiom of law.

Perhaps the perfect meal is one that's been fully paid for, that leaves no debt outstanding. This is almost impossible ever to do, which is why I said there was nothing very realistic or applicable about this meal. But as a sometimes thing, as a kind of ritual, a meal that is eaten in full consciousness of what it took to make it is worth preparing every now and again, if only as a way to remind us of the true costs of the things we take for granted. The reason I didn't open a can of stock was because stock doesn't come from a can; it comes from the bones of animals. And the yeast that leavens our bread comes not from a packet but from the air we breathe. The meal was more ritual than realistic because it dwelled on such things, reminding us how very much nature offers to the omnivore, the forests as much as the fields, the oceans as much as the meadows. If I had to give this dinner a name, it would have to be the Omnivore's Thanksgiving.

It's impossible to prepare and eat a meal quite so physically, intellectually, and emotionally costly without thinking about the incalculably larger debts we incur when we eat industrially—which is to say, when we eat without a thought to what we're doing. To compare my transcendently slow meal to the fast-food meal I "served" my family at that McDonald's in Marin, the one that set me back fourteen bucks for the three of us and was consumed in ten minutes at sixty-five miles per hour, is to marvel at the multiplicity of a world that could produce two such different methods of accomplishing the same thing: feeding ourselves, I mean.

The two meals stand at the far extreme ends of the spectrum of human eating—of the different ways we have to engage the world that sustains us. The pleasures of the one are based on a nearly perfect knowledge; the pleasures of the other on an equally perfect ignorance. The diversity of the one mirrors the diversity of nature, especially the forest; the variety of the other more accurately reflects the ingenuity of industry, especially its ability to tease a passing resemblance of diversity from a single species growing in a single landscape: a monoculture of corn. The cost of the first meal is steep, yet it is acknowledged and paid for; by comparison the price of the second seems a bargain but fails to cover its true cost, charging it instead to nature, to the public health and purse, and to the future.

Let us stipulate that both of these meals are equally unreal and equally unsustainable. Which is perhaps why we should do what a responsible social scientist would do under the circumstances: discard them both as anomalies or outliers—outliers of a real life. Or better yet, preserve them but purely as ritual, for the lessons they have to teach us about the different uses to which the world can be put. Going to McDonald's would be something that happens once a year, a kind of Thanksgiving in reverse, and so would a meal like mine, as slow and storied as the Passover seder.

Without such a thing as fast food there would be no need for slow food, and the stories we tell at such meals would lose much of their interest. Food would be . . . well, what it always was, neither slow nor fast, just food: this particular plant or that particular animal, grown here or there, prepared this way or that. For countless generations eating was something that took place in the steadying context of a family and a culture, where the full consciousness of what was involved did not need to be rehearsed at every meal because it was stored away, like the good silver, in a set of rituals and habits, manners and recipes. I wonder if it isn't because so much of that context has been lost that I felt the need, this one time, to start again from scratch.

This is not the way I want to eat every day. I like to be able to open a can of stock and I like to talk about politics, or the movies, at the dinner table sometimes instead of food. But imagine for a moment if we once again knew, strictly as a matter of course, these few unremarkable things: What it is we're eating. Where it came from. How it found its way to our table. And what, in a true accounting, it really cost. We could then talk about some other things at dinner. For we would no longer need any reminding that however we choose to feed ourselves, we eat by the grace of nature, not industry, and what we're eating is never anything more or less than the body of the world.

ACKNOWLEDGMENTS

I HAD A LOT OF HELP in the kitchen with this one.

First to Gerry Marzorati, my longtime friend and editor at the *New York Times Magazine*, who first suggested five years ago that I spend some time writing about food for the magazine. Unbeknownst to either of us, he was pointing me down the path that led to this book.

I am especially grateful to the farmers and the foragers I write about here. George Naylor in Iowa, Joel Salatin in Virginia, and Angelo Garro in California were my food-chain Virgils, helping me to follow the food from earth to plate and to navigate the omnivore's dilemma. All three gave unstintingly of their time, their wisdom, and their always excellent company. Thanks, too, to the hunters and gatherers who graciously welcomed so rank an amateur on their expeditions: Anothony Tassinello, Bob Baily, Bob Carrou, Richard Hylton, Jean-Pierre Moulle, Sue Moore, and David Evans.

In educating myself on food and agriculture, I've incurred a great many debts. Among my most generous and influential teachers have been: Joan Gussow, Marion Nestle, Fred Kirschenmann, Alice Waters, Todd Dawson, Paul Rozin, Wes Jackson, and Wendell Berry. Thanks also, for information and insight, to Bob Scowcroft, Allan Nation, Kelly Brownell, Ricardo Salvador, Carlo Petrini, Jo Robinson, David Arora, Ignacio Chapela, Miguel Altieri, Peter Hoffman, Dan Barber, Drew and Myra Goodman, Bill Niman, Gene Kahn, and Eliot Coleman.

Many people supported the writing of this book in other ways. In

California, Michael Schwarz generously read the manuscript and offered timely encouragement and helpful suggestions, reminding me what a good editor he was before he forsook print for television. In Berkeley, the faculty, staff, and students of the Graduate School of Journalism, and in particular Dean Orville Schell, have created a stimulating and supportive community in which to do this work. Mark Danner, an old friend and once again a colleague, has, as ever, provided a valuable sounding board. The students in my food chain class have taught me more than they probably realize about these issues over the past few years. Mesa Refuge, in Point Reyes Station, provided the perfect setting in which to write and research a key chapter. And the John S. and James L. Knight Foundation has supported my research in crucial ways.

I'm especially grateful to Chad Heeter, for his dogged research and fact-checking, not to mention his willingness to accompany me on a futile quest to gather salt in San Francisco Bay. Nathanael Johnson, Felicia Mello, and Elena Conis nailed down several elusive facts just when it looked like they might get away. My assistant, Jaime Gross, contributed to this project in many ways, but I'm particularly grateful for her superb research and fact-checking.

In New York, I'm grateful for the excellent work and good cheer of Liza Darnton, Kate Griggs, Rachel Burd, Sarah Hutson, and Tracy Locke at the Penguin Press, my new publishing home. Thanks to Liz Farrell at ICM. At the New York Times Magazine, where some of the material in this book first appeared, I've profited handsomely from the superb editing of Paul Tough and Alex Star and (before they moved on to other magazines) Adam Moss and Dan Zalewski.

In a publishing industry not known for loyalty or continuity, I've been blessed by the constancy of both my editor and agent. This is the fourth book of mine that Ann Godoff has edited, albeit at three different houses. At this point I can't imagine doing a book with anyone else, which is probably why I keep following her around Manhattan. Her moral, intellectual, emotional, and financial support is a critical ingredient in the making of this book. This is also the fourth book of mine represented by Amanda Urban, a verb that doesn't come close to capturing everything she does to keep me whole and on the proper path.

Speaking of constancy, this is also the fourth time I've relied on Mark Edmundson to read and comment on a book manuscript; as ever, his editorial and reading suggestions, as well as his literary judgment, have been invaluable. This time around, he (and his family) contributed in another way as well, by joining me for one of the meals chronicled in these pages. Thanks to Liz, Willie, and Matthew for their gameness, good appetite, and hospitality.

But the prize for gameness in the pursuit of a book chapter must go to Judith, who shared the two meals that bookend the book—the McDonald's cheeseburger at one end and the wild boar at the other— and so much more. A book becomes a sometimes disagreeable member of the family for a period of years, but Judith treated this one with patience, understanding, and good humor. Far more crucial to the book, though, has been her editing. Since I first began publishing, Judith has been my indispensable first reader, and there's no one whose instincts about writing I trust more.

Last but no longer least is Isaac. This is the first book Isaac has been old enough and sufficiently interested in to actually help me with. His own approach to food—Isaac is the pickiest eater I know—has taught me a great deal about the omnivore's dilemma. Though he declined to taste the boar, Isaac's contribution to this book—coming in the form of smart suggestions, stimulating conversations at the dinner table, and, on the bad days, the best comfort a father could wish for—has been more precious than he can know. Thank you.

SOURCES

Listed below, by chapter, are the principal works referred to in the text, as well as others that supplied me with facts or influenced my thinking. Web site URLs are current as of November 2005. All cited articles by me are available at www.michaelpollan.com.

INTRODUCTION: OUR NATIONAL EATING DISORDER

Berry, Wendell. "The Pleasures of Eating," in *What Are People For?* (New York: North Point Press, 1990), pp.145–52.

Kass, Leon. *The Hungry Soul* (New York: The Free Press, 1994). I found the William Ralph Inge quote in this endlessly suggestive philosophical inquiry into how the particular nature of human eating defines us.

Levy, Ariel. "Carb Panic," *New York*, December 12, 2002.

Nestle, Marion. *Food Politics* (Berkeley: University of California Press, 2002).

Rozin, Paul. "The Selection of Foods by Rats, Humans, and Other Animals" in *Advances in the Study of Behavior*, Volume 6, eds. J. Rosenblatt, R. A. Hide, C. Beer, and E. Shaw (New York: Academic Press, 1976), pp. 21–76.

————. "Food Is Fundamental, Fun, Frightening, and Far-Reaching," *Social Research* 66, no. 1 (Spring 1999). This is a special issue on food with many excellent essays.

Taubes, Gary. "What If Fat Doesn't Make You Fat?" *New York Times Magazine*, July 7, 2002.

PART I

CHAPTER 1: THE PLANT: CORN'S CONQUEST

In addition to the printed sources below, I learned a great deal about the natural and social history of *Zea mays* from my conversations with Ricardo Salvador at Iowa State (www.public.iastate.edu/~rjsalvad/home.html) and Ignacio Chapela at the University of California at Berkeley. Ignacio introduced me to his colleague Todd Dawson, who not only helped me understand what a C-4 plant is, but generously tested various foods and hair samples for corn content using his department's mass spectrometer.

The two indispensable books on the history of corn are:

Fussell, Betty. *The Story of Corn* (New York: Knopf, 1994). Columbus's quote on corn is on page 17. The statistics on wheat versus corn consumption are on page 215.

Warman, Arturo. *Corn & Capitalism: How a Botanical Bastard Grew to Global Dominance.* Trans. Nancy L. Westrate (Chapel Hill: University of North Carolina Press, 2003).

Other helpful works touching on the history of corn include:

Anderson, Edgar. *Plants, Man and Life* (Berkeley: University of California Press, 1952).

Crosby, Alfred W. *Germs, Seeds & Animals: Studies in Ecological History* (Armonk, NY: M. E. Sharpe, 1994).

———. *Ecological Imperialism: The Biological Expansion of Europe, 900–1900* (Cambridge, U.K.: Cambridge University Press, 1986).

Diamond, Jared. *Guns, Germs, and Steel* (New York: W. W. Norton, 1997).

Eisenberg, Evan. *The Ecology of Eden* (New York: Alfred A. Knopf, 1998). Very good on the coevolutionary relationship of grasses and humankind.

Iltis, Hugh H. "From Teosinte to Maize: The Catastrophic Sexual Mutation," *Science* 222, no. 4626 (November 25, 1983).

Mann, Charles C. *1491: New Revelations of the Americas Before Columbus* (New York: Alfred A. Knopf, 2005). Excellent on the evolutionary origins of the plant and pre-Columbian maize agriculture.

Nabhan, G. P. *Enduring Seeds: Native American Agriculture and Wild Plant Conservation* (San Francisco: North Point Press, 1989).

Rifkin, Jeremy. *Beyond Beef: The Rise and Fall of the Cattle Culture* (New York: Plume, 1993). The quote from General Sheridan is on page 78.

Sargent, Frederick. *Corn Plants: Their Uses and Ways of Life* (Boston: Houghton Mifflin, 1901).

Wallace, H. A., and E. N. Bressman. *Corn and Corn Growing* (New York: John Wiley & Sons, 1949).

Weatherford, Jack. *Indian Givers: How the Indians of the Americas Transformed the World* (New York: Crown, 1988).

Will, George F., and George E. Hyde. *Corn Among the Indians of the Upper Missouri* (Lincoln: University of Nebraska Press, 1917).

CHAPTER 2: THE FARM

The best accounts of the history and workings of the commodity corn complex in the United States are a series of studies by Richard Manning and

C. Ford Runge commissioned by the Midwest Commodities and Conservation Initiative, a joint project of the World Wildlife Fund, the American Farmland Trust, and the Henry A. Wallace Center for Agricultural & Environmental Policy: All three studies are available online at: www.worldwildlife.org/commerce.

Manning, Richard. *Commodities, Consensus, and Conservation: A Search for Opportunities and The Framework of a Commodities System* (April 2001).

Runge, C. Ford. *King Corn: The History, Trade, and Environmental Consequences of Corn (Maize) Production in the United States* (September 2002).

In writing about the rise of industrial agriculture I also drew on the following works:

Kimbrell, Andrew. *The Fatal Harvest Reader: The Tragedy of Industrial Agriculture* (Washington, D.C.: Island Press, 2002).

Manning, Richard. *Against the Grain* (New York: North Point Press, 2004).

Morgan, Dan. *Merchants of Grain* (New York: Viking, 1979).

Russell, Edmund. *War and Nature: Fighting Humans and Insects with Chemicals from World War I to Silent Spring* (Cambridge, U.K.: Cambridge University Press, 2001).

Schwab, Jim. *Raising Less Corn and More Hell: Midwestern Farmers Speak Out* (Urbana: University of Illinois Press, 1988). See the interview with George Naylor beginning on page 111.

Scott, James. *Seeing Like a State: How Certain Schemes to Improve the Human Condition Have Failed* (New Haven: Yale University Press, 1998). Scott, an anthropologist and political scientist, puts industrial agriculture in the illuminating context of other modernist schemes, including architecture and Soviet collectivization.

Smil, Vaclav. *Enriching the Earth: Fritz Haber, Carl Bosch, and the Transformation of World Food Production* (Cambridge, MA: M.I.T. Press, 2001). This indispensable book tells the story of Fritz Haber's life and work, explains the technology of synthesizing nitrogen, and explores its impact on the environment and world population.

———. *Feeding the World* (Cambridge, MA: M.I.T. Press, 2000).

Wargo, John. *Our Children's Toxic Legacy* (New Haven: Yale University Press, 1996). An important work on the regulation and biology of pesticides.

For detailed information on individual pesticides, see the Web site of the Pesticide Action Network (www.panna.org). On atrazine, the herbicide most widely applied to U.S. cornfields, see Hayes, Tyrone, et al. "Atrazine-Induced Hermaphroditism at 0.1 PPB in American Frogs (*Rana pipiens*): Laboratory and Field Evidence," *Environmental Health Perspectives* 3, no. 4 (April 2003), and Hayes, Tyrone B. "There Is No Denying This: Defusing the Confusion about Atrazine," *BioScience* 54, no. 12 (December 2004).

On the question of industrial agriculture's dependence on fossil fuel, there is a rich and somewhat daunting literature. Marty Bender, at the Land Institute, helped me to navigate a great many complexities as did David Pimentel at Cornell. The figure of 0.25 gallons of oil per bushel of corn comes from unpublished research by Ricardo Salvador (see his Web site, cited earlier); David Pimentel, et al., offers a figure of 0.33 gallons in "Environmental, Energetic, and Economic Comparisons of Organic and Conventional Farming Systems," *BioScience* 55, no. 7 (July 2005). For more on the general subject of energy use in agriculture, see chapter 9 following.

On the equally vexing topic of federal agriculture policy, I have had many fine tutors, foremost among them George Naylor himself, as well as the staff of the National Family Farms Coalition (www.nffc.net), of which he is president. Other sources for this material (which figures in chapter 3 as well) included:

Michael Duffy, Iowa State (www.sust.ag.iastate.edu/gpsa/faculty/duffy.html).

Daryll Ray, University of Tennesse Institute of Agriculture (www.agpolicy.org). See especially his report "Rethinking US Agricultural Policy: Changing Course to Secure Farmer Livelihoods Worldwide" (issued by the Institute's Agricultural Policy Analysis Center in September 2003 and available at www.agpolicy.org/blueprint.html).

Dan McGuire, American Corngrower's Association (www.acga.org). McGuire generously shared his archive of documents on the history of U.S. agricultural policy since the 1930s.

Mark Ritchie, Institute for Agriculture and Trade Policy (www.iatp.org).

Other sources on the history of farm policy:

Critser, Greg. *Fat Land: How Americans Became the Fattest People in the World* (Boston: Houghton Mifflin, 2003). Critser summarizes the history of farm policy since the 1970s, linking it to the current surplus of food and the consequent epidemic of obesity.

Duscha, Julius. "Up, Up, Up: Butz Makes Hay Down on the Farm," *New York Times Magazine*, April 16, 1972.

Rasmussen, Wayne D., and Gladys L. Baker. *Price Support and Adjustment Programs from 1933 through 1978: A Short History* (Washington, D.C.: USDA Economics, Statistics and Cooperatives Service, 1978).

Ritchie, Mark. *The Loss of Our Family Farms: Inevitable Results or Conscious Policies? A Look at the Origins of Government Policies for Agriculture* (Minneapolis: League of Rural Voters, 1979). Ritchie also shared with me his archive of policy statements by the Committee for Economic Development. The CED, an influential business group from the 1950s through the 1970s, led the campaign to dismantle New Deal farm policy. See their "Toward a Realistic Farm Program" (1967) and "A New U.S. Farm Policy for Changing World Food Needs" (1974).

————, et al. *United States Dumping on World Agricultural Markets* (Minneapolis: Institute for Agriculture and Trade Policy, 2003).

CHAPTER 3: THE ELEVATOR

My estimate of the portion of the U.S. corn crop that passes through the corporate hands of Cargill and ADM is based on Richard Manning's reporting in *Against the Grain* (New York: North Point Press, 2004, p. 128) that ADM buys 12 percent of the nation's corn crop, and on a 1999 estimate by Alexander Cockburn and Jeffrey St. Clair (*Counterpunch*, November 20, 1999) that Cargill buys 23 percent of the corn crop.

Cronon, William. *Nature's Metropolis: Chicago and the Great West* (New York: W. W. Norton, 1991).

Kneen, Brewster. *Invisible Giant: Cargill and Its Transnational Strategies* (London: Pluto Press, 2002).

Manning, Richard. *Against the Grain* (New York: North Point Press, 2004). Manning uses the metaphor of biomass to describe the surplus of commodity grain on page 137.

Sahagún, B. de (*Historia general de las cosas de Nueva Espana, 1558–69*) *Florentine Codex: A General History of the Things of New Spain.* 12 vols. Trans. A. J. O. Anderson and C. E. Dibble (Santa Fe, NM: School of American Research and University of Utah, 1950–69).

Michael Duffy and George Naylor helped me to sort out exactly what a farmer receives for a bushel of corn from the market and the government. That said, the various formulae and contingencies involved, not to mention the nomenclature, are dauntingly complex, and neither Naylor nor Duffy bears responsibility for any oversimplifications or errors in my computations. What I call the county "target price" is technically a "marketing loan rate," but since the program is structured in such a way as to make taking out loans unattractive (unlike the old nonrecourse loan program), the wording is confusing. However, it's important to understand that this price level is not a target price in the sense that it once was, when the USDA set a floor for commodity prices that it then supported by offering farmers nonrecourse loans.

CHAPTER 4: THE FEEDLOT: MAKING MEAT

This chapter had its origins in a piece I wrote for the *New York Times* called "Power Steer" (March 31, 2002). In researching cattle and the U.S. cattle industry, I learned a great deal from Bill Niman of Niman Ranch in Oakland; Kansas feedlot operator Mike Callicrate; Colorado rancher Dale Lassiter; animal-handling expert Temple Grandin (www.grandin.com); South Dakota bison rancher and writer Dan O'Brien; Cornell microbiologist James Russell; and Rich and Ed Blair, the South Dakota ranchers profiled in this chapter. Valuable published sources include:

Carlson, Laurie Winn. *Cattle: An Informal Social History* (Chicago: Ivan R. Dee, 2001).

Durning, Alan B., and Holly B. Brough. *Taking Stock: Animal Farming and the Environment* (Washington, D.C.: World Watch Institute, 1991).

Engel, Cindy. *Wild Health: How Animals Keep Themselves Well and What We Can Learn from Them* (Boston: Houghton Mifflin, 2002).

Frazier, Ian. *Great Plains* (New York: Picador, 1989).

Grandin, Temple. *Animal Handling in Meat Plants* (video: Grandin Livestock Handling System, www.grandin.com, undated).

Johnson, James R., and Gary E. Larson. *Grassland Plants of South Dakota and the Northern Great Plains* (Brookings, SD: South Dakota State University, 1999).

Hamilton, Doug. *Modern Meat* (a documentary for *Frontline*; aired on PBS, April 18, 2002).

Lappé, Frances Moore. *Diet for a Small Planet* (New York: Ballantine Books, 1991). Still the strongest case against eating beef, though in making it Lappé assumes a production system based on grain.

Luttwak, Edward. "Sane Cows, or BSE Isn't the Worst of It," *London Review of Books* 23, no. 3 (February 8, 2001).

Manning, Richard. *Grassland: The History, Biology, and Promise of the American Prairie* (New York: Penguin, 1997).

Nierenberg, Danielle. *Happier Meals: Rethinking the Global Meat Industry* (Washington, D.C.: Worldwatch Institute, 2005).

O'Brien, Dan. *Buffalo for the Broken Heart: Restoring Life to a Black Hills Ranch* (New York: Random House, 2001). This is a rancher's account of the cattle business and a promising alternative to it. O'Brien's ranch happens to share a fence with the Blairs'.

Ozeki, Ruth L. *My Year of Meats* (New York: Penguin, 1999). Very funny, well-researched novel about the U.S. meat industry.

Rampton, Sheldon, and John Stauber. *Mad Cow U.S.A.: Could the Nightmare Happen Here?* (Monroe, ME: Common Courage Press, 1997).

Rifkin, Jeremy. *Beyond Beef* (New York: Plume, 1993).

Russell, James B. *Rumen Microbiology and Its Role in Ruminant Nutrition* (Ithaca, NY: self-published, 2002).

Schell, Orville. *Modern Meat: Antibiotics, Hormones, and the Pharmaceutical Farm* (New York: Vintage, 1985).

Schlosser, Eric. *Fast Food Nation* (Boston: Houghton Mifflin, 2001).

Sinclair, Upton. *The Jungle* (London: Penguin, 1985).

Smil, Vaclav. *Feeding the World: A Challenge for the Twenty-First Century* (Cambridge, MA: M.I.T. Press, 2001).

CHAPTER 5: THE PROCESSING PLANT: MAKING COMPLEX FOODS

I've written about the imperatives behind the processing of food on several occasions (the articles are listed below), and on that subject have profited enormously from my conversations with nutritionists Marion Nestle and Joan Gussow, and my readings of industry trade magazines, especially *Food Technology* (Institute of Food Technologists, Chicago). Larry Johnson at the Center for Crops Utilization Research at Iowa State was generous with his time and expertise, showing and telling me all I wanted to know about the wet-milling of corn and soybeans. The Corn Refiners Association (www.corn.org) is an invaluable resource on the history, technology, and products of corn refining; see especially their annual reports, a trove of interesting statistics and history.

Ford, Brian J. *The Future of Food* (New York: Thames & Hudson, 2000).

Goodman, Michael, and Michael Redclift. *Refashioning Nature: Food, Ecology, and Culture* (London: Routledge, 1991).

Gussow, Joan Dye, ed. *The Feeding Web: Issues in Nutritional Ecology* (Palo Alto, CA: Bull Publishing, 1978). This remains an invaluable anthology (unfortunately out of print) on the entire range of food issues, and serves as a reminder that much of the discussion our culture is having about the politics and ecology of food today is a reprise of a discussion that took place in the 1970s. The quote about the relationship between a food's identity and its raw materials and the excerpt from the IFF annual report appear in an essay by Gussow titled "Whatever Happened to Food? Or Does It Pay to Fool with Mother Nature?" pp. 200–4.

Levenstein, Harvey. *Paradox of Plenty* (Berkeley: University of California Press, 2003).

———. *Revolution at the Table: The Transformation of the American Diet* (Berkeley: University of California Press, 2003).

Nestle, Marion. *Food Politics* (Berkeley: University of California Press, 2002).

Pollan, Michael. "Naturally," *New York Times Magazine*, May 13, 2001.

———. "The Futures of Food," *New York Times Magazine*, May 4, 2003.

———. "The (Agri)cultural Contradictions of Obesity," *New York Times Magazine*, October 12, 2003.

Schlosser, Eric. *Fast Food Nation* (Boston: Houghton Mifflin, 2001).

Tannahill, Reay. *Food in History* (New York: Stein and Day, 1973). The quote on manufacturing steak from petroleum is on page 394.

Tisdale, Sally. *The Best Thing I Ever Tasted: The Secret of Food* (New York: Riverhead, 2001). The quote from Massimo Montanari, the Italian food historian, about how processing food frees us from the vicissitudes of nature appears on page 66.

CHAPTER 6: THE CONSUMER: A REPUBLIC OF FAT

Bray, George, et al. "Consumption of High-fructose Corn Syrup in Beverages May Play a Role in Epidemic of Obesity," *American Journal of Clinical Nutrition* 79 (2004), 537–43.

Brownell, Kelly D., and Katherine Battle Horgen. *Food Fight: The Inside Story of the Food Industry, America's Obesity Crisis, and What We Can Do About It* (Chicago: Contemporary Books, 2004).

Critser, Greg. *Fat Land: How Americans Became the Fattest People in the World* (Boston: Houghton Mifflin, 2003).

Drewnowski, Adam, and S. E. Specter. "Poverty and Obesity: The Role of Energy Density and Energy Costs in the American," *American Journal of Clinical Nutrition* 79 (January 2004), 6–16. For this important article, Drewnowski and Specter studied how many and what kind of calories a dollar can buy in various parts of the supermarket.

Kroc, Ray. *Grinding It Out: The Making of McDonald's* (Chicago: Contemporary Books, 1977).

Lender, Mark E., and James Kirby Martin. *Drinking in America: A History* (New York: The Free Press, 1982).

Logsdon, Gene. *Good Spirits: A New Look at Ol' Demon Alcohol* (White River Junction, VT: Chelsea Green, 1999).

Love, John F. *McDonald's: Behind the Arches* (New York: Bantam, 1986). Love tells the story of David Wallerstein, pages 296–97.

Narayan, K. M. Venkat, et al. "Lifetime Risk for Diabetes Mellitus in the United States," *Journal of the American Medical Association* 290 (2003), 1884–90.

Nestle, Marion. *Food Politics* (Berkeley: University of California Press, 2002).

Pollan, Michael. "The (Agri)cultural Contradictions of Obesity," *New York Times Magazine*, October 12, 2003. This chapter extends and elaborates the argument I made in this article.

———. *The Botany of Desire* (New York: Random House, 2001). See the material on sweetness in the chapter on apples, as well as the bibliography on sweetness.

Rorabaugh, W. J. *The Alcoholic Republic: An American Tradition* (Oxford: Oxford University Press, 1979). An eye-opening account of American drinking habits from the Revolution through the temperance movement. The book is the main source for my information on early American alcohol consumption; the quote from William Cobbett appears on page 59.

Satcher, David. "The Surgeon General's Call to Action to Prevent and Decrease Overweight and Obesity," (Washington, D.C.: U.S. Department of Health and Human Services, 2001); available on the Web at www.surgeon general.gov.

Winson, Anthony. "Bringing Political Economy into the Debate on the Obesity Epidemic," *Agriculture and Human Values* 21 (2004), 299–312.

CHAPTER 7: THE MEAL: FAST FOOD

"A Full Serving of Nutrition Facts," pamphlet published by McDonald's (2003).

Schlosser, Eric. *Fast Food Nation* (Boston: Houghton Mifflin, 2001).

On ethanol and air pollution see Libecap, Gary D. "Environmental Phantasm: Political Forces Keep Dreams of Ethanol Alive," Property and Environment Research Center (PERC) (June 2003); www.perc.org/publications/perc reports/june 2003/phantasm.php and the Web site of the Sierra Club, www.sierraclub.org.

PART II
CHAPTER 8: ALL FLESH IS GRASS

On the pastoral tradition, Leo Marx is invaluable. I learned a great deal about farming, grass, animals, and Joel Salatin from Salatin's books, all of which are worth reading, even if you aren't planning to raise chickens; he's a consistently entertaining writer. *Stockman Grass Farmer*, Allan Nation's monthly tabloid for grass farmers, is the indispensable media outlet for the movement.

Klinkenborg, Verlyn. *Making Hay* (Guilford, CT: Lyons Press, 1997).

Marx, Leo. *The Machine in the Garden* (Oxford: Oxford University Press, 2000). The quote from Henry James is on page 352.

Pollan, Michael. "Sustaining Vision," *Gourmet* (September 2002).

Salatin, Joel. *Family Friendly Farming* (Swoope, VA: Polyface, 2001).

———. *Holy Cows & Hog Heaven: The Food Buyer's Guide to Farm Friendly Food* (Swoope, VA: Polyface, 2004).

———. *Pastured Poultry Profit$: Net $25,000 in 6 Months on 20 Acres* (Swoope, VA: Polyface, 1996).

———. *Polyface Farm* (video: Moonstar Films, www.moonstarfilms.com, undated).

———. *$alad Bar Beef* (Swoope, VA: Polyface, 1995).

———. *You Can Farm: The Entrepreneur's Guide to Start and $ucceed in a Farming Enterprise* (Swoope, VA: Polyface, 1998).

Virgil. *Eclogues, Georgics, Aeneid 1–6*, Volume 1. Trans. H. Rushton Fairclough (Cambridge, MA: Harvard University Press, 1986).

Williams, Raymond. *The Country and the City* (New York: Oxford University Press, 1973).

CHAPTER 9: BIG ORGANIC

Parts of this chapter are based on an article on the industrialization of organic food I published in the *New York Times Magazine* (May 13, 2001). Among the sources in the organic movement who have done the most to educate me are: Joan Gussow; Fred Kirschenmann at the Leopold Center at Iowa State (www.leopold.iastate.edu); Bob Scowcroft at the Organic Farming Research

Foundation; Michael Sligh and Hope Shand at ETC (www.etcgroup.org); the late Betsy Lydon; farmer and author Eliot Coleman; farmer Woody Derycks; farmers Tom and Denesse Willy; farmer Warren Weber; farmer and author Michael Ableman; Drew and Myra Goodman and Mark Merino at Earthbound Farm; George Siemens at Organic Valley; John Diener at Greenways Organic; Gene Kahn at General Mills; Miguel Altieri; Julie Guthman; Peter Rosset; Charles Benbrook; Roger Blobaum; and Maria Rodale. Several of the scientific articles comparing organic and conventional produce are included in the list of printed sources following; others are available at the Organic Center (www.organic-center.org).

Altieri, Miguel. *Agroecology:The Science of Sustainable Agriculture* (Boulder, CO: Westview Press, 1995).

————. "The Ecological Role of Biodiversity in Agroecosystems," *Agric. Ecosyst. and Env.* 74 (1999), 19–31.

Asami, Danny K., et al. "Comparison of the Total Phenolic and Ascorbic Acid Content of Free-Dried and Air-Dried Marionberry, Strawberry, and Corn Using Conventional, Organic, and Sustainable Agricultural Practices," *Journal of Agricultural and Food Chemistry* 51 (2003), 1237–41.This is the study I discuss at some length.

Barron, R. C. ed. *The Garden and Farm Books of Thomas Jefferson* (Golden, CO: Fulcrum, 1987). In a letter to his daughter, Jefferson suggests that the problems she's having with insects could be the result of exhausted soil; see page 156. Eliot Coleman first told me about this passage.

Belasco, Warren. *Appetite for Change: How the Counterculture Took on the Food Industry 1966–1988* (New York: Pantheon, 1989). Belasco persuasively traces organic food's roots to the sixties counterculture.The contemporary accounts of People's Park and the "People's Garden" are on pages 19–22.

Benbrook, Charles M. *Elevating Antioxidant Levels in Food Through Organic Farming and Food Processing:An Organic Center State of Science Review* (Foster, RI: Organic Center, 2005).

Berry, Wendell. *The Gift of Good Land* (San Francisco: North Point Press, 1981).

————. *Home Economics* (San Francisco: North Point Press, 1987).

————. *The Unsettling of America: Culture and Agriculture* (San Francisco: Sierra Club Books, 1977).The quote from Sir Albert Howard about soil and health appears on page 46.

Carbonaro, Marina, and Maria Mattera. "Polyphenoloxidase Activity and Polyphenol Levels in Organically and Conventionally Grown Peaches," *Food Chemistry* 72 (2001), 419–24.

Coleman, Eliot. "Can Organics Save the Family Farm?" *The Rake* (September 2004).

Curl, Cynthia L., et al. "Organophosphorus Pesticide Exposure of Urban and Suburban Pre-school Children with Organic and Conventional Diets," *Environmental Health Perspectives* 3, no. 3 (March 2003).

Davis, Donald R., et al. "Changes in USDA Food Composition Data for 43 Garden Crops, 1950 to 1999," *Journal of the American College of Nutrition* 23, no. 6 (2004), 669–82.

———. "Trade-Offs in Agriculture and Nutrition," *Food Technology* 59, no. 3, 120.

Dewhurst, R. J., et al. "Comparison of Grass and Legume Silages for Milk Production," *Journal of Dairy Science* 86, no. 8 (2003), 2598–2611.

Diamond, Jared. *Collapse: How Societies Choose to Fail or Succeed* (New York: Viking, 2005).

Freyfogle, Eric T., ed. *The New Agrarianism: Land, Culture, and the Community of Life* (Washington, D.C.: Island Press, 2001).

Guthman, Julie. *Agrarian Dreams* (Berkeley: University of California Press, 2004).

Harvey, Graham. *The Forgiveness of Nature: The Story of Grass* (London: Jonathan Cape/Random House, 2001). For the great humus controversy, see chapter 17, pages 300–19.

Hayes, Tyrone, et al. "Atrazine-Induced Hermaphroditism at 0.1 PPB in American Frogs (*Rana pipiens*): Laboratory and Field Evidence," *Environmental Health Perspectives* 3, no. 4 (April 2003).

———. "There Is No Denying This: Defusing the Confusion about Atrazine," *BioScience* 54, no. 12 (December 2004).

Howard, Sir Albert. *An Agricultural Testament* (New York: Oxford University Press, 1943).

———. *The Soil and Health* (New York: Schocken, 1972).

Lewis, W. J., et al. "A Total System Approach to Sustainable Pest Management," *The Proceedings of the National Academy of Sciences* 84 (1997).

Manning, Richard. *Commodities, Consensus and Conservation* (April 2001). In his study of commodity agriculture, Manning quotes Plato on agriculture's impact on the environment, and the importance of healthy soils (page 2):

> What now remains of the formerly rich land is like the skeleton of a sick man . . . : Formerly, many of the mountains were arable. The plains that were full of rich soil are now marshes. Hills that were once covered with forests and produced abundant pasture now produce only food for bees. Once the land was enriched by yearly rains, which were not lost, as they are now, by flowing from the bare land into the sea. The soil was deep, it absorbed and kept the water in loamy soil, and the water that soaked into the hills fed springs and running streams everywhere. Now the abandoned shrines at spots where formerly there were springs attest that our description of the land is true.

Marx, Leo. *The Machine in the Garden* (Oxford: Oxford University Press, 2000).

Rosset, Peter M. *The Multiple Functions and Benefits of Small Farm Agriculture* (Oakland: Food First, 1999). Rosset documents the ways in which small diversified farms are actually more efficient than large ones.

Sligh, Michael, and Carolyn Christman. *Who Owns Organic?* (Pittsboro, NC: RAFI-USA, 2003).

Stoll, Steven. *The Fruits of Natural Advantage: Making the Industrial Countryside in California* (Berkeley: University of California Press, 1998).

Tilman, David. "The Greening of the Green Revolution," *Nature*, 396 (November 19, 1998).

Wargo, John. *Our Children's Toxic Legacy* (New Haven: Yale University Press, 1996).

Wirzba, Norman, ed. *The Essential Agrarian Reader* (Lexington, KY: University Press of Kentucky, 2003).

Wolfe, M. S. "Crop Strength Through Diversity," *Nature* 406, no. 17 (August 2000).

On the complex and contentious subject of energy use in conventional and organic agriculture, I relied on many sources, including David Pimental, Rich Pirog at the Leopold Center, Marty Bender at the Land Institute, and Karen Klonsky and Peter Livingston at the University of California at Davis, as well as the indefatigable work of my researcher Chad Heeter. Pimental helped us calculate the energy required to grow, pack, wash, cool, and ship across country a pound of organic lettuce, using his data and additional information graciously provided by Earthbound Farm. Pimentel's numbers are sometimes criticized as high because he includes "embodied energy," i.e., the fossil fuel required to manufacture things like tractors. His numbers remain the most comprehensive, however, and whenever a specific figure is in dispute I've used the more conservative number or stated the range. On energetics in agriculture see also:

Carlsson-Kanyama, Annika, and Mireille Faist. *Energy Use in the Food Sector: A Data Survey.* AFN-report 291 (Swedish Environmental Protection Agency: Stockholm, Sweden, 2000).

Heller, Martin C., and Gregory A. Keoleian. *Life Cycle–Based Sustainability Indicators for Assessment of the U.S. Food System*, Report No. CSS00-04. (Center for Sustainable Systems, University of Michigan, 2000). This study is the source for my figures on the portion of U.S. energy consumption devoted to the food system (one-fifth) and the portion of that amount (one-fifth) accounted for by farming (as opposed to packing, cooling, or shipping).

Livingston, Peter. "A Comparison of Economic Viability and Measured Energy Required for Conventional, Low Input, and Organic Farming Systems over a Rotational Period." Unpublished thesis. California State University, Chico, CA, 1995.

Lovins, Amory, L. Hunter Lovins, and Marty Bender. "Agriculture and Energy," *Encyclopedia of Energy Technology and the Environment* (New York: John Wiley & Sons, 1995).

Pimentel, David, ed. *Handbook of Energy Utilization in Agriculture* (Boca Raton, FL: CRC Press, 1980).

Pimentel, David, and Marcia Pimentel, eds. *Food, Energy, and Society* (Niwot, CO: University Press of Colorado, 1996).

Pimentel, David, et al. "Environmental, Energetic, and Economic Comparisons of Organic and Conventional Farming Systems," *BioScience* 55, no. 7 (July 2005), 573–82. The statistic on the energy savings of organic production (30 percent) comes from this study, though as Pimentel acknowledges, if the farm's fertility is not generated on the farm or nearby, this savings quickly disappears.

Tourte, Laura, et al. "Sample Costs to Produce Organic Leaf Lettuce." University of California Cooperative Extension, 2004.

CHAPTER 10: GRASS: THIRTEEN WAYS OF LOOKING AT A PASTURE

Benyus, Janine M. *Biomimicry: Innovation Inspired by Nature* (New York: Perennial, 2002). Offers a fine account of the Land Institute's project to perennialize agriculture.

Eisenberg, Evan. *The Ecology of Eden* (New York: Knopf, 1998).

Farb, Peter. *Living Earth* (New York: Pyramid Publications, 1959).

Harvey, Graham. *The Forgiveness of Nature: The Story of Grass* (London: Jonathan Cape/Random House, 2001).

Hawken, Paul, Amory Lovins, and L. Hunter Lovins. *Natural Capitalism* (New York: Bay Books, 2000). Another good account of the Land Institute's work.

Jackson, Wes, et al., eds. *Meeting the Expectations of the Land: Essays in Sustainable Agriculture and Stewardship* (San Francisco: North Point Press, 1984).

———. *New Roots for Agriculture* (Lincoln, NE: University of Nebraska Press, 1985).

Judy, Greg. *No Risk Ranching: Custom Grazing on Leased Land* (Ridgeland, MS: Green Park Press, 2003).

Logsdon, Gene. *All Flesh Is Grass: The Pleasures and Promises of Pasture Farming* (Athens, OH: Swallow Press/Ohio University, 2004).

Nation, Allan. *Knowledge Rich Ranching* (Ridgeland, MS: Green Park Press, 2002).

Savory, Allan. *Holistic Management: A New Framework for Decision Making* (Washington, D.C.: Island Press, 1999). Savory is a pioneer in using intensive grazing to restore arid grasslands, and is changing the way environmentalists regard the role of grazing in ecosystem health.

The Stockman Grass Farmer, published monthly.

Voisin, André. *Grass Productivity* (Washington, D.C.: Island Press, 1989).

CHAPTER 11: THE ANIMALS: PRACTICING COMPLEXITY

For further reading on the advantages of polyculture, see *Permaculture* magazine (www.permaculture.co.uk); *Permaculture Activist* (www.permacultureactivist.net); and the works of Bill Mollison. Also see:

Furuno, Takao. *The Power of Duck: Integrated Rice and Duck Farming* (Tasmania, Australia: Tagari Publications, 2001). This is another example from another tradition of a polyculture farm. Furuno is the Joel Salatin of Japan.

Imhoff, Dan. *Farming with the Wild: Enhancing Biodiversity on Farms and Ranches* (San Francisco: Sierra Club Books, 2003).

Rosset, Peter. *The Multiple Functions and Benefits of Small Farm Agriculture* (Oakland: Food First, 1999).

CHAPTER 12: SLAUGHTER: IN A GLASS ABATTOIR

Joel explains exactly how to kill a chicken and compost slaughter waste in chapters 15 and 16 of *Pastured Poultry Profit$* (Swoope, VA: Polyface, 1996).

On slaughter practices, humane and otherwise, see Temple Grandin's Web site (www.grandin.com).

CHAPTER 13: THE MARKET: "GREETINGS FROM THE NON-BARCODE PEOPLE"

To find local producers of meat, eggs, poultry, and milk in your area, go to www.eatwellguide.org and www.eatwild.com. The Web site for Slow Food USA is www.slowfood.com.

Berry, Wendell. *Citizenship Papers* (Washington, D.C.: Shoemaker & Hoard, 2003). See especially the essays "The Total Economy" (pp. 63–76) and "The Whole Horse" (pp. 113–26), where the Berry quotes in this chapter are found.

Blank, Steven. *The End of Agriculture in the American Portfolio* (Westport, CT: Quorum Books, 1998).

Fallon, Sally. *Nourishing Traditions* (Washington, D.C.: New Trends Publishing, 2001). Fallon is the president of the Weston A. Price Foundation: www.westonaprice.org.

Fernald, Anya, et al. *A World of Presidia: Food, Culture, and Community* (Bra, Italy: Slow Food Editore, 2004).

Gussow, Joan Dye. *This Organic Life: Confessions of a Suburban Homesteader* (White River Junction, VT: Chelsea Green Publishing, 2001).

Halweil, Brian. *Eat Here: Reclaiming Homegrown Pleasures in a Global Supermarket* (New York: W. W. Norton & Company, 2004).

———. *Home Grown: The Case for Local Food in a Global Market* (Washington, D.C.: Worldwatch Institute, 2002).

Kloppenberg, J., Jr., et al. "Coming into the Foodshed," *Agriculture and Human Values* 13, no. 3 (1996), 33–41. This article appears to be the first use of the term "foodshed": "The concept of a 'foodshed' (a term that elicits images of food flowing into a place) has been developed to promote discussion and action about the disempowerment and destructive nature of this current system with regards to the community and the environment."

Lyson, Thomas A. *Civic Agriculture: Reconnecting Farm, Food, and Community* (Medford, MA: Tufts University Press, 2004).

McKibben, Bill. "Small World: Why One Town Stays Unplugged," *Harper's* 307, no. 1843 (December 2003), 46–54.

Nabhan, Gary Paul. *Coming Home to Eat: The Pleasures and Politics of Local Foods* (New York: W. W. Norton, 2001).

Norberg-Hodge, Helena, et al. *Bringing the Food Economy Home: Local Alternatives to Global Agribusiness* (London: Zed Books, 2002).

Petrini, Carlo, ed. *Slow Food: Collected Thoughts on Taste, Tradition, and the Honest Pleasures of Food* (White River Junction, VT: Chelsea Green Publishing, 2001). See also Petrini's speeches on the Slow Food Web site.

Pollan, Michael. "Cruising on the Ark of Taste," *Mother Jones* (May 2003). An essay on the politics of Slow Food.

Porter, Michael E. *The Competitive Advantage of Nations* (New York: The Free Press, 1990).

CHAPTER 14: THE MEAL: GRASS-FED

For a digest of research on the health benefits of grass-fed meat, milk, and eggs, see www.eatwild.com.

Brillat-Savarin, Jean-Anthelme. *The Physiology of Taste*. Trans. Anne Drayton (London: Penguin, 1994).

Child, Julia. *Mastering the Art of French Cooking* (New York: Alfred A. Knopf, 2001).

McGee, Harold. *On Food and Cooking: The Science and Lore of the Kitchen* (New York: Charles Scribner, 2004).

Robinson, Jo. *Pasture Perfect: The Far-Reaching Benefits of Choosing Meat, Eggs, and Dairy from Grass-Fed Animals* (Vashon, WA: Vashon Island Press, 2004).

———. *Why Grassfed Is Best! The Surprising Benefits of Grassfed Meat, Eggs, and Dairy Products* (Vashon, WA: Vashon Island Press, 2000).

For recent research on the role of omega-3s and other fats in the diet, see the proceedings of the 2004 meeting of the International Society for the Study of Fatty Acids and Lipids (www.issfal.org.uk). The research on the benefits of omega-3s cited in my chapter came from the following articles:

de Groot, R. H. M., et al. *Correlation Between Plasma (N-3) Fatty Acid Levels and Cognitive Performance in Women.* Report. Department of Psychiatry and Neuropsychology, Nutrition and Toxicology Research Institute Maastricht (Maastricht University, The Netherlands, 2004).

Kelley, R. L., et al. *Effect of Dietary Fish Oil on Puppy Trainability.* Report. The Iams Company Technical Centre (Lewisburg, OH: 2004).

Smuts, C. M., et al. *The Effect of Omega-3 Rich Spread on the Cognitive Function of Learners 6–9 Years Old from a Low Socio-Economic Community.* Nutritional Intervention Research Unit, MRC. Report (Parow Valley, Stellenbosch, South Africa, 2004).

PART III

CHAPTER 15: THE FORAGER

Allport, Susan. *The Primal Feast: Food, Sex, Foraging, and Love* (Lincoln, NE: Universe, 2003).

Budiansky, Stephen. *The Covenant of the Wild: Why Animals Chose Domestication* (New Haven: Yale University Press, 1999). Thoreau's quote on hunting is on page 157.

Leopold, Aldo. *A Sand County Almanac* (New York: Ballantine, 1986). The Leopold quotes are from page 177.

Nelson, Davia, and Nikki Silva. *Hidden Kitchens: Stories, Recipes, and More from NPR's The Kitchen Sisters* (New York: Rodale, 2005). See especially the chapter on Angelo Garro, pages 172–89.

CHAPTER 16: THE OMNIVORE'S DILEMMA

Allport, Susan. *The Primal Feast: Food, Sex, Foraging and Love* (Lincoln, NY: Writers Club Press, 2003).

Fernández-Armesto, Felipe. *Near a Thousand Tables: A History of Food* (New York: The Free Press, 2002).

Harris, Marvin. *The Sacred Cow and the Abominable Pig: Riddles of Food and Culture* (New York: Simon & Schuster, 1987).

Kass, Leon. *The Hungry Soul* (New York: The Free Press, 1994).

Katz, Solomon H. "Food and Biocultural Evolution: A Model for the Investigation of Modern Nutritional Problems," *Nutritional Anthropology*, ed. Francis E. Johnston (New York: Alan R. Liss, 1987), 41–63.

Lévi-Strauss, Claude. *The Origin of Table Manners: Introduction to a Science of Mythology*, Volume 3. Trans. John and Doreen Weightman (New York: Harper & Row, 1978).

———. *The Raw and the Cooked: Introduction to a Science of Mythology*, Volume 1. Trans. John and Doreen Weightman (Chicago: University of Chicago Press, 1983).

Mooallem, Jon. "The Last Supper: Living by One-handed Food Alone," *Harper's* (July 2005). My source for the statistic that 19 percent of American meals are eaten in the car.

Pinker, Steven. *How the Mind Works* (New York: W. W. Norton, 1997). Valuable on hunting and gathering; visual perception; the cognitive niche; and the evolution of disgust; the quote on "intuitive microbiology" is on page 383.

Pollan, Michael. "Our National Eating Disorder," *New York Times Magazine*, October 17, 2004.

Rozin, Paul, et al. "Attitudes to Food and the Role of Food in Life: Comparisons of Flemish Belgian, France, Japan and the United States," *Appetite* (1999).

———, et al. "The Borders of the Self: Contamination Sensitivity and Potency of the Mouth, Other Apertures and Body Parts," *Journal of Research in Personality* 29 (1995), 318–40.

————, et al. "The Cultural Evolution of Disgust," in *Food Preferences and Taste: Continuity and Change,* ed. H. M. Macbeth (Oxford: Berghahn, 1997).

————, et al. "Disgust," in *Handbook of Emotions,* 2nd ed., eds. Lewis M. and J. Haviland (New York: Guilford, 1999).

————, et al. "Lay American Conceptions of Nutrition: Dose Insensitivity, Categorical Thinking, Contagion, and the Monotonic Mind," *Health Psychology* 15 (1996), 438–47.

————, and A. E. Fallon. "A Perspective on Disgust," *Psychological Review* 94, no. 1 (1987), 23–41.

————, and J. Schulkin. "Food Selection," in *Handbook of Behavioral Neurobiology, Food and Water Intake,* Volume 10, ed. E. M. Stricker (New York: Plenum, 1990), 297–328.

Wrangham, Richard, et al. "The Raw and the Stolen: Cooking and the Ecology of Human Origins," *Current Anthropology* 40, no. 5 (December 1999). Wrangham argues persuasively here and elsewhere that it is cooking that made us human.

CHAPTER 17: THE ETHICS OF EATING ANIMALS

Berger, John. *About Looking* (New York: Vintage International, 1991).

Budiansky, Stephen. *The Covenant of the Wild: Why Animals Choose Domestication* (New York: Morrow, 1992). A valuable book on the evolution of domestication in animals.

————. *If a Lion Could Talk: Animal Intelligence and the Evolution of Consciousness* (New York: The Free Press, 1998).

Coetzee, J. M. *The Lives of Animals* (Princeton: Princeton University Press, 1999).

Dennett, Daniel C. *Kinds of Minds: Toward an Understanding of Consciousness* (New York: Basic Books, 1996).

Ehrenfeld, David. *Beginning Again: People and Nature in the New Millenium* (New York: Oxford University Press, 1995).

Ovid. *Metamorphoses.* Trans. A. D. Melville. (Oxford: Oxford University Press, 1998).

Flannery, Tim. *The Eternal Frontier: An Ecological History of North America and Its Peoples* (New York: Atlantic Monthly Press, 2001). Flannery's account of how the Plains bison evolved under the pressure of hunting by Indians is on pages 223–29; the quote is on page 227.

Regan, Tom. *The Case for Animal Rights* (Berkeley: University of California Press, 1983).

————, and Peter Singer, eds. *Animal Rights and Human Obligations* (Englewood Cliffs, NJ: Prentice Hall, 1989).

Scully, Matthew. *Dominion: The Power of Man, the Suffering of Animals, and the Call to Mercy* (New York: St. Martin's Press, 2002). An eloquent defense of animals, and an indictment of factory farming, from the right.

Singer, Peter. *Animal Liberation* (New York: Ecco, 2002).

————. *Practical Ethics* (Cambridge, U.K.: Cambridge University Press, 1999).

————, ed. *In Defense of Animals* (New York: Basil Blackwell, 1985).

Thomas, Keith. *Man and the Natural World: A History of the Modern Sensibility* (New York: Pantheon, 1983).

Williams, Joy. *Ill Nature: Rants and Reflections on Humanity and Other Animals* (New York: Vintage, 2001).

Wise, Steven M. *Drawing the Line: Science and the Case for Animal Rights* (Cambridge, MA: Perseus, 2002).

CHAPTER 18: HUNTING: THE MEAT

Nelson, Richard. *The Island Within* (New York: Vintage, 1991). "The Gifts of the Deer" is one of the great accounts of hunting.

Ortega y Gasset, José. *Meditations on Hunting*. Trans. Howard B. Westcott (New York: Scribner's, 1972). A remarkable book, brilliant and more than a little mad. My own meditations on hunting owe a large debt to Ortega's.

Shepard, Paul. *Coming Home to the Pleistocene* (Washington, D.C.: Island Press, 1998).

————. *Nature and Madness* (Athens, GA: University of Georgia Press, 1998). Writing in the tradition of Ortega, Shepard's work offers a bracing reevaluation of Paleolithic culture and psychology.

————. *The Tender Carnivore and the Sacred Game* (Athens, GA: University of Georgia Press, 1998).

CHAPTER 19: GATHERING: THE FUNGI

My education in the mysteries of the fungal kingdom profited from time spent in the field with Ignacio Chapela and David Arora, as well as with mushroom hunters Anthony Tassinello, Bob Baily, Sue Moore, and Angelo Garro. The following books and articles were also valuable:

Arora, David. *Mushrooms Demystified* (Berkeley: Ten Speed Press, 1986).

Hudler, George W. *Magical Mushrooms, Mischievous Molds* (Princeton: Princeton University Press, 2000).

Krieger, Louis C. C. *The Mushroom Handbook* (New York: Dover Publications, 1967).

Lincoff, Gary H. *National Audubon Society Field Guide to North American Mushrooms* (New York: Alfred A. Knopf, 2003).

McKenna, Terence. *Food of the Gods: The Search for the Original Tree of Knowledge* (New York: Bantam, 1993).

Rommelmann, Nancy. "The Great Alaskan Morel Rush of '05," *Los Angeles Times Magazine*, July 10, 2005.

Schaechter, Elio. *In the Company of Mushrooms: A Biologist's Tale* (Cambridge, MA: Harvard University Press, 1998).

Stamets, Paul. *Growing Gourmet and Medicinal Mushrooms* (Berkeley: Ten Speed Press, 2000).

————. *Mycelium Running: How Mushrooms Can Help Save the World* (Berkeley: Ten Speed Press, 2005).

Treisman, Ann. "Features and Objects in Visual Processing," *Scientific American* 254, no. 11 (November 1986), 114–25. Treisman, a research psychologist, developed the concept of the "pop-out effect" in human visual processing.

Weil, Andrew. *The Marriage of the Sun and Moon: Dispatches from the Frontiers of Consciousness* (Boston: Houghton Mifflin, 2004). See chapters 7 to 9; the quotes come from chapter 8.

CHAPTER 20: THE PERFECT MEAL

Brillat-Savarin, Jean-Anthelme. *The Physiology of Taste.* Trans. Anne Drayton (London: Penguin, 1994).

Leader, Daniel, and Judith Blahnik. *Bread Alone: Bold Fresh Loaves from Your Own Hands* (New York: Morrow, 1993). See chapter 13 on gathering wild yeast for the chef and baking a levain. I also learned about baking with wild yeasts from Robbie Barnett.

McGee, Harold. *On Food and Cooking: The Science and Lore of the Kitchen* (New York: Charles Scribner, 2004).

Waters, Alice. *The Chez Panisse Café Cookbook* (New York: Morrow, 1999). The recipe for the galette dough is on page 227.

INDEX

A NOTE ON THE AUTHOR

Michael Pollan is the author of three previous books: *Second Nature*, *A Place of My Own*, and *The Botany of Desire*, a *New York Times* bestseller that was named a best book of the year by Borders, Amazon, and the American Booksellers Association. Pollan is a longtime contributing writer at the *New York Times Magazine* and teaches journalism at the University of California–Berkeley. He lives in Berkeley with his wife, the painter Judith Belzer, and their son, Isaac. To read more of his work, go to www.michael pollan.com.